☀ The Doctrine of Triangles

The Doctrine of Triangles

A History of Modern Trigonometry

GLEN VAN BRUMMELEN

PRINCETON UNIVERSITY PRESS Princeton and Oxford

Published by Princeton University Press
41 William Street, Princeton, New Jersey 08540
6 Oxford Street, Woodstock, Oxfordshire OX20 1TR

press.princeton.edu

Library of Congress Cataloging-in-Publication Data
Names: Van Brummelen, Glen, author.
Title: The doctrine of triangles : a history of modern trigonometry / Glen Van Brummelen.
Description: Princeton : Princeton University Press, 2021. | Includes bibliographical
  references and index.
Identifiers: LCCN 2020047691 (print) | LCCN 2020047692 (ebook) |
  ISBN 9780691179414 (hardback) | ISBN 9780691219875 (ebook)
Subjects: LCSH: Trigonometry—History.
Classification: LCC QA24 .V35 2021 (print) | LCC QA24 (ebook) |
  DDC 516.2409/03—dc23
LC record available at https://lccn.loc.gov/2020047691
LC ebook record available at https://lccn.loc.gov/2020047692

British Library Cataloging-in-Publication Data is available

Editorial: Susannah Shoemaker, Kristen Hop
Jacket Design: Chris Ferrante
Production: Jacqueline Poirier
Publicity: Matthew Taylor, Amy Stewart

Jacket image: From *Trigonometrie: Or, the Doctrine of Triangles*, by Richard Norwood, 1685 ed. Norwood's book classifies various types of spherical triangle, here postulated as astronomical entities, and shows how to solve problems in spherical astronomy using logarithms

This book has been composed in Times LT Std

Printed on acid-free paper. ∞

Printed in the United States of America

10 9 8 7 6 5 4 3 2 1

To the memory of

Joel Silverberg: Gentleman, scholar, friend

Contents ⚝

Preface   xi

1. **European Trigonometry Comes of Age**   1

   What's in a Name?   3

   ▪ Text 1.1 Regiomontanus, Defining the Basic Trigonometric
   Functions   4

   ▪ Text 1.2 Reinhold, a Calculation in a Planetary Model Using
   Sines and Tangents   6

   Trigonometric Tables Evolving   16

   Algebraic Gems by Viète   25

   ▪ Text 1.3 Viète, Finding a Recurrence Relation for $\sin n\theta$ 25

   New Theorems, Plane and Spherical   30

   ▪ Text 1.4 Snell on Reciprocal Triangles   37

   Consolidating the Solutions of Triangles   39

   Widening Applications   45

   ▪ Text 1.5 Clavius on a Problem in Surveying   49

   ▪ Text 1.6 Gunter on Solving a Right-Angled Spherical Triangle
   with His Sector   56

2. **Logarithms**   62

   Napier, Briggs, and the Birth of Logarithms   62

   ▪ Text 2.1 Napier, Solving a Problem in Spherical Trigonometry
   with His Logarithms   65

   Interlude: Joost Bürgi's Surprising Method of Calculating a
   Sine Table   69

   The Explosion of Tables of Logarithms   71

   Computing Tables Effectively: Logarithms   76

   Computing Tables Effectively: Interpolation   78

   ▪ Text 2.2 Briggs, Completing a Table Using Finite
   Difference Interpolation   81

   Napier on Spherical Trigonometry   84

   Further Theoretical Developments   91

   Developments in Notation   97

Practical and Scientific Applications   99

⬛ Text 2.3 John Newton, Determining the Declination of an Arc of the Ecliptic with Logarithms   100

3. Calculus   110

Quadratures in Trigonometry Before Newton and Leibniz   110

⬛ Text 3.1 Pascal, Finding the Integral of the Sine   118

Tangents in Trigonometry Before Newton and Leibniz   120

⬛ Text 3.2 Barrow, Finding the Derivative of the Tangent   122

Infinite Sequences and Series in Trigonometry   126

⬛ Text 3.3 Newton, Finding a Series for the Arc Sine   129

Transforming the Construction of Trigonometric Tables with Series   135

Geometric Derivatives and Integrals of Trigonometric Functions   143

A Transition to Analytical Conceptions   145

⬛ Text 3.4 Cotes, Estimating Errors in Triangles   149

⬛ Text 3.5 Jakob Kresa, Relations Between the Sine and the Other Trigonometric Quantities   155

Euler on the Analysis of Trigonometric Functions   161

⬛ Text 3.6 Leonhard Euler, On Transcendental Quantities Which Arise from the Circle   165

⬛ Text 3.7 Leonhard Euler, On the Derivative of the Sine   175

Euler on Spherical Trigonometry   177

4. China   185

Indian and Islamic Trigonometry in China   185

⬛ Text 4.1 Yixing, Description of a Table of Gnomon Shadow Lengths   188

Indigenous Chinese Geometry   191

⬛ Text 4.2 Liu Hui, Finding the Dimensions of an Inaccessible Walled City   192

Indigenous Chinese Trigonometry   198

The Jesuits Arrive   202

Trigonometry in the *Chongzhen lishu*   204

Logarithms in China   208

The Kangxi Period and Mei Wending   213

Dai Zhen: Philology Encounters Mathematics   222

Infinite Series   227

▪ Text 4.3 Mei Juecheng, On Calculating the Circumference of a
Circle from Its Diameter   228

▪ Text 4.4 Minggatu, On Calculating the Chord of a Given Arc   231

5. **Europe After Euler**   243

Normal Science: Gap Filling in Spherical Trigonometry   244

▪ Text 5.1 Pingré, Extending Napier's Rules to Oblique
Spherical Triangles   245

Symmetry and Unity   253

The Return of Stereographic Projection   255

Surveying and Legendre's Theorem   260

Trigonometry in Navigation   264

▪ Text 5.2 James Andrew, Solving the *PZX* Triangle Using
Haversines   268

Tables   273

Fourier Series   281

▪ Text 5.3 Jean Baptiste Joseph Fourier, A Trigonometric
Series as a Function   287

Concerns About Negativity   290

Hyperbolic Trigonometry   294

▪ Text 5.4 Vincenzo Riccati, The Invention of the
Hyperbolic Functions   294

Education   303

Concluding Remarks   314

Bibliography   317

Index   363

Preface ☀

I am grateful to the readers of the first volume of my history of trigonometry, *The Mathematics of the Heavens and the Earth: The Early History of Trigonometry* (2009), for their warm reception of my work. This second volume, 12 years later, has revealed to me only well after I was committed to the project the magnitude of this endeavor. The effort required to research the dizzying network of topics and subdisciplines involved in the trigonometric enterprise was at times overwhelming. I stand amazed at the virtuosity of my predecessor Anton von Braunmühl, who was able to compile what was truly a comprehensive masterwork (*Vorlesungen zur Geschichte der Trigonometrie*) in his 1900 and 1903 volumes on the subject.

Von Braunmühl's accomplishment made it possible to establish a framework for the history of trigonometry for many decades, and his work is still quoted occasionally 120 years later. However, a great deal has changed in that time. The history of mathematics as a discipline has gone through sea changes, often relegating excellent work by past historians to history itself. Prominent among those changes with respect to trigonometry is our greatly increased knowledge of the extensive contributions from outside the Western canon. Although there is a lifetime or more of work to do before these cultures are understood as well as the West, it is a salutary sign of our growth in these areas that von Braunmühl devoted 55 pages to India, Persia, and Islam (still a remarkable piece of work, given the limited literature to which he had access at his time) while *The Mathematics of the Heavens and the Earth* takes almost 130 pages.

As significant as the measure of attention that we now pay to non-Western cultures is the way that we write about them. The past 50 years have seen a fundamental shift in our view of the mathematics of "the other," and this ground is still controversial today. The question at the heart of the matter: to what extent does a representation of historical mathematics in modern terms distort how the historical actors saw themselves and their subject? It is safe to say that today, most historians would respond with more feeling than they did half a century ago. As a result, it has become almost an axiom for many authors to portray historical mathematics as much as possible in the terms and language of the originating culture.

Finally, what precisely one means by "mathematics" has broadened considerably. We have recognized that the naïve definition of mathematics as a formal science reflects the diversity of historical mathematical practices rather poorly. The boundary between mathematics on the one hand, and so-called "applications" like science and the trades on the other, has become blurred—

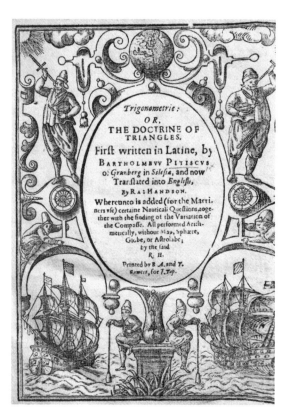

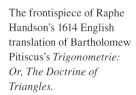

The frontispiece of Raphe Handson's 1614 English translation of Bartholomew Pitiscus's *Trigonometrie: Or, The Doctrine of Triangles.*

admittedly, more so in some subdisciplines than in others. In addition, mathematical activities outside of traditional academic practice (whatever that means) are being recognized increasingly as valid subjects of historical discourse and brought into the light. This has enriched the discipline to include, for instance, ethnomathematics and the mathematics of practitioners and students, not just the theoretical pinnacles. A positive effect of this broadening is the inclusion of historically disadvantaged groups in the dialogue.

Each of these developments has had an impact on this book: some more than others, and some perhaps less than I would have liked. As with *The Mathematics of the Heavens and the Earth*, my first goal is reportorial rather than polemical. That is, I hope to bring the existing research together, returning it to accessibility so that historians may once again use it in their narratives. Some of that research, often up to a century old, has not been fashionable for some while. Nevertheless, especially when gathered together, it opens up an important part of our history that deserves more attention than it has recently received.

Since von Braunmühl's volumes, the history of trigonometry (especially in the West) has moved gradually to the scholarly background. One might speculate on reasons for this: other themes in Western mathematics have attracted more attention; trigonometry itself has receded into the mathematical background; or perhaps trigonometric developments frequently came about in practitioner circles rather than academic ones. Part of the issue may be that much of the trigonometry practiced in the West between 1550 and 1900 is simply no longer known to many historians or mathematicians. How many of us today know anything about advanced spherical trigonometry? Whatever the reason, it seems valuable to reintroduce this history to the current generation of scholars and learners. As a reintroduction, then, this book reflects the existing research and does not intend to argue for a specific thesis or advance new historiographies.

That is not to say that historiographic concerns are not reflected here. For instance, our narrative often strays outside the bounds of mathematics, questioning the extent to which the history of mathematics may be said to have boundaries at all. Surveyors and navigators in particular play leading roles in this story, and their practical concerns reflected their own nonmathematical goals. The fact that some of their accomplishments are not well connected to recent publications illustrates a second concern. The bias of the modern author appears not only in how they write the history but also in what they choose to write. In this regard, much press has been devoted, and with good reason, to the unfortunate effects of Eurocentrism. However, as I researched this book I also found large parts of Western trigonometry that have been forgotten or never noticed by historians in the first place—especially from times before 1600, from disciplines allied with but not part of mathematics and from mathematical topics outside the modern curriculum. These huge gaps illustrate another way in which, subconsciously, we tend to write royal roads to us. To address this problem, I have given space, for example, to developments in spherical trigonometry that likely are outside the direct knowledge of many readers. My goal is to focus on what the historical actors found important rather than what we recognize today.

The last choice was whether to write a scientific or, more popular these days, a social history. *The Mathematics of the Heavens and the Earth* was primarily scientific history. The main reason for this was a lack of documentary evidence to support a detailed social narrative over the cultures and times being described. That evidence is more available for some of the periods covered in this book, although not much research of this sort has been done as yet. I have moved a little in the direction of social history, especially with respect to the teaching of trigonometry, but have continued to lean into the science for two reasons. The first is to retain narrative coherence with *The Mathematics of the Heavens and the Earth*; the second is that most of the existing secondary literature is scientific rather than social. There is plenty of room to write a social history of trigonometry; the history of mathematics is more than wide enough to permit multiple approaches, and I would be delighted to see another take on the subject. This is the reason for a change in the first word of the subtitle of this book, as opposed to the first volume: "*A History of Modern Trigonometry.*"

A few words on the boundaries of this study. One could end writing on one's favorite topic forever; however, eventually one must stop. Some of the most interesting current research in the history of trigonometry concerns developments in Islam and in India after the time period of *The Mathematics of the Heavens and the Earth*, including newly discovered indigenous contributions to both planar and spherical trigonometry. Especially fascinating is a wealth of material gradually becoming available on Indian appropriations of both Islamic trigonometrical methods and European logarithms of trigonometric functions up to the eighteenth century.[1] This includes writings on the construction of trigonometric tables in 1639 by Mughal astronomer Nityānanda[2] and two anonymous texts on the determination of the Sine of 1° in the court of Jai Singh in the early eighteenth century,[3] among other developments. At the moment this story is still emerging from the manuscripts, and scholars are not yet settled on how one should properly embed these events in the various cultural histories that they intersect. It seems very likely that more manuscripts will bring to light a much richer account than we know now. Soon it may be possible to write another chapter, perhaps as a supplement to this volume, that deals thoroughly with this and other topics that I regretfully have had to pass over. In the meantime, I eagerly await the results of research in this area.

Boundaries had to be drawn also in the chapters focusing on European trigonometry. It would have been possible to go into greater depth on applications like astronomy and navigation and into mathematical topics like cyclometry. Chapters 2 (Logarithms) and 3 (Calculus) are not about logarithms and calculus themselves but restrict their attention to how trigonometry interacted with the new conceptions and applications provoked by these new ideas. Some purely trigonometric events occur within these chapters as well,

---

[1] On logarithms in early eighteenth-century India, see among others [Pingree 1999], [Pingree 2002a], and [Pingree 2002b].
[2] See [Montelle/Ramasubramanian/Dhammaloka 2016] and [Montelle/Ramasubramanian 2018] as well as work to appear by A. J. Misra.
[3] An announcement of this content may be found in [Plofker 2017]; an edition and analysis of the text is being prepared by Plofker, Montelle, and Van Brummelen.

captured in our text by the time periods in which they occurred. This account ends around the year 1900, more or less, with a couple of incursions into the twentieth century.

Finally, the fourth chapter of this book fulfills a promise I made in the preface of *The Mathematics of the Heavens and the Earth* (p. xiv): to treat the history of Chinese trigonometry with the attention and thoroughness that it deserves. The placement of this chapter was a problem with no solution; it would disrupt the narrative wherever I included it. Since it deals occasionally with the introduction of concepts from the European trigonometry of the first three chapters, I placed it where it is now. Of course, the reader may feel free to read these chapters in any order.

This is the end of a long and rewarding journey that turned out to be much deeper and broader than I had expected. I hope that bringing it back to the public eye will provoke others to explore directions suggested here and help to enrich the history of mathematics and allied disciplines.

A project of this magnitude does not reach completion without the indispensable assistance of countless people and organizations. Financial support came primarily from two sources: Parker Collier, whose inspiring interest in the history of mathematics was a breath of fresh air from well outside the world of academia; and the Smithsonian Institution, which supported me through a fellowship in the Dibner Library Resident Scholar Program. My time at the Dibner Library in the National Museum of American History was where the book really took shape. Several colleagues welcomed me warmly and provided a collegial environment that helped make the absence from home tolerable, including Division of Medicine and Science Curator Deborah Warner and her husband Jack, Peggy Aldrich Kidwell (also a curator in the Division of Medicine and Science), Amy Shell-Gellasch, and Amy Ackerberg-Hastings. The staff of the Dibner Library, Lilla Vekerdy and Kirsten van der Veen, made the biggest impact; they were daily companions as I explored the incredible resources of the Dibner collection. I also had the opportunity to spend two formative months working at the history department of the California Institute of Technology where regular lunches and meetings with the faculty were invaluable. Jed Buchwald and Mordechai Feingold, and especially Noel Swerdlow, provided a rich scholarly community and crucial input in an early stage of this work. I was also able to use the resources of the nearby Huntington Library, for which I am most grateful. Various people have selflessly provided commentaries and critiques that have helped to improve the manuscript immeasurably, including Rob Bradley, Jeff Chen, Dominic Klyve, Arnaud Michel, Fred Rickey, Fei Shi, Joel Silverberg, and three anonymous reviewers. My original editor Vickie Kearn, who retired while I was completing this manuscript, was as supportive as one could hope as well as being a good friend. Her successor Susannah Shoemaker had big shoes to fill and has done so more than admirably. In the end, as always, all remaining short-comings fall on my shoulders. In the book's production, Melody Negron was outstanding, dealing marvelously with a book that must have been an unusually difficult challenge to typeset properly.

This project has been an enormous effort but has been worth every minute. I apologize to readers who hoped to see it much sooner. Here, however belatedly, is volume 2.

✦ The Doctrine of Triangles

1 ✼ European Trigonometry Comes of Age (1552–1613)

The subject we know today as trigonometry has a long, complex history that weaves through several major cultures and more than two millennia. Perhaps more than any other subject in the modern mathematics curriculum, trigonometry has been shaped, has been reconfigured, and gone through metamorphoses several times. Born of needs in ancient astronomy, it has been repurposed by many scientific disciplines and worked to serve several cultural and religious perspectives. It has been a participant, active or passive, in many of humanity's most significant scientific pursuits. The tidy, polished package found in today's high school and university textbooks camouflages a tangled story that interacts with many themes in the history of science, often with implications for some of the most transformative moments in our and other cultures.

I told the first half of this story in *The Mathematics of the Heavens and the Earth: The Early History of Trigonometry*.[1] This volume narrates the second half, but we begin with a brief summary of what went before. Trigonometry began with Greek astronomers such as Hipparchus of Rhodes, who had constructed geometric models of the motions of the sun and moon that reproduced qualitatively the phenomena he witnessed in the sky. Converting these models into tools for prediction of events like eclipses required the translation of their geometric components into numerical measures. Since these components were lines and circles, it quickly became necessary to convert the magnitudes of circular arcs into lengths of line segments and vice versa. Hence the chord function was formulated,[2] giving the astronomer the ability to compute the length of a chord within a circle given the magnitude of the arc that it spans. The earliest table of chords of which we are aware was constructed by Hipparchus; the earliest account of the construction of chord tables is in Claudius Ptolemy's *Almagest*. The mathematical preparation for astronomy began with these chords and grew from there. However, since the geometric arena was often the celestial sphere rather than a flat surface, plane trigonometry was only the beginning. Perhaps already from the time of

---

[1] [Van Brummelen 2009].

[2] The term "function" has a long and complicated history. Properly speaking, according to the term's modern usage, it is an anachronism to refer to functions at all before the modern period. However, there is an affinity at least between ancient numerical tables and our use of the term: ancient astronomers found the length of the chord of a given arc by inputting the numerical value of that arc into a table and treating the value obtained as an output. In this book the word "function" is used in this loose sense, unless stated otherwise.

Hipparchus, astronomers quickly moved from the plane to the sphere, where much of the most important work was done.

The first major transformation occurred with the complicated and controversial transmission of mathematical astronomy from Greece to India. The early Indian astronomers' appropriation of the geometric models of the planets, much more than a simple transmission of knowledge (but a topic for another book), also extended to many new ways of thinking in trigonometry. The most obvious effect of the transformation of trigonometry in India is the introduction of the sine function: a slightly less intuitive quantity from a geometric point of view but a more efficient tool for astronomical computation. The versed sine followed quickly afterward. The inventions of new mathematical methods to work with these functions, such as iterative solutions to equations and higher-order interpolation within numerical tables, greatly enriched mathematical astronomy. In the fourteenth and fifteenth centuries, astronomers even employed infinitesimal arguments that we recognize today as related to calculus to derive several powerful results beneficial to astronomy, most famously the Taylor series for the sine and cosine.

The reception and naturalization of trigonometry in medieval Islam is no less complicated. In the eighth and ninth centuries Indian astronomy found its way through Persia to Baghdad. As interest grew, a translation movement brought a fresh crop of Greek texts to Islamic scholars. This produced the curious circumstance that two approaches to astronomy, both of which contained at least some trace of Greek origin, were in opposition to each other. The Greek texts gradually took precedence during the ninth and tenth centuries, but many of the Indian advances (including the sine and iterative methods) were retained. Around the end of the tenth century several advances streamlined eastern Islamic trigonometry. The tangent, invented in the process of sundial construction, became part of the trigonometric toolkit. New theorems reformulated the foundations of spherical trigonometry and delivered greater power to both astronomy and astrology. Trigonometry was also applied to new contexts, including ritual needs like determining the beginning of the month of Ramadan and the direction of prayer toward Mecca. Some of the work done on the latter problem became a standard tool in mathematical geography, bringing trigonometry down from the heavens to the earth for the first time.

From the tenth century onward, Islamic science gradually diversified according to cultural subgroups spread across its vast geographical area. The most prominent division was between eastern Islam and al-Andalus, in what is now Spain. Andalusian mathematical astronomy retained Indian and Greek influences, but after AD 1000 it developed without much conversation with the East. Rather, their knowledge spread northward into Europe, especially through the Toledan and Alfonsine Tables. Some innovations in trigonometry occurred in medieval Europe, sometimes through interactions with practical geometry and with astronomical instruments. However, the fifteenth century saw the beginning of tremendous growth through the theoretical astronomy of people such as Giovanni Bianchini (ca. 1410–1469) and Regiomontanus (1436–1476). This period set in motion the events that we shall survey in this chapter.

It is a reflection of the richness of the history of trigonometry that after more than one and a half millennia of years of progress, in the year 1550 the word itself was still 50 years away from being coined. Indeed, triangles did not really emerge as the primitive objects of study until Regiomontanus's *De triangulis omnimodis* ("Concerning Triangles of Every Kind") became popular in the mid-sixteenth century. This volume's title, *The Doctrine of Triangles*, is taken from one of the names that was given to trigonometry in the sixteenth and seventeenth centuries.

▨ What's in a Name?

By 1550, the central problem of trigonometry—determining lengths in geometric diagrams from corresponding circular arcs and vice versa—had long been solved. European astronomers had within their grasp a somewhat compact theory that allowed them to solve every problem that they needed to solve, both on the plane and on the sphere. Regiomontanus's *De triangulis omnimodis*, written in the fifteenth century but published in 1533,[3] provided a unified source for the mathematical methods and most (although not quite all) of the fundamental theorems. Sine tables composed by Regiomontanus and others provided a straightforward tool for working out the practical calculations. Seemingly, there was not much left to do.

However, there was a great deal left to do. Over the next 50 years, the mathematical structure and even the basic notions of trigonometry were overhauled. New theorems were discovered, and more elegant and efficient ways of organizing the material were found. By the beginning of the seventeenth century, new ways to employ the subject, both within science and outside of it, were being devised with regularity. Even the basic functions, the fundamental building blocks of trigonometry, went through multiple reinventions. By 1613, the subject no longer looked much like Regiomontanus's *De triangulis omnimodis*.

We may begin to get a sense of the contrast by comparing basic definitions in the works of two of the dominant figures in the mid-sixteenth century, Regiomontanus and Rheticus. We start with Regiomontanus's *De triangulis omnimodis*.

---

[3] [Regiomontanus 1533]; see also the edition [Regiomontanus 1561]. *De triangulis* has been translated in [Regiomontanus (Hughes) 1967]. Finally, see [de Siebenthal 1993, chapter 5, 268–352] for an account of the mathematics in French.

Text 1.1
Regiomontanus, Defining the Basic Trigonometric Functions
(from *De triangulis omnimodis*)

Definitions:

. . .

An *arc* is a part of the circumference of a circle.

The straight line coterminous with the arc is usually called its *chord*.

When the arc and its chord are bisected, we call that half-chord the *right sine*[4] of the half-arc.

Furthermore, the *complement* of any *arc* is the difference between [the arc] itself and a quadrant.

The *complement* of an *angle* is the difference between [the angle] itself and a right angle.

Book I, Theorem 20: In every right triangle, one of whose acute vertices becomes the center of a circle and whose [hypotenuse] its radius, the side subtending this acute angle is the right sine of the arc adjacent to that [side and] opposite the given angle, and the third side of the triangle is equal to the sine of the complement of the arc.[5]

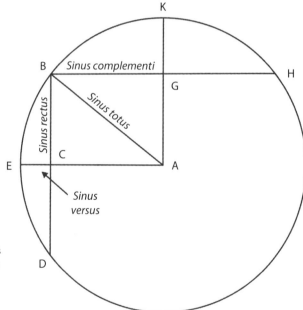

Figure 1.1
Regiomontanus's definitions of the primitive trigonometric functions.

---

[4] In Latin, *sinum rectum*.
[5] [Regiomontanus (Hughes) 1967, 31, 59].

Explanation: In right-angled triangle *ABC* (figure 1.1), draw a circle centered at *A* with radius *AB*. Draw *AK* vertically, *BC* parallel to *AK*, and *BH* parallel to *AC*; and extend *AC* to *E* and *BC* to *D*. Several differences between Regiomontanus's structure and the modern definition are apparent. Firstly, following his predecessors, he defines the trigonometric functions as lengths of line segments in the diagrams, not as ratios. Secondly, again following convention, he relies on the ancient Greek chord function by defining the sine *BC* (*sinus rectus*) as half the length of the chord *BD*. Thirdly, he allows the radius *R* of the base circle to be any chosen value. In the *De triangulis* Regiomontanus at times uses $R = 60{,}000$ but at other times uses $R = 10{,}000{,}000$. Such large radii were chosen to avoid having to work with decimal fractions.

Regiomontanus calls the circle's radius $R = AB$ the *sinus totus*, a term used already in medieval Islam that represents the greatest possible sine value. The right sine of $\widehat{BE}$ is *BC*; in modern terms, $\mathrm{Sin}(\widehat{BE}) = R\sin\widehat{BE} = BC$.[6] This is the only function used in most of the *De triangulis*. What we call the cosine is called simply the *sinus complementi*, the sine of the complement of the given arc. Near the end of the book Regiomontanus uses the versed sine, the *sinus versus EC*, the difference between the *sinus totus* and the *sinus rectus*. This function originated in India.

Just like Ptolemy's *Almagest* a millennium and a half earlier, the *De triangulis* lacks an equivalent to the tangent function. In I.28, Regiomontanus describes how to find an angle in a right triangle if the ratio between two sides is known, a simple but nontrivial process if one does not have a tangent. But Regiomontanus did not have long to wait. In his popular collection of tables for spherical astronomy, the *Tabulae directionum* ("Tables of directions"),[7] he borrowed several tables from his predecessor Giovanni Bianchini to solve stellar coordinate conversion problems.[8] One of these tables, repeatedly borrowed in turn by various successors, was recognized as useful in many other calculations, hence the name bestowed on it by Regiomontanus, the *tabula fecunda* ("fruitful table"). Mathematically equivalent to the tangent, it would become accepted gradually as a full-fledged trigonometric function on its own.

Regiomontanus was the most frequently quoted trigonometer of the sixteenth century, and we shall see more of his influence later in this chapter. His definitions and terms, most of them not original to him but spread by him, became the foundation of the field. One of his early adopters was Erasmus

[6] Here and throughout, we capitalize a trigonometric function if it is used with a circle with $R \neq 1$.
[7] See [Van Brummelen 2009, 261–263], as well as [Delambre 1819, 292–293] and [Folkerts 1977, 234–236].
[8] [Van Brummelen 2018].

Reinhold (1511–1553), one of the best quantitative astronomers of his generation. A colleague of Georg Rheticus at the University of Wittenberg, Reinhold was one of the first to receive a copy of Copernicus's work. Reinhold is most known for his very successful astronomical *Prutenic Tables*, but more relevant to us is his posthumous 1554 *Tabularum directionum.*[9] This collection of tables is an expansion of Regiomontanus's work of the same name and includes a tangent table ("*canon fecundus*") greatly expanded from Regiomontanus's. This table gives values to at least seven places for every minute of arc from 0° to 89° and for every 10 seconds of arc between 89° and 90° where the values change rapidly from entry to entry.[10] To give the reader a sense of calculations in typical astronomical work of the time, we provide a short passage of his commentary on Copernicus, one of many where Reinhold uses his tangent table.

Text 1.2
Reinhold, a Calculation in a Planetary Model Using Sines and Tangents
(from Reinhold's commentary on Copernicus's *De revolutionibus*)

Likewise, because angle *FEN* is 39°37′38″, therefore in right [triangle] *EPL* the remaining angle of *LEP* [that is, angle *ELP*] is 50°22′22″; and when *EL* is 100,000, then *LP* is 63,779 and *PE* is 77,021. And now when *EL* is taken to be 5,943, such that it is half the eccentricity, then *LP* is 3,790 and *EP* is 4,577. And from here, their doubles are *DQ* = 7,580 and *EQ* = 9,154, when *EN* . . . is 100,000. Therefore, the whole of these, *QEN*, is 109,154. And with *QN* taken to be 10,000,000, then *QD* is 694,432. And from our table, angle *DNQ* is 3°58′21″.[11]

Explanation: (See figure 1.2.) In the figure, *D* is the center of the universe and *E* is the center of the topmost eccentric deferent circle.

  Reinhold knows that $\angle FEN = 39°37′38″$ and wants to find $\angle QND$. Firstly, since $\angle FEN = \angle PEL$ and $\angle EPL$ is a right angle, $\angle ELP = 90° - 39°37′38″ = 50°22′22″$. Next, in right-angled triangle *EPL*, Reinhold sets the hypotenuse $R = 100,000$. This allows him to use his Sine table; he finds $LP = \text{Cos } \angle ELP = R \cos \angle ELP = 63779$ and $PE = \text{Sin } \angle ELP = R \sin \angle ELP = 63779$. But *EL* is a known parameter with value 5,943, so *LP* and *EP* are scaled downward to 3,790 and 4,577, adjusting from the hypotenuse of 100,000 assumed by the Sine table to a hypotenuse of 5,943. Now, the astronomical model assumes that $EL = LD$, so the sides of triangle *DQE* are double those of $\triangle LPE$, which

---

[9] [Reinhold 1554]. The "canon fecundus" may be found on folios 17 through 51.

[10] The values in the table stray significantly away from the correct ones as the argument approaches 90°, a problem that plagued both medieval Islamic and especially early European table makers. See the account of Rheticus, Romanus, and Pitiscus in [Van Brummelen 2009, 280–282]. See also the analysis of early European tangent tables in [Pritchard, forthcoming].

[11] [Nobis/Pastori 2002, 246–247]. Translated from the Latin.

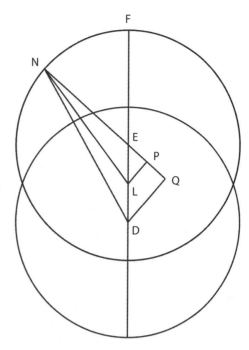

Figure 1.2
Reinhold's calculation with a
planetary model using the
tangent.

make $DQ = 2LP = 7,580$ and $EQ = 2EP = 9,154$. But EN is the radius of the circle, previously set to 100,000; therefore $QEN = 109,154$. Finally, consider right triangle NQD. Reinhold's *Canon fecundus* uses a radius of 10,000,000, so he sets QEN (the side adjacent to the angle we seek) equal to that value rather than 109,154. This requires him to adjust DQ's value accordingly, from 7,580 upward to 694,432. He can now look up this value in the *Canon fecundus* (figure 1.3); we can see for ourselves that $\angle QND$ is between 3°58′ and 3°59′.

Clearly the tangent has come a long way from its initial role as a helper to Bianchini and Regiomontanus in solving stellar coordinate problems. Reinhold is now using his *Canon fecundus* as a general purpose tool for dealing with arbitrary right triangles.

The approach shared by Regiomontanus and Reinhold, dominant in the sixteenth century, was opposed by Georg Rheticus (1514–1574). Known as the man who discovered Copernicus and convinced him to publish his heliocentric theory, Rheticus hailed from the region of Rhaetia, which overlaps Austria, Switzerland, and Germany.[12] In his mid-twenties he visited Copernicus and became his student; he announced the heliocentric theory in his

---

[12] We have already discussed Rheticus and Copernicus in [Van Brummelen 2009, 273–282]. For more on Rheticus, see [Burmeister 1967–1968] and [Danielson 2006].

FŒCVNDVS.   18

	O	1	2	3	
30	87168	261859	436609	611625	30
31	90177	264770	439523	614544	29
32	93086	267681	442438	617464	28
33	95995	270592	445353	620384	27
34	98904	273503	448267	623304	26
35	101814	276414	451182 (2915)	626225	25
36	104723	279325	454097	629145	24
37	107632	282237	457012	632066	23
38	110541	285148	459927	634986 (2921)	22
39	113450	288059 (2912)	462842	637907	21
40	116360 (2910)	290970	465757	640818	20
41	119269	293882	468672	643749	19
42	122178	296794	471588	646671	18
43	125088	299705	474503	649592	17
44	127997	302617	477419	652514	16
45	130906	305528	480335	655435 (2922)	15
46	133816	308439	483251	658357	14
47	136725	311351	486166	661278	13
48	139635	314262	489082	664200	12
49	141544	317174	491997 (2916)	667121	11
50	145454	320085	494013	670043	10
51	148363	322997	497829	672565	9
52	151173	325909	500745	675888 (2923)	8
53	154182	328821	503662	678810	7
54	159092	331733	506578	681733	6
55	160001	334645	509495	684656	5
56	162911	337558	512411	687578	4
57	165820	340470	515328	690501	3
58	168730	343382	518244	693423	2
59	171630	346295 (2913)	521161	696346	1
60	174550	349207	524078	699269	0
	89	88	87	86	

Figure 1.3
A page from Reinhold's *Canon fecundus*. This section gives tangents from 0° to 4°, and cotangents from 86° to 90°. This page includes tangent values for arcs with minute values between 30′ and 60′; the grid on the facing page gives values for arcs with minute values between 0′ and 30′.

Narratio prima and helped Copernicus bring his *De revolutionibus* (and separately its trigonometry under the title *De lateribus et angulis triangulorum*, "On the Sides and Angles of Triangles")[13] to press.

Rheticus's accomplishments after Copernicus's death in 1543 are primarily trigonometric, especially in the design and production of tables. His short 1551 tract *Canon doctrinae triangulorum* ("Table of the Doctrines of Triangles"),[14] consisting of nothing more than a short introductory poem, 14 pages of tables, and a six-page dialogue, seems at first glance unassuming. But within its pages one finds not only tables of all six trigonometric func-

---

[13] [Copernicus 1542]. For an account of the trigonometry in this treatise (which is not very original), see [Swerdlow/Neugebauer 1984, part 1, 99–104]. See also [Rosińska 1983], which argues that the sine table in this work was computed by Copernicus himself but corrected by Rheticus based on Regiomontanus's tables.

[14] [Rheticus 1551]. This treatise has an unusual history. Since it was placed on the *Index expurgatorius* (and since Rheticus's later work, the *Opus palatinum*, rendered it obsolete), it disappeared from view after the sixteenth century. It was rediscovered by Augustus De Morgan in the mid-nineteenth century. See [De Morgan 1845], [Hunrath 1899], [Archibald 1949b], and [Archibald 1953]. [Roegel 2011d] contains a recomputation of all of its tables.

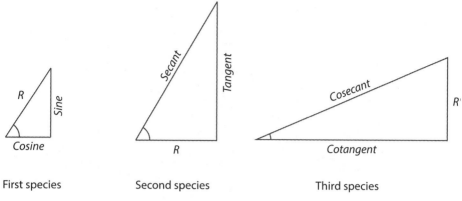

Figure 1.4
Rheticus's six trigonometric functions.

tions now considered standard (sine, cosine, secant, tangent, cosecant, and cotangent) but also a completely new and elegant set of terminology to describe them. Consider three "species" of right triangles (figure 1.4) described with respect to a given radius R.[15] In the first species the hypotenuse is set equal to R; in the second, the base; and in the third, the perpendicular. Then:

- in the first species, we have two functions, the perpendicular and the base (equivalent to Sine and Cosine respectively);
- in the second species, we have the hypotenuse and the perpendicular (Secant and Tangent); and
- in the third species, we have the hypotenuse and the base (Cosecant and Cotangent).

When Rheticus solves triangles, circles play no role. Thus, Rheticus's system not only defines all six trigonometric functions compactly but also divorces them from circular arcs: the arguments are now simply angles within the triangles, as they are today.

Rheticus found posthumous support for his design in the writings of the possibly the most well-known mathematician of the sixteenth century, François Viète (1540–1603). Viète's career was in the French civil service—not mathematics, on which he worked in his spare time. As a Huguenot during a time of unrest between Catholics and Protestants in France, his position was often hardly stable. He lived through an authorized massacre of Huguenots (which claimed the life of his older colleague Peter Ramus) and five years of

[15] In the *Canon doctrinae triangulorum* Rheticus sets $R = 10,000,000$; in the *Opus palatinum*, $R = 10,000,000,000$.

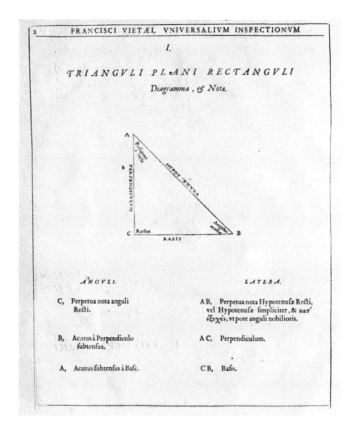

Figure 1.5
A page from Viète's *Canon mathematicus seu ad triangula* (1579),
naming the sides and angles of a right triangle.

banishment from Paris, during which he worked on his mathematics. His interests were diverse, including astronomy and cryptography; but today he is recognized most for his contributions to the revolution of symbolic algebra, especially his *In artem analyticam isagoge*.[16]

While Viète's role in transforming algebra was fundamental, he was also deeply involved in the evolution of trigonometry. His first mathematical work, *Canon mathematicus seu ad triangula* ("Mathematical Canon, or On Trian-

---

[16] For editions and translations of Viète's mathematical treatises, see [Viète 1646; 1983] and [Viète/Girard/de Beaune 1986]; [van Egmond 1985] is a catalog of his works. None of these books contains *Canon mathematicus seu ad triangula* [Viète 1579], which occupies our attention here. See also [Ritter 1895] and [Reich/Gericke 1973]; the latter contains accounts of several of Viète's works in algebra. The secondary literature on Viète's role in the transformation of algebra is too large to be summarized here.

	Hypotenusa	Perpendiculum	Basis
I.	Totus	Sinus Anguli, vel Peripheriae (*sine*)	Sinus anguli Reliqui, seu Residuae peripheriae (*cosine*)
II.	Hypotenusa Faecundi Anguli, vel Peripheriae (*secant*)	Faecundus Anguli, vel Peripheriae (*tangent*)	Totus
III.	Hypotenusa Faecundi anguli Reliqui, vel Residuae peripheriae (*cosecant*)	Totus	Faecundus anguli Reliqui, vel Residuae peripheriae (*cotangent*)

Figure 1.6
Viète's nomenclature for the six trigonometric functions, taken from page 16 of *Universalium inspectionum* of his *Canon mathematicus seu ad triangula*. The Roman numerals on the left refer to Rheticus's triangle species.

gles," 1579),[17] is an unusual volume—as close as it comes to being a coffee table book on trigonometry. For instance, the first page of text (figure 1.5) lays out the names of the sides and angles of a right-angled triangle with an eye to filling the page in a pleasing way. The book begins with a set of trigonometric tables designed according to the methods of Rheticus's *Canon doctrinae triangulorum*, with all six functions grouped according to the three triangle species we saw in figure 1.4. Although his names for the various functions often vary (see figure 1.6) and borrow the term *fecunda* from Regiomontanus, the structure clearly imitates that of Rheticus.[18]

Most of Viète's colleagues and contemporaries, however, were content to stick with the language of Regiomontanus.[19] For instance, only eight years after the *De triangulis omnimodis* was published, the great German astronomical

---

[17] See [Viète 1579], [Hunrath 1899], and [Rosenfeld 1988, 24–27]. See also [Roegel 2011g] for a recomputation of the tables.

[18] See page 16 of the *Universalium inspectionum* within [Viète 1579], and [Ritter 1895, 40]. Viète applies the term *fecunda* to several quantities.

[19] [Von Braunmühl 1900/1903, vol. 1, 183] suggests that Viète's unique notation here and elsewhere, brilliant as it was, may have contributed to his colleagues' lack of appetite for his trigonometric inventions. But Rheticus and Viète were not without followers; Adrianus Romanus's *Canon triangulorum* [Romanus 1609], for instance, adopts some of Viète's structure and terminology, including the terms "transsinuousae" for the secant and "prosinus" for the tangent (even though the standard terms are on the title page).

and geographical instrument maker Peter Apian (1495–1552)[20] had followed with his 1541 *Instrumentum sinuum seu primi mobilis*, a well-known treatise on trigonometric instruments and their use in solving various astronomical problems, which we shall consider later. Apian uses names that would have been familiar to Regiomontanus and his colleagues: the *sinus rectus primus* for the sine and the *sinus rectus secundus* for the cosine.[21] There is no reference to Regiomontanus's *tabula fecunda* or indeed to anything resembling a tangent function.

Apian's traditional names for the sine and cosine are found again in the 1558 collection of works on spherical astronomy[22] by Francesco Maurolico (1494–1575). A Sicilian priest, Maurolico held a variety of civil positions over the course of his life, including master of the mint, and was eventually appointed professor at the University of Messina. He was active in a wide variety of areas of mathematics and science, including optics and music; within astronomy he was especially prolific in spherical astronomy and edited several Greek works on the subject. Although he does not define the tangent and cotangent directly in his book on spherics, they do appear as *umbra versa* and *umbra recta* in Book II, Proposition 30,[23] as they often had before. These terms derive from ancient and medieval references to "shadows" in sundials, and Maurolico himself defines the *umbra versa* and *umbra recta* in this way in his astronomical treatise *De sphaera,* a work infamous for his vicious condemnation of Copernicus.[24] However, as we noted earlier, it was not from the *umbra versa* and *umbra recta* that the modern tangent and cotangent evolved.

We do find one innovation in Maurolico's work on spherics. Near the end he describes a new table as follows: "In imitation of the *tabula fecunda* of Johannes Regiomontanus, we made another table which we have named *benefica*, because certain calculations become easy by means of this table."[25]

---

[20] For a general introduction to Apian's mathematics see [Kaunzner 1997]; for his trigonometry see [Folkerts 1997].

[21] See the third page of the first section of [Apian 1541], *Instrumentum hoc primi mobilis componere.*

[22] [Maurolico 1558] (on which see [Moscheo 1992] on editorial issues) includes Latin editions of Theodosius's *Spherics*, Menelaus's *Spherics*, Autolycus's *Spherics*, Theodosius's *De habilitationibus*, and Euclid's *Phenomena* as well as several small trigonometric tables (sine, *tabula fecunda, tabula benefica,* and declinations and ascensions) and a *Compendium mathematicae.* On Maurolico's sources for his edition of Menelaus, see [Taha/Pinel 1997] or [Taha/Pinel 2001]. See also [Napoli 1876] for an edition of Maurolico's *Geometricarum quaestionum.* [Rose 1975, 159–184] is a good account of Maurolico's life and work.

[23] [Maurolico 1558, f. 58].

[24] *De sphaera* is the first of a number of short treatises in *Opuscula mathematica,* [Maurolico 1575]; the definitions of *umbra versa* and *umbra recta* may be found on page 13. For Maurolico's attack on Copernicus, see [Rosen 1957].

[25] [Maurolico 1558, f. 60], *Demonstratio tabulae beneficae.*

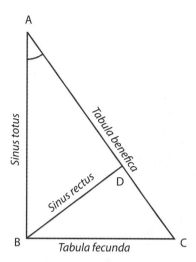

Figure 1.7
Maurolico's trigonometric functions.

Maurolico's new table introduces what today we call the secant.[26] A short table at the end of the book[27] gives secant values, with $R = 100,000$, for integer arguments from 1° to 89°. Rheticus, of course, had already published tables of all six trigonometric functions seven years earlier in his *Canon doctrinae triangulorum*. But he had used his own unique terms and definitions, which make no appearance in Maurolico's work.[28] Instead, consider Maurolico's figure 1.7: within right triangle *ABC*, segment *BD* is perpendicular to *AC*. Set *AB* equal to $R=100,000$. Then, given ∡*A* at the top of the diagram, we may find *BD* from a table of sines, *BC* from the *tabula fecunda*, and *AC* from the *tabula benefica*.

It took another quarter century for the *tabula fecunda* and *tabula benefica* to take on their modern names of tangent and secant in Danish scholar Thomas Fincke's (1561–1656) *Geometriae rotundi* ("Geometry of Circles and Spheres").[29] Still a 22-year-old student in 1583 at its publication, Fincke switched to the study of medicine that same year. Over the course of his very long career, he held professorial positions in medicine, rhetoric, and mathematics and held a number of senior administrative posts (including rector and

---

[26] Copernicus composed a table of secants by hand, but it was never published. See [Glowatzki/ Göttsche 1990, 190–192]. For an analysis of Maurolico's table, see [Van Brummelen/Byrne, forthcoming].

[27] Folio 66. As we shall see later, a controversy arose over whether Maurolico's table owed an unpaid debt to Rheticus.

[28] Here we differ from von Braunmuhl's opinion that Maurolico was following Rheticus; see [von Braunmühl 1900/1903, I, 150–151].

[29] [Fincke 1583]. De Morgan first makes this identification in [De Morgan 1846]. See [Schönbeck 2004] for a detailed account of Fincke's life and a summary of the *Geometriae rotundi*.

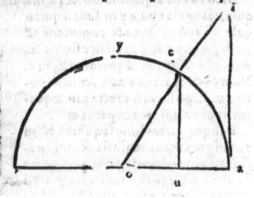

22. Tangens est à termino peripheriæ altero perpendicularis in radium extra per reliquum terminum continuatum.

Esto peripheria a e.sitq; ab ejus termino a perpendicularis a i terminata in radio o e continuato in i. per reliquum datæ peripheriæ terminum. erit a i tangens datæ peripheriæ. Sic vocare pla cuit quia sit perpendicularis extremæ diametro. hoc dato per 1.e.2 erit tangens:itaq; Geometria ipsa commodum suppeditavit nomen : nec aliunde adferri comodius poterit. Nam quod qui-

Figure 1.8
Defining the tangent in Fincke's *Geometria rotundi.*

dean of the medical school for over half a century) at the University of Copenhagen. But the *Geometriae rotundi* remains his most enduring legacy. Inspired by Peter Ramus's 1569 *Geometria*, in a way the book is a step back to an older time, with its emphasis on the ancient spherical Menelaus's theorem.[30] However, it was found to be extremely clear and readable, and it was spoken of highly for several decades.

One of the *Geometriae rotundi*'s most lasting contributions was its creative use of language to simplify the presentation. Among his innovations were the inventions of the names "*tangens*" and "*secans*" for the tangent and secant functions respectively. In Proposition V.22 (figure 1.8), Fincke takes a semicircle of given radius, draws a vertical tangent from its rightmost point, and extends a diagonal at a given angle from center *O* until it touches the tangent line at *I*. Then the length of *AI*, naturally, is the "tangent" of that angle. A few propositions later (V.27), Fincke calls *OI* the secant since it crosses the circle's edge.[31]

The new names were instantly popular among Fincke's colleagues; they are found already three years later in Christoph Clavius's 1586 edition of The-

[30] See [Van Brummelen 2009, 56–61].
[31] [Fincke 1583, 73–74, 76].

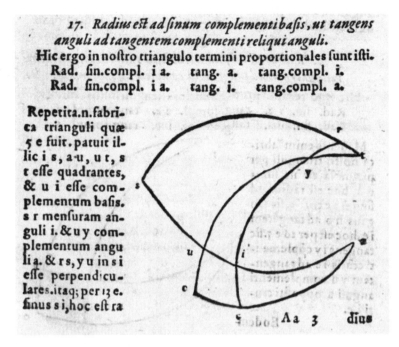

17. Radius eſt ad ſinum complementi baſis, ut tangens
anguli ad tangentem complementi reliqui anguli.
Hic ergo in noſtro triangulo termini proportionales ſunt iſti.
Rad. ſin.compl. i a.   tang. a.   tang.compl. i.
Rad. ſin.compl. i a.   tang. i.   tang.compl. a.

Repetita.n.fabri-
ca trianguli quæ
5 e fuit. patuit il-
lic i s, a u, u t, s
t eſſe quadrantes,
& u i eſſe com-
plementum baſis.
s r menſuram an-
guli i. & u y com-
plementum angu
li a. & rs, y u in s i
eſſe perpend'cu-
lares. itaq; per 13 e.
ſinus s i, hoc eſt ra

Aa 3   dius

Figure 1.9
Fincke's expression of the relation cos c = cot A cot B for a right-angled spherical triangle, *Geometria rotundi* XIV.17. Book XIV contains the earliest appearances of the abbreviations "sin," "tang," and "sec"; the first two can be seen here.

odosius's *Spherics*[32] as well as in Antonio Magini's 1592 *De planis triangulis* (which also contains the terms *tangens secunda* and *secans secunda* for cotangent and cosecant, paralleling the earlier usages of *sinus primus* and *sinus secundus* for sine and cosine), among other works.[33] The abbreviations for the words varied from one author to the next; well into the seventeenth century they had not yet become standardized. François Viète himself objected to the new terms, arguing that they could be too easily confused with other ways that the terms are used in geometry.[34] But in this case at least, Viète's opinion did not transform into practice.

---

[32] [Clavius 1586]. In addition to Theodosius's *Spherics*, the book contains tables of tangents and secants (in which the name *benefica* also appears) and trigonometric treatises by Clavius himself.

[33] [Magini 1592]. [Cajori 1928–1929, vol. 2, 150–151] also refers to the use of these terms by Brahe, Lansberg, Blundeville, and Pitiscus.

[34] [Viète 1593, the third folio numbered 38] ("Immo vero artem confundunt, cum his vocibus necessae habeat uti Geometra abs relatione"); see also [Cajori 1928–1929, vol. 2, 150].

Book XIV, concluding Fincke's *Geometriae rotundi* with some spherical trigonometric results, contains a significant notational development. Perhaps due to the length of text that would otherwise be required to state these theorems, Fincke abbreviates the trigonometric functions in ways that we recognize today. Here we find for the first time "sin." for sine; "tan." and "tang." for tangent; "sec." for secant; and "sin. comp." or "sin. compl." for cosine (and similarly for cotangent and cosecant). In figure 1.9, for instance, we see Fincke's expression of the relation "*R* is to Cos *ia* as Tan *a* is to Cot *i*" in the right-angled spherical triangle at the bottom of the diagram, equivalent to our $\cos c = \cot A \cot B$.

▓ Trigonometric Tables Evolving

Until machines took over the world of computation, numerical tables were how trigonometry was used in the sciences, surveying, and navigation. Hipparchus's invention of the trigonometric table to convert geometric statements into quantitative results was to extend far beyond his predictions of eclipses. In turn, the need for easily computed, yet accurate tables was the motive behind many of the theorems that are now taught in school. The basic formulas of plane trigonometry—for instance, the sine and cosine sum and difference laws and the half-angle formulas—were invented to simplify computations of tables.[35] And as we just saw, the tangent and the secant functions were introduced in Europe not as functions but as tables (the *tabula fecunda* and *tabula benefica*).

The late sixteenth century saw a spectacular rise in the production of trigonometric tables in terms of both the industry required to generate them and the quality of the results.[36] Almost every author participated in the table-making process (see figure 1.10); composing a table was a major part of what it meant to be a practitioner of the doctrine of triangles. Dealing with fractional quantities outside of the astronomers' traditional sexagesimal (base 60) arithmetic was not in the standard toolbox until late in the sixteenth century; table makers usually got around this problem by using a base circle radius equal to some large power of ten.[37] Then, they could represent Sines, Cosines, and so on as large whole numbers.

---

[35] See [Van Brummelen 2009, 41–46, 70–77] for descriptions of trigonometric tables in ancient Greece and in multiple places elsewhere in the book for discussions of tables in medieval cultures.

[36] See [Glowatzki/Göttsche 1990] for a study of Regiomontanus's trigonometric tables and those of his successors.

[37] At least one astronomer of the fifteenth century (Giovanni Bianchini) took some early steps toward decimal fractional notation, including the invention of the decimal point, which we shall describe shortly.

Author	Work	sin	tan	sec	R	Step size	Worst case error
Regiomontanus	*Tabulae directionum* (1490)	✓	✓		60,000 (sine) 100,000 (tangent)	1' (sine) 1° (tangent)	4th of 7 decimal places
Apian	*Introductio geographica* (1541)	✓			100,000	1'	
Regiomontanus	*Tractatus Georgii Peurbachii...* (1541)	✓			6,000,000; 10,000,000	1'	
Copernicus	*De lateribus triangulorum* (1542)	✓			10,000,000	1'	
Rheticus	*Canon doctrinae triangulorum* (1551)	✓	✓	✓	10,000,000	10'	5th of 10 decimal places
Reinhold	*Tabularum directionum* (1554)		✓		10,000,000	1' (10" after 89°)	4th of 12 decimal places (for 89°59':5th of 11 places)
Maurolico	*Theodosii sphaericorum* (1558)	✓	✓	✓	100,000	1°	7th of 7 decimal places (for 89°59': 6th of 9 places)
Viète	*Canon mathematicus seu ad triangula* (1579)	✓	✓	✓	100,000.000	1'	9th of 9 decimal places
Bressieu	*Metrices astronomicae* (1581)	✓	✓	✓	60 (three sexagesimal places)	1°	3rd of 4 sexagesimal places
Fincke	*Geometriae rotundi* (1583)	✓	✓	✓	10,000,000	1'	5th of 11 decimal places
Rheticus/Otho	*Opus palatinum* (1596)	✓	✓	✓	10,000,000,000	10"	7th of 15 decimal places (for 89°59':9th of 14 places)
Pitiscus	*Trigonometriae* (1600)	✓	✓	✓	100,000	1'	5th of 9 decimal places
Van Roomen	*Canon triangulorum sphaericorum* (1607)	✓	✓	✓	1,000,000,000	10'	6th of 12 decimal places
Pitiscus (Rheticus)	*Thesaurus mathematicus* (1613)	✓			1,000,000,000,000,000	10"	

Figure 1.10
Trigonometric tables from Regiomontanus to the eve of logarithms.

A quick examination of figure 1.10 reveals several noteworthy facts. Firstly, it took almost no time for the tangent and the secant functions, under various names, to be accepted and tabulated along with the sine.[38] Secondly, the increments between the arguments became smaller and smaller, achieving more accuracy at the cost of increased labor; the standard increment soon became 1' or even smaller. Finally, often unaware of it, all authors struggled with the entries of a trigonometric table that are most difficult to compute accurately: namely, values for the tangent and secant where the argument approaches 90°. These values were often calculated by dividing by a very small quantity such as the cosine of an angle near 90°.[39] Small rounding errors in

---

[38] In spherical trigonometry the function arcsin (sin x sin y). had currency through the sixteenth century and was often tabulated; see [Van Brummelen 2009, 263] on Regiomontanus's table and [Glowatski/Göttsche 1990, 197–207] for a summary.

[39] See [Pritchard, forthcoming].

the cosine values were thus magnified and became much larger errors in the tangent and secant values.[40]

Several sixteenth-century European authors discussed their methods for computing sines.[41] Usually their methods did not go much beyond what one finds already in the chord table in Ptolemy's *Almagest* along with those developed in early Islam and transmitted to Europe through al-Andalus. A typical early sixteenth-century text is Regiomontanus's *Compositio tabularum sinuum rectorum*, published 65 years after his death in 1541.[42] Regiomontanus begins this work simply by stating that one can find the Sine of the complement of an arc whose Sine is known, using the Pythagorean Theorem:

$$\sin(90° - \theta) = \sqrt{R^2 - Sin^2\theta}. \tag{1.1}$$

He then determines the Sines of the *kardajas*, namely, the multiples of 15°, which can be obtained from the Sines of 30°, 45°, and 60°, a simple geometric argument deriving the Sine of 15°, and (1.1).[43] This results in a small table of sines, listed in the order of their computation rather than in increasing order, with $R = 600,000,000$:

Arcus	Sinus
90	600000000
30	300000000
60	519615242
45	424264069
15	155291427
75	579555496

[40] We have already discussed this problem with respect to Rheticus's tables in the *Opus palatinum*, their identification by Adriaan van Roomen, and the repairs to the table made by Pitiscus; see [Van Brummelen 2009, 280–282]. For the secant function, the alternative method $sec^2\theta = 1 + tan^2\theta$ was much less prone to error (assuming one has an accurate tangent table) and used occasionally; see [Van Brummelen/Byrne, forthcoming].

[41] Occasionally they also discussed the computation of tangents and secants but usually only briefly and simply.

[42] Published as an appendix to [Peurbach 1541]; [Glowatzki/Göttsche 1990, 11–24] contains a reproduction of the manuscript and a translation to German. This is not the earliest sixteenth-century publication describing the calculation of a sine table; Peter Apian's *Introductio geographica* (1533) contains both a sine table (reprinted a year later in his *Instrumentum sinuum seu primi mobilis*) and a description. See [Folkerts 1997, 225–226] for a brief account. The *Instrumentum sinuum seu primi mobilis* also contains a small table of arc sines, the earliest such table of which I am aware with clearly trigonometric intent. An early description of the construction of a sine table, using similar methods and almost contemporaneous with Regiomontanus, may be found in Oronce Fine's 1542 *De sinibus*; see [Ross 1977].

[43] The *kardajas*, from the Persian for "sections," are found in medieval India, Islam, and Europe. For a modern account of this and the following proposition, see [Zeller 1944, 33–34].

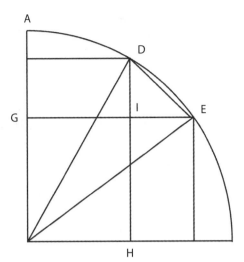

Figure 1.11
Regiomontanus's calculation of the
length of a side of a regular 15-gon.

Proposition 3 gives

$$\frac{\frac{1}{2}R}{\text{Sin}\,\theta} = \frac{\text{Sin}\,\theta}{\text{Vers}\,2\theta}, \tag{1.2}$$

an equivalent to the Sine half-angle formula. This gives Regiomontanus all
the tools he needs to find the Sines of all the multiples of 3°45′, which he
promptly does, in a table similar in form to the above. Proposition 4 uses
constructions of the regular pentagon and decagon inscribed in a circle, as
Ptolemy and many others had done, to determine the values of a couple of
more difficult Sines.[44] For instance, since the side of the inscribed pentagon is equal to the chord of a 72° arc, half of the side of the pentagon is the
Sine of 36°. Once these values are known, proposition 5 allows Regiomontanus to find the length of a side of a regular 15-gon inscribed in a circle,
as follows (figure 1.11): in a quadrant of radius R, let $AD = 30°$ and $AE = 54°$.
Then $EI = EG - IG = \text{Sin}\,54° - \text{Sin}\,30° = \text{Sin}(90° - 36°) - \text{Sin}\,30°$ and $ID = DH - HI = \text{Sin}\,60° - \text{Sin}\,36°$, so we can calculate a value for $ED = \sqrt{EI^2 + ID^2}$.
But $\overset{\frown}{ED} = 24°$ is one side of the regular 15-gon, so $\frac{1}{2}ED = \text{Sin}\,12°$. Now that
we have a value for Sin 12°, apply the half-angle formula four times to get
Sin 45′. Once Regiomontanus has this value under his belt, he needs only
time, patience, and the Sine sum and difference laws to find the Sines of all
multiples of 45′.

But all of this work is only a precursor to the most challenging problem
in the calculation of Sine tables, namely that of finding the Sine values of
multiples not of 45′ but of 1° (or 1′). The only Sines that can be found using

---

[44] See [Van Brummelen 2009, 72–74].

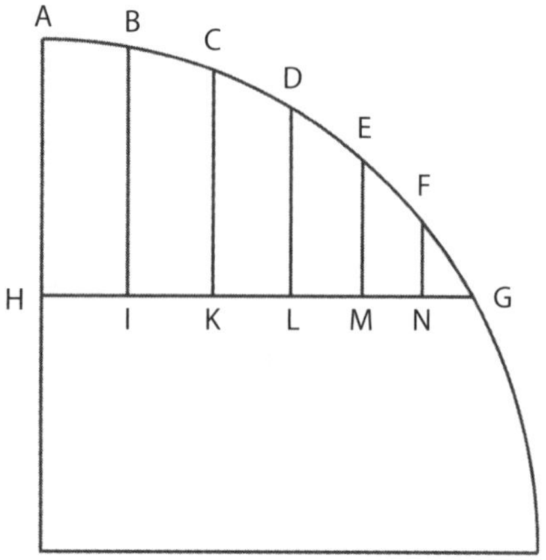

Figure 1.12
Regiomontanus's method to calculate Sin 1°.

geometry alone are those whose arcs can be written in the form $3m/2^{n\circ}$ for whole numbers m and n. To go beyond this set, mathematicians since Ptolemy had had to find a way somehow to break the bounds of the methods available to geometry. Regiomontanus proceeds as follows. Within the quadrant, cut six arcs of ¼° each: AB, BC, . . . , FG (figure 1.12); and drop a perpendicular from G onto AH. Then drop perpendiculars from B, C, D, E, and F to HG. HI, HK, . . . are then the Sines of the successive multiples of ¼°, up to $HG = \text{Sin} 1\frac{1}{2}°$. By a lemma (omitted here, although one can see it is true by inspection), Regiomontanus knows that $HI > IK > \ldots > NG$. Since he already knows from his table calculations that $HL = \text{Sin}\frac{3°}{4} = 7,853,773$ (in a circle of radius 600,000,000), he determines

$$\text{Sin } 1° = HM = HL + LM < \frac{4}{3} HL = \frac{4}{3} \text{Sin} \frac{3°}{4} = \frac{4}{3}(7853773)$$
$$= 10,471,697. \tag{1.3}$$

Similarly, knowing also that $HG = \text{Sin } 1\frac{1°}{2} = 15,706,169$, he finds

$$\text{Sin } 1° = HM = HL + LM > HL + \frac{1}{3}LG =$$
$$\text{Sin}\frac{3°}{4} + \frac{1}{3}\left(\text{Sin } 1\frac{1°}{2} - \text{Sin}\frac{3°}{4}\right) = 10,471,238. \tag{1.4}$$

The result is a narrow interval containing Sin 1°:

$$10,471,238 < \text{Sin } 1° < 10,471,697. \tag{1.5}$$

From here Regiomontanus uses his half-angle formula to obtain[45]

$$5,235,818 < \operatorname{Sin} \frac{1^\circ}{2} < 5,236,044. \tag{1.6}$$

Since he wishes to compute a sine table with $R = 6,000,000$ rather than 600,000,000, Regiomontanus divides by 100, leaving

$$52,358 < \operatorname{Sin} \frac{1^\circ}{2} < 52,360, \tag{1.7}$$

from which he concludes that $\operatorname{Sin} \frac{1^\circ}{2} = 52,359$. Armed with this approximation, the half-angle formula, the Sine sum and difference laws, and a lot of patience, he is able to fill in the Sines of all the multiples of $\frac{1}{4}^\circ$.[46]

This technique is an enhancement on the approach used by Ptolemy in the *Almagest*, but it is essentially the same idea. Various eastern Arabic enhancements of Ptolemy's procedure from the tenth and eleventh centuries had generated similar results.[47] Curiously, only a few decades before Regiomontanus wrote this treatise but far to the East in Samarqand, Jāmshīd al-Kāshī had overturned the rules of this problem by introducing algebra and an iterative procedure that allows the determination of Sin 1° to as many places as one has the patience to calculate. However, his solution was not to find its way to Europe.[48] Even more curiously and much closer to Regiomontanus's home, his older colleague Giovanni Bianchini had done something similar, also with a method capable of generating arbitrary levels of precision, and we know that Regiomontanus became aware of it at some point.[49] However, there is no trace of anything new on this topic in this work.

The divide over terminology that we saw in the previous section was about to make a reappearance in the context of tables. Rheticus's new structure and his tables for all six trigonometric functions appeared only a decade after the publication of Regiomontanus's book, in the 1551 *Canon doctrinae triangulorum*. While this latter work eventually became very difficult to find, clearly the word about it spread through the mathematical community; his name is mentioned frequently in the late sixteenth century in conjunction with the new trigonometric functions well before his massive *Opus palatinum*,

---

[45] These two values are in error in the last two places, but this is about to become irrelevant.

[46] Regiomontanus goes on to describe how to enhance the process to work one's way down to Sin 1′, which would allow him to build a table with an increment of 1′, but he does not provide the calculations.

[47] See [Van Brummelen 2009, 140–145].

[48] See [Van Brummelen 2009, 146–149].

[49] See [Gerl 1989, 265–268]. A marginal note by Regiomontanus in the margin of the manuscript Cracow BJ 558 (f.22v) states that Bianchini's method is superior to Ptolemy's.

a full treatment of his trigonometry with gigantic tables, was published in 1596. In fact, although Maurolico published his table of secants under a different name (*tabula benefica*) imitating the style of Regiomontanus in 1558, Thomas Fincke asserted in his 1583 *Geometriae rotundi* that Maurolico had simply taken over Rheticus's secant table. Magini, in his 1592 *De planis triangulis*, defended Maurolico, arguing that he had worked independently of Rheticus.

The question may be resolved by a closer inspection of Maurolico's table, which gives the secant for $R = 100{,}000$ and for every degree up to 89°.[50] Since the secant grows without bound as the argument approaches 90°, the last few values in any secant table are difficult to compute and are highly sensitive to rounding errors. For instance, the correct value of Sec 89° is 5,729,869. Maurolico's value is 5,729,868 while Rheticus's is 5,729,838.[51] Another example: immediately below Maurolico's table, he gives a few values of Sec θ for arguments greater than 89°, one of which (89°30′) has the same argument as an entry in Rheticus's table. The correct value of Sec 89°30′ is 11,459,301; Maurolico's is 11,459,309; Rheticus's value is 11,459,348. In both cases (and in others) Maurolico's value is much more accurate than Rheticus's. Therefore, he did not appropriate Rheticus's table.[52]

François Viète dealt with the problem of finding Sine values for arguments where geometry alone does not suffice, both early and late in his career. In his 1579 *Canon mathematicus seu ad triangula*, he determines sin 1′ as follows.[53] Beginning with sin 30° = 0.5, he applies the sine half-angle formula (in the form $\sin^2(\theta/2) = \frac{1}{2}\text{vers}\,\theta$) 11 times in a row. In his last two iterations he finds

$$\sin^2\left(\frac{450'}{256}\right) = 0.000000261455205834 \text{ and}$$

$$\sin^2\left(\frac{225'}{256}\right) = 0.000000065363805733. \tag{1.8}$$

---

[50] [Von Braunmühl 1900/1903, vol. 1, 151–152] reports on the controversy and mentions a table of secants by Maurolico with arguments up to 45°; this table is mentioned by several later writers, apparently taking their information from von Braunmühl. The manuscript in fact does contain a secant table as described by von Braunmühl but in two columns, the first of which ends at 45°. Perhaps von Braunmühl did not notice the second column and thus did not have the opportunity to compare the values in the two secant tables for arguments near 90°.

[51] This entry cannot be a typographical error since Rheticus's interpolation column confirms this value. Since Rheticus's value for R is larger, it contains two more decimal places, suppressed here; likewise for the entry for sec 89°30′.

[52] For a full analysis and the background to the controversy, see [Van Brummelen/Byrne, forthcoming].

[53] See [Viète 1579, 62–67]. For the reader's ease, we have converted Viète's calculations to a base circle of $R = 1$.

From these values Viète derives two estimates for sin 1′ as follows:

$$\sin 1' > \sqrt{\left(\frac{256}{450}\right)^2 \cdot 0.00000026 1455205834} = 0.0002908881959 \qquad (1.9)$$

and

$$\sin 1' < \sqrt{\left(\frac{256}{225}\right)^2 \cdot 0.000000065363805733} = 0.0002908882056. \qquad (1.10)$$

The former comes from the assertion that $\dfrac{\sin\left(\frac{450'}{256}\right)}{\sin 1'} < \dfrac{450/256}{1}$; the latter comes from $\dfrac{\sin 1'}{\sin\left(\frac{225'}{256}\right)} < \dfrac{1}{225/256}$. As impressive as these calculations are, this inequality—the heart of Viète's method—goes all the way back to Ptolemy's *Almagest*. Now, since $\frac{225'}{256}$ is closer to 1′ than $\frac{450'}{256}$ is, Viète proposes (but does not carry out in the text) that the final value for Sin 1′ should be a weighted average favoring (1.10) over (1.9). This would result in Sin 1′ ≈ 0.0002908882042, a value that is completely accurate except for the last decimal place. Decades later, Viète would invent (but not carry out) a method that applies algebra to the problem in the spirit of al-Kāshī; we shall examine it later in this chapter.

Also in the *Canon mathematicus*, we find a very large and rather odd table, the *Canonion triangulorum laterum rationalium*.[54] Within it, Viète provides 45 pages of over 1,400 Pythagorean triples, scaled so that one of the three sides of the triangle is exactly equal to 100,000. These triples are ordered sequentially so that they can be used as a trigonometric table. Their values can be quite complicated. For instance, the first entry is

$$\frac{19,988,480,000}{49,942,416,589} \text{ and } 99,999\frac{49,942,376,589}{49,942,416,589};$$

and in fact, the square root of the sum of the squares of these two numbers is precisely 100,000. Viète himself states at the end of the *Canon mathematicus* that the *Canonion* "is of very little use."[55] One wonders, then, why he put so much effort into it. Perhaps he was concerned about issues of roundoff error in conventional tables, or he wished not to stray from the realm of pure geometry into approximation, or he thought of this work more as number theory

---

[54] [Viète 1579], pages numbered separately as pp. 1–45. See also [Tanner 1977] for offshoots of this work by Torporley and Harriot, [Hutton 1811b, 5–6], [Zeller 1944, 73–74], and [Roegel 2011h] for a reconstruction of Viète's table.

[55] [Viète 1579, 75].

than as support for astronomy. We shall encounter this "rational trigonometry" again in chapter 5.

Before we move on, it is also worth mentioning an unusual small treatise by Nicolaus Raymarus Ursus (1551–1600), a German astronomer known primarily for his rivalry with Tycho Brahe over priority to the geoheliocentric system for the motions of the planets. The work in which he propounded this model, his 1588 *Fundamentum astronomicum*,[56] also contains some computational mathematics, including discussions of the computation of sine tables. Here he refers, not entirely clearly, to a method developed by his teacher Joost Bürgi involving finite differences, which we shall discuss later.[57] The method Ursus describes for finding sin 1′ is similar to those we have seen before. However, once he has it, he uses the identity

$$2 \sin(A - x) \cos x - \sin A = \sin(A - 2x) \tag{1.11}$$

cleverly to fill in the remaining entries: starting with $A = 90°$ and $x = 1'$ and the knowledge of sin 90° and sin 89°59′, he uses it to calculate sin 89°58′; and by decreasing A again and again by one minute, he is able to calculate the sines of 89°57′, 89°56′, and so forth.[58] We shall see identities used in this way again, in chapter 3.

Meanwhile, Rheticus had died in 1574, but the massive tables of the *Opus palatinum* were finally published in 1596 by Valentin Otho. We have already described these tables elsewhere.[59] The 700-page tables, the largest ever compiled up to that time, contain all six of the standard trigonometric functions. Computed for every 10″ of arc to ten decimal places, they constitute one of the most intensive computational efforts in human history. However, the methods Rheticus used, although inventive, did not extend beyond the approximation methods we have seen in this section. In fact, in figure 1.10 we see that Rheticus encountered the same difficulties with numerically sensitive trigonometric values that plagued almost all of his colleagues. The errors in Rheticus's tables were noticed by Romanus[60] and repaired by Pitiscus in 1607. Six years later Pitiscus would release *Thesaurus mathematicus*, an even more precise set of tables based on some of Rheticus's unpublished calculations.[61]

---

[56] [Ursus 1588]. On sine tables, see especially the second of the seven chapters.

[57] See [Delambre 1821, vol. 1, 289–291, 299–301].

[58] See an account in [Delambre 1821, vol. 1, 306–307].

[59] See [Van Brummelen 2009, 273–282]. Since then a recomputation of the entire set of tables has appeared ([Roegel 2011e]).

[60] See [Bockstaele 1992] for a Latin edition of the passage and a modern account of Romanus's criticism.

[61] See the description in [Van Brummelen 2009, 281–282]. Since then [Roegel 2011c] has given a recomputation.

▨ Algebraic Gems by Viète

A tantalizing hint suggests that Rheticus was dissatisfied with existing methods for the construction of sine tables; he may have been aware that the $3m/2^{no}$ barrier could be broken by solving an appropriate cubic equation as al-Kāshī had done (unbeknownst to Rheticus) just over a century earlier. Rheticus visited Gerolamo Cardano in 1545, the year Cardano published his solution to the cubic in his *Ars Magna*, "hoping it would be of some use to me in grappling with the science of triangles."[62] But he was sent away empty handed, and the *Opus palatinum* contains no hint of the use of a cubic equation. Its accomplishment, then, owes as much to industry as it does to creativity.

On the other hand, François Viète managed to make the transition to the algebraic problem, showed how to solve the relevant equations, and described how they could be used to generate sine tables—but he seems never to have implemented the solution. His methods appear in *Ad angularium sectionum analyticen*, published by Alexander Anderson in 1615 more than a decade after Viète's death.[63] The key to the solution comes early in this work where Viète determines recurrence relations for sin $n\theta$ and cos $n\theta$.

Text 1.3
Viète, Finding a Recurrence Relation for sin $n\theta$
(from *Ad angularium sectionum analyticen*)

Theorem IIII: If beginning as a point on the circumference of a circle any number of equal segments are laid off and straight lines are drawn [from the beginning point] to the individual points marking the segments, as the shortest is to the one next to it, so any of the others above the shortest will be [to] the sum of the two nearest to it.

[A geometric proof follows.]

(After Theorem VII:) Cut the circumference of a circle into a number of equal parts beginning at any assumed point and from it draw straight lines to the ends of the equal arcs. Let the shortest of these lines be Z and the next shortest B. Hence, from Theorem IIII, the first is to the second as the second is to the sum of the first and the third. The third, therefore, will be $(B^2 - Z^2)/Z$. By the same method used in the preceding [theorem],

the fourth will be $\dfrac{B^3 - 2Z^2B}{Z^2}$

---

[62] [Danielson 2006, 121].
[63] See [Viète 1615]; it also appears as "Theoremata ad sectiones angulares" in [Viète 1646]. See [Viète (Witmer) 1983, 418–450] for a translation.

the fifth will be $\dfrac{B^4 - 3Z^2B^2 + Z^4}{Z^3}$

. . .

the tenth will be $\dfrac{B^9 - 8Z^2 + 21Z^4B^5 - 20Z^6B^3 + 5Z^8B}{Z^8}$.[64]

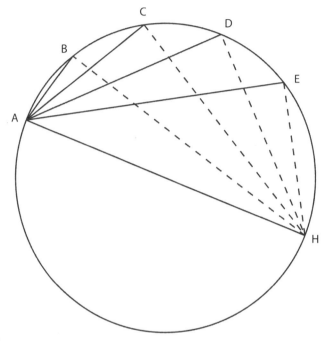

Figure 1.13
Viète's diagram for the sine multiple-angle recurrence relation (simplified). The dashed lines are used in our explanation but do not appear in Viète's figure.

Explanation: (See figure 1.13.) First, we must understand Viète's notation. Arcs $\widehat{AB}, \widehat{BC}, \widehat{CD},$ and $\widehat{DE}$ are all equal; it is understood that AH has been divided into arbitrarily many arcs. AH is a diameter, which implies that the triangles ABH, ACH, and so forth are all right angled. Let θ be the angles $\angle AHB, \angle BHC, \angle CHD,$ and $\angle DHE$; by *Elements* III.20, they are equal to half the posited arcs $\widehat{AB}, \widehat{BC},$ and so on. Then (assuming we are in a unit circle)

[64] [Viète (Witmer) 1983, 426, 435–436]. Viète's algebraic notation in the original differs somewhat

from Witmer's transcription; for instance, $(B^2 - Z^2)/Z$ is rendered as $\dfrac{\begin{array}{c} \text{Bq.} \\ -\text{Zq.} \end{array}}{\text{Z}}$.

chord $Z = AB$ is equal to $2 \sin \theta$ while chord $B = AC$ is equal to $2 \sin 2\theta$. Viète asserts that

$$\frac{Z}{B} = \frac{D}{C + E},\tag{1.12}$$

where D is the second-longest chord in the diagram, C is the third longest, and E is the longest. In modern notation, this turns out to be equivalent to the recurrence relation

$$\frac{\sin \theta}{\sin 2\theta} = \frac{\sin(n-1)\,\theta}{\sin(n-2)\,\theta + \sin n\theta}.\tag{1.13}$$

Viète also determines a recurrence relation for cosines:

$$\frac{1}{2\cos\theta} = \frac{\cos(n-1)\theta}{\cos(n-2)\theta + \cos n\theta}.\tag{1.14}$$

By increasing n successively by one and solving for $\sin n\theta$ each time, Viète is able to generate formulas for $\sin n\theta$ for any n, including an equivalent to the sine triple-angle formula used by al-Kāshī.[65]

Viète compiles a table of the coefficients in the formulas for $\cos n\theta$, going as far as $n = 21$.[66] Clearly, this would have been virtually impossible without his symbolic notation.

Was Viète simply showing off by deriving higher and higher multiple-angle formulas in this way? Perhaps. Certainly, he could hardly have illustrated more effectively the power of combining symbolic algebra with trigonometry; higher-order formulas beyond the triple-angle formula had not been discovered anywhere else, even in the Islamic world. But there was more to it than demonstrating his prowess. He reveals at least part of his intent at the end of *Ad angularium sectionum analyticen*: to find a precise value for sin 1′ in order to construct a table of sines. He begins with a value for sin 18°, which is a value that one can compute using geometric theorems. From it, Viète applies his sine quintuple-angle formula, generating sin 3°36′. This requires solving a quintic equation, which Viète does not explain how to do; however,

---

[65] It came to light in the nineteenth century that Joost Bürgi had followed a similar algebraic path; see [Wolf 1872–1876, 7–28; 1890, vol. 1, 169–175] and [von Braunmühl 1900/1903, vol. 1, 205–208] for accounts and [Roegel 2010a, 5–7] for a discussion of his sine table. Unfortunately, Bürgi's failure to publish rendered his work a dead end.

[66] Viète also derives equivalents to multiple-angle sine and cosine formulas up to $n = 5$ in Propositions 48–51 of his *Ad logisticem speciosam notae priores*, published in 1631 with notes by Jean de Beaugrand; it is the second treatise in [Viète (van Schooten) 1646]. For an English translation see [Viète (Witmer) 1983, 72–74]; for a French translation see [Ritter 1868, 245–276]. Witmer remarks (pp. 6–7) that Viète comes close to, but does not quite arrive at, general expressions for $\cos n\theta$ and $\sin n\theta$.

in another work he had shown how to approximate solutions to polynomial equations.[67] Likewise, using the sine triple-angle formula (and solving a cubic), we may move from sin 60° to sin 20°. Trisect again to get sin 6°40′; then bisect to get sin 3°20′. Apply the sine difference law to 3°36′ and 3°20′ to get sin 16′; finally, bisect four times, and we have sin 1′.[68] Viète never did implement this method, but three decades later Henry Briggs would exploit it in the construction of massive trigonometric tables in his *Trigonometria Britannica*.

We are not yet finished with Viète's algebra. Before applying his multiple-angle formulas to sine tables in the *Ad angularium*, Viète shows how one may work sometimes in the other direction using trigonometry to solve problems in algebra. His most spectacular example is his 1595 *Ad problema quod omnibus mathematicis totius orbis construendum proposuit Adrianus Romanus*.[69] This dramatic story begins two years earlier. In 1593 Romanus had proposed to the world an apparently unsolvable problem, to find roots of the 45th-degree equation

$$45x - 3795x^3 + 95634x^5 - 1138500x^7 + 7811375x^9 - 34512075x^{11}$$

$$+ 105306075x^{13} - 232676280x^{15} + 384942375x^{17} - 488494125x^{19}$$

$$+ 483841800x^{21} - 378658800x^{23} + 236030652x^{25} - 117679100x^{27}$$

$$+ 46955700x^{29} - 14945040x^{31} + 3764565x^{33} - 740259x^{35} + 111150x^{37}$$

$$- 12300x^{39} + 945x^{41} - 41x^{43} + x^{45} = K. \tag{1.15}$$

A quick examination reveals that this is no ordinary 45th-degree polynomial; for instance, all the powers of x are odd. However, at first glance it is a mystery how, when presented this problem by a Dutch ambassador through the king of France, Viète was able to come up with one solution almost immediately, and 22 others by the next day.

---

[67] *De numerosa potestatum purarum* [Viète 1600]; also available in [Viète 1646, 163–228]. The method for the extraction of roots is based on finding an initial approximation a to the solution x of the polynomial, substituting $a + b$ for x in the polynomial, and applying the binomial theorem to expand the result. See also [Goldstine 1977, 66–68].

[68] [Viète 1615, 47]; an English translation is in [Viète (Witmer) 1983, 450].

[69] [Viète 1595]; also available in [Viète 1646, 305–324]. Our account is based on [Viète (Witmer) 1983, 445n46], a translation of [Viète 1595, folio 12]. Viète deals with these issues in other treatises as well, including *De aequationum recognitione* and *Supplementum geometriae*, both available in [Viète 1646]. Viète's calculus of triangles, appearing also in *Ad logisticen speciosam notae priores* and *Zeteticorum*, has drawn attention; some of its calculations are isomorphic to the use of arithmetic with complex numbers, although [Glushkov 1977] is careful to point out the danger of such "unhistorical analysis"; see also [Itard 1968], [Bekken 2001], and [Reich 1973, chapter 3]. Also, [Bachmakova/Slavutin 1977] argue that Viète's calculations with triangles are dedicated to the solution of indeterminate equations.

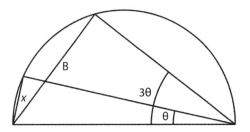

Figure 1.14
Viète's solution of the irreducible cubic equation.

We illustrate with a (thankfully) simpler case, an example of the first "Theoremation" of *Ad problema*: the equation $3x - x^3 = 1$, an example of the irreducible (sometimes called "depressed") cubic $ax - x^3 = b$ that Scipione del Ferro, Tartaglia, and Gerolamo Cardano had solved several decades earlier. Viète recognizes that the form of this cubic equation is related to the sine triple-angle formula that he expresses as $3R^2x - x^3 = R^2B$, where R is the base circle radius, x is the chord subtending angle θ in figure 1.14, and B is the chord subtending 3θ. If we are in a unit circle, then we may verify that $B = 2\sin 3\theta$ and $x = 2\sin\theta$. For our example we have $B = 1$. This implies that $\sin 3\theta = \frac{1}{2}$. Thus $3\theta = 30°$ or $150°$, so $\theta = 10°$ or $50°$. Hence $x = 2\sin 10° = 0.347296$ or $x = 2\sin 50° = 1.53208$, and Viète has found two of the three roots of the cubic equation. (Since Viète can consider only angles between 0 and 180° he cannot find the third root, which is negative.)

This remarkable use of trigonometry to solve the irreducible cubic can be extended to certain polynomials of higher powers using higher multiple-angle formulas, thereby extending beyond Cardano's solutions of the cubic and quartic equations. Of course, bringing in a sine table to solve a polynomial alters the problem by expanding the set of tools permitted to generate a solution. Nevertheless, it is ingenious and, within its parameters, successful. One can see now how Viète upheld the honor of French mathematics by solving the 45th-degree polynomial so quickly: he recognized that it is the result of two angle trisections and a quintisection $(3 \times 3 \times 5)$. He was able to generate only 23 of the 45 solutions for the same reason that we generated only two of the three solutions in our cubic; the other solutions are negative.[70]

Through this tour de force, Viète had clearly demonstrated the power of the new algebra. He ends the treatise, and we end our treatment of Viète's contributions to trigonometry, as follows: "Embrace the new, lovers of knowledge; farewell, and consult the just and the good."[71]

---

[70] [Hollingdale 1984, 135–136] contains an account of how Viète might have gone about solving Romanus's equation.
[71] [Viète 1595, unnumbered folio after folio 13].

▨ New Theorems, Plane and Spherical

Complete solutions to all conceivable triangles, both plane and spherical, had existed in Europe since Regiomontanus's *De triangulis omnimodis*, which remained the dominant textbook for most of the sixteenth century. One might wonder, then, what there was left to do. But Regiomontanus's book was written before advances in the mid-sixteenth century made possible certain ways to streamline the theory. Primary among these was the advent of the new functions, especially the tangent and the secant. Regiomontanus, restricted to the sine, cosine (expressed as the sine of the complement of the angle), and the versed sine, naturally approached solutions of triangles with only these three functions in mind. As the tangent and secant (and their complements) gradually established themselves as members of an expanded set of primitive functions, new and more attractive options for solving triangles became readily available.

Today, the most well known of the new sixteenth-century formulas is the planar Law of Tangents,[72]

$$\frac{a-b}{a+b} = \frac{\tan \frac{1}{2}(A-B)}{\tan \frac{1}{2}(A+B)}. \tag{1.16}$$

Most modern sources assign the first European appearance of this formula to Thomas Fincke in proposition X.15 of his 1583 *Geometriae rotundi*.[73] He introduces the law to solve triangles where two sides and the included angle are known. His first example illustrates how it works. Let $a=21$, $b=13$, and $\angle C = 67°22'49''$; then $\frac{1}{2}(a+b) = 17$ and $\frac{1}{2}(a-b) = 4$.[74] We also know that $\frac{1}{2}(A+B) = \frac{1}{2}(180° - C) = 56°18'35''$, so by the Law of Tangents, $\frac{1}{2}(A-B) = 19°26'24''$. Finally, A and B may be found as the sum and difference of $\frac{1}{2}(A+B)$ and $\frac{1}{2}(A-B)$ respectively, namely, $75°45'$ and $36°52'11''$.

Many other authors picked up the Law of Tangents shortly after its appearance in Fincke's book.[75] We find it used for the same purpose in, for instance, Christoph Clavius's 1586 *Triangula rectilinea*,[76] Philip van Lansberge's 1591 *Triangulorum geometriae*,[77] and Viète's 1593 *Variorum de*

---

[72] The theorem was known in medieval Islam, but (as far we know) it was not transmitted to Europe.

[73] [Fincke 1583, 292–293].

[74] Fincke expresses the left side of the Law of Tangents as $\frac{1}{2}(a-b) / \frac{1}{2}(a+b)$, which simplifies the calculations slightly.

[75] See [Tropfke 1903, vol. 2, 238] for a short discussion.

[76] In an appendix to his edition of Theodosius's *Spherics* [Clavius 1586, 328–329].

[77] [Van Lansberge 1591, 162].

rebus mathematicis responsorum, liber VIII.[78] So the theorem was integrated quickly into the standard corpus of plane trigonometry and has remained there ever since.

It comes as a mild surprise that the Law of Tangents does not appear directly in Viète's earlier *Canon mathematicus seu ad triangula* (1579), for that work is full of new identities, most of which have fallen out of common use today.[79] Some of the more interesting of Viète's new theorems are equivalents in his notation to

$$\tan(45° + \theta / 2) = 2 \tan \theta + \tan(45° - \theta / 2) \qquad (1.17)$$

and

$$\sec \theta = \frac{1}{2} \tan(45° + \theta / 2) + \frac{1}{2} \tan(45° - \theta / 2). \qquad (1.18)$$

The first allows a tangent table to be computed quickly (using only additions) once the entries up to an argument of 45° have been found; the second allows the easy completion of a secant table once a tangent table has been completed. Others of Viète's theorems include

$$\cot \frac{\alpha + \beta}{2} = -\frac{\sin \alpha - \sin \beta}{\cos \alpha - \cos \beta} \qquad (1.19)$$

and

$$\frac{\tan \frac{\alpha + \beta}{2}}{\tan \frac{\alpha - \beta}{2}} = \frac{\sin \alpha + \sin \beta}{\sin \alpha - \sin \beta}, \qquad (1.20)$$

with the latter being related to the Law of Tangents. As part of his work on solving planar oblique triangles, Viète also presents the sine and cosine *difference-to-product* identities,[80]

$$\sin \alpha - \sin \beta = 2 \cos \frac{\alpha + \beta}{2} \sin \frac{\alpha - \beta}{2} \qquad (1.21)$$

and

$$\cos \alpha - \cos \beta = -2 \sin \frac{\alpha + \beta}{2} \sin \frac{\alpha - \beta}{2}. \qquad (1.22)$$

---

[78] [Viète 1593, 32].

[79] See [Delambre 1821, vol. 2, 19] on the identities useful for computing tables. For a survey of the new identities in the *Canon mathematicus seu ad triangula*, see [Ritter 1895, 48–53].

[80] There are corresponding formulas for the sums of sines and cosines.

These two equations are close cousins of the *product-to-difference* (or just *product*) identities

$$\sin\alpha\sin\beta = \frac{1}{2}[\cos(\alpha-\beta) - \cos(\alpha+\beta)] \qquad (1.23)$$

and

$$\cos\alpha\cos\beta = \frac{1}{2}[\cos(\alpha-\beta) + \cos(\alpha+\beta)], \qquad (1.24)$$

which were of considerable interest. They were studied intensely, first by Johann Werner in the early sixteenth century and then in the 1580s by Nicolai Ursus and the group led by Tycho Brahe.[81] Their attraction lay in the fact that they could be used to transform the need to multiply two trigonometric quantities, a tedious process common in spherical trigonometry and astronomy, into the much easier task of adding or subtracting—essentially the same benefit that would be associated later with logarithms. This became known as *prosthaphairesis*; we already discussed its history in the previous volume.[82]

Spherical trigonometry also saw its share of new theorems; in fact, the subject underwent a metamorphosis during the sixteenth century. We begin where the theory itself begins, with right-angled triangles. The modern treatment reduces to these ten identities:

$$\sin b = \tan a \cot A \qquad \sin a = \sin A \sin c$$
$$\cos c = \cot A \cot B \qquad \cos A = \sin B \cos a$$
$$\sin a = \cot B \tan b \qquad \cos B = \cos b \sin A$$
$$\cos A = \tan b \cot c \qquad \sin b = \sin c \sin B$$
$$\cos B = \cot c \tan a \qquad \cos c = \cos a \cos b$$

Many of these results had been known already to ancient and medieval astronomers, especially those in the right column consisting entirely of sines and cosines. In various forms, some of them may be found buried in texts as old as Ptolemy's *Almagest*, embedded in the language of chords and often presented within solutions to problems in spherical astronomy. The second and third identities on the right are known as Geber's theorem, named after the twelfth-century Andalusian astronomer. But neither the ancient Greek nor the medieval eastern Islamic astronomers dealt solely with the triangle as the fundamental figure of spherical trigonometry; the Greeks worked with Menelaus's theorem, and in eastern Islam after the tenth century the emphasis

---

[81] We do not suggest that the later interest in these formulas came from Viète.
[82] [Van Brummelen 2009, 264–265].

was on the Rule of Four Quantities.[83] Our ordered list of identities would not have been familiar to either culture.

The idea of gathering the ten fundamental identities into a unified whole is first hinted at by Georg Rheticus in a six-page dialogue at the end of his 1551 *Canon doctrinae triangulorum*.[84] Explicitly rejecting both Ptolemy and Geber, Rheticus claims to have a new approach to spherical trigonometry that requires knowledge of only ten identities applied to a right triangle. One can hardly imagine what else he may have meant, other than these. But in this dialogue, he does not elaborate or even state what they are.

Rheticus's comprehensive theory of spherical trigonometry would not appear until 22 years after his death in the 1596 *Opus palatinum* with Valentin Otho. In the meantime, several authors had beaten him to publication. The first was François Viète in his 1579 *Canon mathematicus seu ad triangula*. Viète lists all ten of the basic identities in a table as follows:[85]

	Totus	Sinus	Sinus	Sinus		Totus	Fecundus	Fecundus	Sinus
I	C	A B	A	C B	VI	C	A C	~~B~~	C B
II	C	A B	B	A C	VII	C	C B	~~A~~	A C
III	C	~~A~~~~C~~	A	~~B~~	VIII	C	~~A~~~~B~~	C B	~~B~~
IIII	C	~~C~~~~B~~	B	~~A~~	IX	C	~~A~~~~B~~	A C	~~A~~
V	C	~~A~~~~C~~	~~C~~~~B~~	~~A~~~~B~~	X	C	~~B~~	~~A~~	~~A~~~~B~~

The table may be read as follows. Under "Totus" the C represents the sine of the right angle at C, in other words, the radius of the base circle. "Sinus" represents the sine; "fecundus" represents the tangent. A pair of letters represents the side we would represent by the missing letter (i.e., AB represents c). A strikethrough represents the complementary function of that quantity. Each row expresses an equality of ratios. Thus the first row represents $\frac{\sin 90°}{\sin c} = \frac{\sin A}{\sin a}$ or $\sin a = \sin A \sin c$. The other rows give the remaining nine identities; for instance, rows III and IIII are Geber's Theorem, and row V is the spherical Pythagorean Theorem. As we shall see, the ten identities exhibit an extraordinary structure when arranged appropriately, but Viète's arrangement does not reflect this structure. Viète proceeds to rearrange the identities in various ways corresponding to his version of Rheticus's scheme for

---

[83] In two nested right-angled triangles sharing the angles on the bases, the ratio of the sines of the altitudes is equal to the ratio of the sines of the hypotenuses.

[84] [Rheticus 1551, third and fourth pages of the dialogue].

[85] [Viète 1579, 36–37].

grouping right-angled triangles in three species. This results in another 50 mathematically trivial variations of the ten identities. He does not prove any of them; his interest here (and elsewhere in the *Canon*) is to present the theorems compactly and systematically so that the reader may apply them easily to any triangle problem—provided that Viète's unique notation is mastered.

A couple of pages later, Viète presents another table of 60 identities.[86] The first ten are as follows:

	Sinus	Sinus	Sinus	Sinus			Sinus	Faecundus	Faecundus	Sinus
I	B	A̸	A̸C̸	A̸B̸		VI	A	B̸	A C	A B
II	A	B̸	C̸B̸	A̸B̸		VII	B	A̸	C B	A B
III	C̸B̸	A̸	A B	A C		VIII	A	C B	A̸B̸	A̸C̸
IIII	A̸C̸	B̸	A B	C B		IX	B	A C	A̸B̸	C̸B̸
V	A	C B	B	A C		X	C̸B̸	A̸	B̸	A̸C̸

The notation is identical to the preceding table, so for instance, the first identity should be read as sin B/cos A = cos b/cos c. Each of these ten identities may be derived by solving for the same term in two of the original ten theorems and setting them equal to each other; for example, this one may be found by solving for cos a in cos A = sin B cos a and cos c = cos a cos b. Hence these new results are not particularly interesting here. But Viète's thoroughness occasionally leads him to stumble upon theorems that had had currency in medieval Islam; for instance, the third identity is cos a/cos b = sin c/sin b, which had appeared three centuries earlier in Naṣīr al-Dīn al-Ṭūsī's thirteenth-century *Treatise on the Quadrilateral*.[87]

Viète seemed to realize that such a surfeit of formulas could be confusing to the reader. Later, in his 1593 *Variorum de rebus mathematicis responsorum*, following the textbook writers of the previous decade, he selected and reported on the identities most useful for solving triangles according to which of the triangle's elements are known and which are to be found.[88] As we shall see, the theory was streamlined between 1580 and 1609; Simon Stevin has been credited with the conclusion that the original ten identities are sufficient for all right triangles in his 1608 book *Driehouckhandel*.[89]

---

[86] [Viète 1579, 40–41]

[87] See [Van Brummelen 2009, 190].

[88] [Viète 1593, folios 32–35].

[89] Within [Stevin 1608a]; a Latin version may be found at the beginning of the first volume of [Stevin 1608b]; the credit is given in [von Braunmühl 1900/1903, vol. 1, 227]. Here and elsewhere, he and some other writers sometimes refer to six rather than ten identities; this reflects

As for oblique spherical triangles, many authors continued to treat them simply by dropping a perpendicular from one of the vertices onto the opposite side and working with the resulting pair of right triangles, an approach that would later pay dividends in the age of logarithms. But others treated oblique triangles directly. The two fundamental results are the Law of Sines,

$$\frac{\sin a}{\sin A} = \frac{\sin b}{\sin B} = \frac{\sin c}{\sin C};$$

(1.25)

and the Law of Cosines,

$$\cos c = \cos a \cos b + \sin a \sin b \cos C.$$

(1.26)

Both had been stated and proved already in Regiomontanus's *De triangulis omnimodis*.[90] However, Regiomontanus's expression of the Law of Cosines is in a form that might not be recognized immediately today. It refers not to cosines but rather to versed sines:

$$\frac{\text{vers } C}{\text{vers } c - \text{vers}(a - b)} = \frac{1}{\sin a \sin b}.$$

(1.27)

The Law of Cosines refers to all three sides of the triangle but only one angle. There is another spherical Law of Cosines, this one referring to three angles and one side:

$$\cos C = -\cos A \cos B + \sin A \sin B \cos c.$$

(1.28)

This theorem did not appear in Regiomontanus or anywhere else for some time; it is stated for the first time in print (but not proven), again in a form that applies the versed sine rather than the cosine, in IV.16 of Phillipp van Lansberge's 1591 *Triangulorum geometricae*.[91] It seems that it was known earlier to Brahe[92] and possibly others. In both van Lansberge's book and in its next appearance in Viète's 1593 *Variorum de rebus mathematicis responsorum* (the latter using cosines rather than versed sines), it is placed in direct

---

the fact that four of the identities are identical to four others, up to switching the *A*s with the *B*s and the *a*s with the *b*s.

[90] The Law of Sines is theorem IV.17, [Regiomontanus (Hughes) 1967, 225–227]; the Law of Cosines is theorem V.2, [Regiomontanus (Hughes) 1967, 271–275].

[91] [Van Lansberge 1591, 196–197]. In later editions it appears as IV.17. Lansberge claims the theorem as his own and inserts a proof, on which we shall comment shortly, in the second edition, [van Lansberge 1631, 158–161].

[92] [Von Braunmühl 1900, vol. 1, 181] notes its appearance in one of Brahe's unpublished manuscripts.

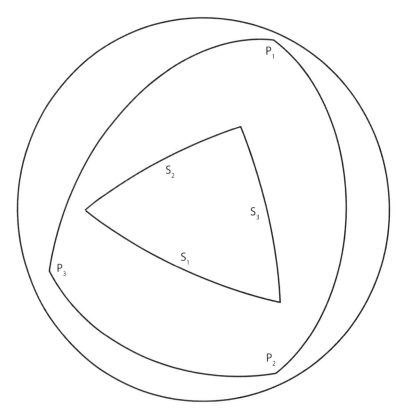

Figure 1.15
The construction of the polar triangle. For each side S_n of the original triangle,
draw the pole P_n on the side of S_n that contains the interior of the triangle; then join
the P_ns.

parallel with the Law of Cosines.[93] The earliest proof may be found a couple
of years later, in Pitiscus's 1595 edition of the *Trigonometriae*.[94]

The correspondence between the two Laws of Cosines is no coincidence;
they are linked by a duality relation. If one considers each side of a given
spherical triangle to be an equator and draws the pole of that equator on the
side of the triangle's interior, then joins the three poles, the resulting *polar
triangle* has some remarkable properties (figure 1.15). In particular, the polar
triangle of the polar triangle is the original triangle, the sides of the polar triangle are the supplements of the angles of the original, and the angles of the
polar triangle are the supplements of the sides of the original. Applying this

latter statement to the Law of Cosines immediately gives the Law of Cosines for Angles and vice versa.

The polar triangle had been discovered centuries earlier by astronomer Abū Naṣr Manṣūr ibn ʿIrāq around the turn of the millennium,[95] but it (along with most other trigonometric innovations from eastern Islam) does not seem to have found its way to Europe. The story of its rediscovery is more complicated. We find something like the polar triangle first in the creative hands of François Viète in his 1593 *Variorum de rebus mathematicis responsorum* where he refers somewhat obscurely to sides and angles of triangles being reciprocal.[96] Later in the same chapter Viète constructs diagrams of triangles with great circles connecting all six poles of the three sides of the original triangle; the polar triangle is one of the triangles in these figures.[97] Later in the same text, Viète gives a series of eight theorems about spherical triangles that happen to be aligned in four pairs, a theorem along with its dual result through the polar triangle. This has been taken as evidence that Viète was in fact using polar triangles as a device to convert theorems to their dual partners. In any case, Viète's presentation is sufficiently vague that it appears not to have spread far; until Viète's work was reexamined much later, credit went instead to Willebrord Snell.[98] The expression of polar triangles in the latter's 1627 *Doctrinae triangulorum canonicae* is certainly much clearer and gives a good sense as to how they can be used.

Text 1.4
Snell on Reciprocal Triangles
(from *Doctrinae triangulorum canonicae*)

Book III: PROPOSITION 8: If from the three given angles of the triangles [taken as] poles, great circles are described, the sides and angles of the triangle will be expressed, [and] the remaining sides and angles are first found reciprocally.[99]

---

[95] See [Van Brummelen 2009, 184–185].

[96] The tenth statement on spherical triangles in [Viète 1593, folio 41], which reads: "Si sub apicibus singulia proposti Tripleuri sphaerici describantur maximi circuli, Tripleurum ita descriptum Tripleuri primum proposti lateribus et angulis est reciprocum."

[97] [Viète 1593, folios 42–45]. See an exposition of one of the examples of this section in [Zeller 1944, 83–84].

[98] [Delambre 1819, 478–479] argues that Viète's words are not sufficiently clear to be certain that he was referring specifically to the polar triangle; [Ritter 1895, 56] disagrees. [Von Braunmühl 1898] and [1900/1903, vol. 1, 182–183] pay special attention to the problem, noting the sequence of theorems in polar pairs as evidence. [Tropfke 1923, vol. 5, 125] notes that Viète's sparse presentation likely led to the public credit passing to Snell.

[99] [Snell 1627, 120].

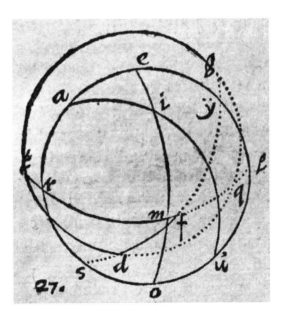

Figure 1.16
The polar triangle in
Snell's *Doctrinae
triangulorum
canonicae.*

Explanation: (See figure 1.16) The diagram represents a sphere; the original triangle is *aei*. Snell instructs us to draw the equator *sdy* with pole *a*; equator *rfl* with pole *e*; and equator *tdqb* with pole *i*. Snell's construction is a little different than how it is usually done today; it begins by considering the poles of the original triangle and constructs equators rather than the other way around. The relation between spherical triangles and their polar duals implies that there is no difference in the final result as long as one selects the correct triangle among those formed by the intersections of the three equators.

There is one other candidate for the discovery of the polar triangle in Europe. As noted above, the Law of Cosines for Angles is stated in Philip van Lansberge's 1591 *Triangulorum geometricae*. It seems a natural inference that he might have used polar triangles to arrive at the statement of this theorem.[100] In his second edition, published four years after Snell's book in 1631, van Lansberge inserts a proof based on the idea of the polar triangle, introducing it as follows: "the second part of the [Law of Cosines for Angles], which we have the right to claim that we were the first to discover, is proved in the same way as the first, if first we describe a new triangle by means of the poles of the sides of the given triangle."[101] This is a claim for the discovery of the Law

---

[100] The suggestion is made in [von Braunmühl 1900/1903, vol. 1, 192–193].
[101] [Van Lansberge 1631, 158]. The description and diagram in [Zeller 1944, 97] are from the 1631 edition and are not found in the 1591 first edition.

of Cosines for Angles but not quite for polar triangles. Nevertheless Simon Stevin credits van Lansberge in his 1608 *Hypomnemata mathematica* (the Latin version of his *Dreihouckhandel*)[102] and provides essentially the same proof of the complementarity of sides and angles. Perhaps van Lansberge had circulated his ideas privately.

▩ Consolidating the Solutions of Triangles

François Viète's 1579 *Canon mathematicus* seems to have triggered a period of about three decades of textbook writing. There were enough new trigonometric functions, theorems, and approaches to solving triangles that a book to replace Regiomontanus's universal triangle solver *De triangulis omnimodis* was sorely needed, and a number of authors attempted to fill the gap. Neither Viète's notation nor the structure of his 1579 *Canon mathematicus* conformed to Regiomontanus's style, which most of his contemporaries were used to reading. Thus, while clearly most mathematicians read Viète and profited by his work, many continued to approach trigonometry within Regiomontanus's tradition (soon to be augmented by Fincke's 1583 introduction of the "tangent" and "secant"). Perhaps the earliest of these textbooks was Maurice Bressieu's 1581 *Metrices astronomicae*,[103] written and titled to position the science of triangles as a computational foundation for astronomy. Bressieu presents various different kinds of triangles and in each case outlines how to solve it, often presenting alternate methods he credits to Ptolemy and sometimes to Regiomontanus; following this he provides a numerical example. Figure 1.17 shows his solution to a plane right triangle where the two sides adjoining the right angle are known and the beginning of a numerical example.[104] Note the hash marks drawn on the given segments; Bressieu seems to have been the first of a number of authors to indicate the givens in the diagram in this way.[105]

One of the most influential of the early texts, Thomas Fincke's 1583 *Geometriae rotundi* appeared two years later. The book itself was not especially innovative mathematically, relying especially on Menelaus's theorem for its spherical results (although, as we saw, it does contain the first appearance of the planar Law of Tangents). However, it came recommended by Clavius, Pitiscus, and Napier for its exceptional clarity. Fincke's presentation (Book X for plane triangles, Book XIV for spherical) is organized around theorems

---

[102] [Stevin 1608b, vol. 1, 223–224].

[103] Little has been written about Bressieu; see [de Merez 1880] for a short biography.

[104] For readers attempting to translate the Latin, the "canone adscriptarum" refers to a tangent table.

[105] [Zeller 1944, 87].

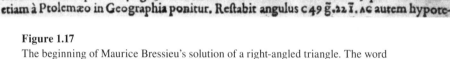

NOSTRA METHODVS.

Repetatur tertium & vltimum antecedentis propofitionis dia-
gramma. Quia igitur AB & BC in iifdem partibus datæ funt, Qualium AB, vt radius, eft ɪ fex. ɪ̄, talium dabitur BC, & ex canone adfcriptarum arcus BD, angulufque BAC. per arcum vero BD dabitur
hypotenufa AC earundem partium atque AB. Ita ratione AB ad AC
data, Qualium AB ex hypothefi data eft partium, talium dabitur
AC per 19 lib. 7 elem. Ex angulo autem BAC dato, dabitur & recti
reliquus ACB.

Exemplum, Cum AB fit ad BC vt 9 ad 8, qualium AB eft ɪ fex. ɪ̄, talium BC erit 53 ḡ. 20 m̄. per 19 lib. 7 elem. & arcus BD, angulúfue BAC
erit ex can. adfcriptarum 41 ḡ. 38 m̄. AC tanta erit latitudo vrbis Romæ, quemadmodum
etiam à Ptolemæo in Geographia ponitur. Reftabit angulus C 49 ḡ. 22 ɪ̄. AC autem hypote-

Figure 1.17
The beginning of Maurice Bressieu's solution of a right-angled triangle. The word
"adscriptum" refers to his version of a tangent.

rather than triangles: that is, he presents a theorem and afterward describes how it may be used to solve a certain kind of triangle rather than the other way around. Christoph Clavius's 1586 text[106] similarly emphasizes theorems and proofs, interspersing them with "problems" that demonstrate how to use the theorems to solve certain triangles. Pitiscus's famous 1595 *Trigonometriae*[107] follows Regiomontanus's model in *De triangulis omnimodis*: he states all the theorems first and then uses them to solve various kinds of triangles. For spherical triangles he begins with four results, calling them "axioms": the Rule of Four Quantities, the Law of Tangents, the Law of Sines, and the Law of Cosines.

However, perhaps driven by the increasing use of trigonometry in applications such as surveying, navigation, and science, some texts started to emphasize an algorithmic approach based on the presentation of triangles rather than theorems: if the triangle has such and such a property, then follow this path; if it does not, then the triangle does not exist; and so forth. Some of the books we have just mentioned had an inkling of such schemes in short indexes that list the various types of triangles in sequence and indicate where one should go in the text to solve them. The index in Phillip van Lansberge's

[106] Published as a supplement to his edition of Theodosius's *Sphaerica*; see [Clavius 1586].

[107] Pitiscus's *Trigonometriae* first appeared at the end of Scultetus's *Sphaericorum* in 1595 and was published separately in a revised edition five years later [Pitiscus 1600]. For Handson's translation, see [Pitiscus (Handson) 1614]; the frontispiece is reproduced in the preface of this book. For a summary of the various editions and translations of the *Trigonometriae* and Pitiscus's other works, see [Archibald 1949a]. See also [Delambre 1821, vol. 2, 28–35]; [Gravelaar 1898] in Dutch, mostly on the computation of tables; [Hellmann 1997] for some discussion of the mathematics; and [Miura 1986] on the applications.

Figure 1.18
A page from van Lansberge's 1591
Triangulorum geometriae, classifying methods
to solve right-angled spherical triangles.

1591 *Triangulorum geometriae* is elaborate; figure 1.18, for instance, shows the first of three pages of his index for right-angled spherical triangles, grouping the various identities according to what element is sought and what elements are known.[108] Antonio Magini's 1609 *Primum mobile* goes further with similar classifications, grouping different types of spherical triangle in a 16-page-long scheme[109] and elsewhere providing grids showing which problem in his treatise solves which type of triangle.[110] Simon Stevin's *Driehouckhandel* (*Trigonometry*), published as part of his 1608 *Wisconstighe Ghedachtenissen* (*Mathematical Memoirs*),[111] divides the discussions of both planar and spherical triangles into three distinct parts: (a) preliminary theorems, (b) identities, and (c) solutions of triangles. This structure endured for hundreds of years; it is found in Todhunter's *Spherical Trigonometry*, the dominant textbook of the late nineteenth and early twentieth centuries.[112]

---

[108] [Van Lansberge 1591, 202].
[109] [Magini 1609, folios 38–45].
[110] [Magini 1609, folios 47, 68]. Several other grids in this work explain how to handle certain cases of problem.
[111] [Stevin 1608a]. The book, written in Dutch, was translated several times. See, for instance, the Latin edition by Snell [Stevin 1608b], and a French translation with supplements by Albert Girard [Stevin (Girard) 1634]. A selection from the treatise appears in Struik's *The Principal Works of Simon Stevin* [Struik 1958, vol. IIB, 757–761].
[112] The original edition is [Todhunter 1859]; it was revised and expanded in [Todhunter/Leathem 1901].

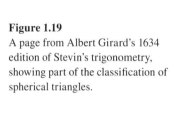

42 Chapter 1

Figure 1.19
A page from Albert Girard's 1634
edition of Stevin's trigonometry,
showing part of the classification of
spherical triangles.

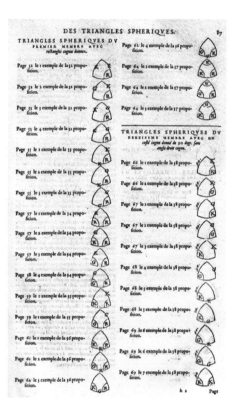

Figure 1.19 shows part of the index from Albert Girard's French edition of the *Driehouckhandel*, illustrating the classification of spherical triangles (including a special category of quadrantal triangles).[113]

But when it came to algorithmic thinking, no one went further than Adrianus Romanus in his 1609 *Canon triangulorum sphaericorum*. Other than its tables and a section describing how to compute them, the entire book is a 270-page-long detailed algorithm for solving spherical triangles with dozens of examples. Book II begins with a detailed nine-page classification of triangles into various genera, followed by 40 pages of examples and diagrams of each genus. The remaining 200 pages are divided into six problems: the first dealing with triangles where two sides are given as well as one of the angles not included between the given sides, the second dealing with two given angles and one of the sides not included between the given angles, and so on. In each case Romanus provides an algorithm for solving the triangle and for handling the various cases that arise. At the bottom of figure 1.20,

---

[113] [Stevin 1634, vol. 2, 87]. A quadrantal triangle has a side (not an angle) equal to 90°.

Figure 1.20
A page from Adrianus Romanus's 1609
Triangulorum sphaericorum.

the beginning of his algorithm for the first problem, Romanus solves it (as many others did) by dropping a perpendicular from a vertex to the opposite side, thereby dividing it into two right triangles. He then applies (but does not prove) the right-angled triangle identities to the two right triangles.[114]

One of the most eccentric, yet remarkable methods ever developed to solve spherical triangles appears in Christoph Clavius's lengthy treatise, the *Astrolabium*.[115] This mostly astronomical treatise works extensively with the technique for spherical geometry known as the ***analemma***.[116] Dating back to ancient Greece, the analemma deals with a problem in spherical geometry by rotating one or more circles on the sphere into the plane of a particular great circle, thereby reducing it to a problem in plane geometry.

The central topic of the *Astrolabium* is ***stereographic projection***, which maps a sphere onto the plane through its equator as follows. In figure 1.21

---

[114] [Romanus 1609, 100]. We should mention Nathaniel Torporley's bizarre 1602 *Diclides Coelo-metricae*, which we saw before. Its unique approach reduces the six cases of right spherical triangles to two, but its obscurity renders it close to impenetrable. See [Delambre 1821, vol. 2, 37–40], [von Braunmühl 1900/1903, vol. 1, 183–186], [Zeller 1944, 106–107], and [Silverberg 2009].

[115] [Clavius 1593]. For an account of Clavius's interactions with Ptolemy's and Copernicus's cosmological theories, see [Lattis 1994]. For a survey of Clavius's mathematics, see [Naux 1983].

[116] See [Van Brummelen 2009, 66–67].

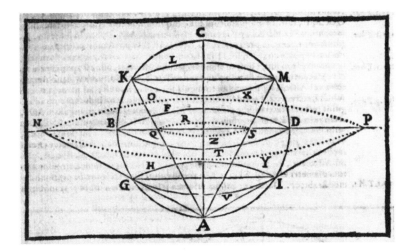

Figure 1.21
From Clavius's *Astrolabium*, illustrating stereographic projection.

the south pole is *A*, and a plane is drawn through equator *BFDT*. For any point *M* on the sphere, a line is drawn from *M* to *A*. Point *S*, where the line crosses the plane through the equator, is considered to be the projection of *M* onto the plane.[117] Stereographic projection has two advantages: circles on the sphere map to circles or lines on the plane and the angle between two great circles is mapped to the same angle on the plane. The ancient astronomical instrument, the astrolabe, is simply a physical realization of a stereographic projection of the celestial sphere.

Clavius's approach to solving spherical triangles begins by positioning the sphere so that some side of the triangle is placed along the equator (called by later authors the *primitive circle*) or so that one vertex is at the north pole. Using the given quantities, as much as possible of the projected triangle is drawn on the primitive circle. Once this is done, the remaining elements are constructed geometrically, if possible. Sometimes other great circles are rotated onto the plane, taking the place of the primitive circle. Once the projected triangle has been drawn, the sought angles and sides are measured with a ruler or protractor. Finally, these data are used as inputs into a mathematical process that reconstructs the values of the sought elements of the original spherical triangle.[118]

---

[117] Points on the sphere below the equator are mapped to points on the plane outside the equator; for instance, *G* maps to *N*.

[118] Clavius's methods, as well as related work by Dutch mathematician Adrian Metius (1571–1635) in Book V of *De astrolabio catholico* (1633), are described in [Haller 1899].

This method, ingenious as it is, was not seen as very practical even by some of its adherents; the famous instrument maker Benjamin Martin, introducing the subject in his 1736 *Young Trigonometer's Compleat Guide*, states that "this way is (generally speaking) more artful than useful"; but he goes on to say that "by a little use, [it] is very practicable and easy."[119] It had currency in some textbooks until as late as the nineteenth century, appearing alongside more conventional solutions as a legitimate alternative.[120]

▨ Widening Applications

Through the fifteenth century and into the sixteenth, trigonometry had been a handmaid to astronomy; Regiomontanus himself called it "the foot of the ladder to the stars."[121] In medieval Islam, spherical trigonometry had come to be applied to finding distances and directions on the surface of the earth, originally through the determination of the direction of Mecca. But even these calculations had taken place on the celestial rather than the terrestrial sphere. This makes the sixteenth century one of the most remarkable periods in the history of mathematics, for it was during the latter part of this century that trigonometry started to become genuinely applicable to the physical world: not just for determining distances and directions in the heavens but also on the earth and sea. Raphe Handson's 1614 translation of Pitiscus's *Trigonometriae* presents a transformed view of trigonometry, liberated from its servanthood to astronomy by linking to many other earthly activities:

> All arts are in themselves so infinite, that the life of man is first consumed before he comes to know; yet, the pleasure is such (especially in the mathematics) that the more a man understandeth, the less he thinks to know; as still covetous of more, and never satisfied. And amongst all the sciences mathematical, this trigonometry, or dimension of triangles, is copious in the contemplation of it, and more profitable in the practice: For thereby all heights, depths, distances, questions of the map, globe, sphere, or astrolabe, may be more truly supputated [calculated], than by any instrument whatsoever; besides

---

[119] [Martin 1736, vol. 2, 150]. See pp. 150–160 for his treatment of the subject and [Van Brummelen 2013, 133–139] for a modern mathematical explanation based on Martin's text. See also [von Braunmühl 1900/1903, vol. 1, 189–191], who expresses admiration but also reserves doubts about its efficacy.

[120] See for instance [Wilson 1720] and [Keith 1826]. Other graphical methods were invented to solve spherical triangles, and interest continued (especially in educational circles) as late as the 1950s, at which point interest in spherical trigonometry itself faded away. [Bradley 1920] contains a useful bibliography of references up to that date.

[121] [Regiomontanus (Hughes) 1967, 28–29].

the infinite use thereof in geometry, astronomy, cosmography, etc. Wherefore I have adventured thereon, as a subject, which generally in its own nature carrieth much reputation amongst the sincere lovers of those sciences.[122]

Modern students may dispute Handson's characterization of the pleasure of the subject but perhaps not its practical value.

The most obvious places for trigonometry to spread its wings were still with mathematics—in particular, to measurement within geometry, for which there was a healthy tradition dating back to ancient times. From the sixteenth century onward, a number of authors were interested in questions of goniometry and cyclometry. These related subjects dealt with measurements of various lengths, angles, and areas of certain geometric figures, especially regular polygons and circles. Trigonometry can of course be applied to such questions, but it can also benefit from such study. For instance, the study of the lengths of regular polygons is related to the determination of the sines of small arcs (such as sin 1°, which is half the length of a side of a 180-gon inscribed in a unit circle). Cyclometry in particular is intimately related to approximations for π. It was at this time that Adrianus Romanus, Ludolph van Ceulen, and Philip van Lansberge derived their values of π accurate to 16, 35, and 28 digits respectively.[123]

Genuine applications of trigonometry outside of mathematics were more difficult to find at first. There was of course no end to the uses of trigonometry in astronomy: they had been present since the birth of the subject, especially models of the motions of the planets, spherical astronomy, and solar timekeeping. However, earthly applications were much rarer.[124] From the thirteenth century, the genre of "practical geometry" had dealt with questions related to altimetry, stereometry, and mensuration. This subject was defined by its interaction with the physical world and often involved the use of measurements made by instruments. Its audience consisted of surveyors, architects, cartographers, observational astronomers, navigators, the military, and artists, among others.[125] A few of these treatises made some small use of trig-

---

[122] [Pitiscus (Handson) 1614, beginning of the dedicatory epistle].

[123] Romanus's text is the incomplete *Ideae mathematicae pars prima* [Romanus 1593]; van Ceulen's is *Arithmetische en Geometrische Fondamenten* [van Ceulen 1615] (which contains 33 digits; his final value appears in [Snell 1621]); van Lansberge's is the *Cycometriae novae* [van Lansberge 1616].

[124] One must not forget the determination of the *qibla* in medieval Islam; but even here, correct solutions relied primarily on spherical astronomy. Some other uses of trigonometry in geography in Islam do exist; see [Van Brummelen 2009, 215–217].

[125] The literature on practical geometry and its history is too extensive to be described exhaustively here; we refer only to a few texts. See [Victor 1979] for a description of its origins in medieval Europe; [Busard 1998, 7–12] for a survey of practical geometry to the mid-sixteenth century; [Taylor 1954] for a history of practical mathematics in England from 1485 to 1715;

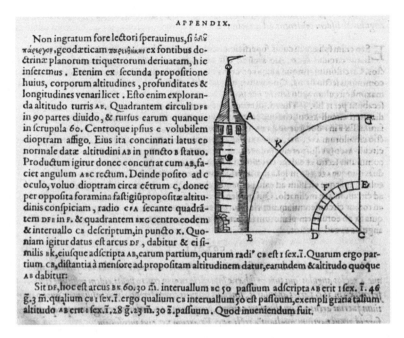

Figure 1.22
Finding the altitude of a tower in Bressieu's *Metrices astronomicae*.

onometry, but most were devoid of it, sticking to basic geometric tools like similar triangles and the Pythagorean theorem.[126]

Prior to 1580, texts devoted to trigonometry had stayed within the confines of mathematics and astronomy. This changed dramatically with the consolidation movement of the 1580s and beyond. What one might consider to be the first practical "story problem" in a trigonometry textbook appears at the end of the chapter on planar trigonometry in Maurice Bressieu's 1581 *Metrices astronomicae*.[127]

Bressieu seems hesitant to introduce the world of practice into his trigonometry, separating the problem from his main text and introducing it with the phrase, "Hoping it will not be unwelcome to the reader." His goal is to find the height of a tower (figure 1.22) where the distance *BC* from the base is given and the angle of altitude from the observer at *C* of the top of the tower is measured. This elementary problem had been solved previously in practical geometry textbooks but not with trigonometry. Bressieu replaces the

---

and several of Jim Bennett's publications, especially [Bennett 1998], a survey of the relation between instruments and practical geometry.

[126] See [Van Brummelen 2009, 224–230, 239–240], where the works of Abraham bar Hiyya, Fibonacci, and John of Murs are considered.

[127] [Bressieu 1581, 49].

Figure 1.23
From Book XI of
Fincke's
*Geometriae
rotundi.*

shadow square and similar triangles with an angle measurement and a tangent table, eventually finding the height as an equivalent to 50 tan θ, where θ is the altitude. In the example calculation Bressieu has an angle of elevation of 60.5° and the distance to the tower of 50 paces, which makes the height of the tower an impressive 88.5 paces.

Bressieu's tentative foray into practical geometry seems not to have had the feared effect of deterring readers, for several texts over the next decade traversed similar ground but with much more commitment. Only two years later, Thomas Fincke's *Geometriae rotundi* devoted the entire 11th book (out of 14) to problems involving altitudes and distances.[128] Its mathematics is straightforward; it consists of applying similar triangles and (often) the tangent function to measurements obtained with a quadrant and other simple instruments to determine various heights, distances, and lengths. Fincke's text and images would have appealed to surveyors, the military, and architects (see figure 1.23). Pitiscus's *Trigonometriae* goes even further; it includes chapters on geodesy, altimetry, geography, gnomometry (sundials), and astronomy and in a later edition another chapter on architecture (especially military). These applications take up over half of his text (aside from the tables).[129]

---

[128] [Fincke 1583, 296–322].
[129] [Miura 1986] contains a brief account of the applications chapters in Pitiscus's *Trigonometriae*.

Trigonometry and practical geometry truly came together in a meaning-ful way with Christopher Clavius's revolutionary new *Geometria practica* in 1604. Near the beginning of this book Clavius presents a full summary of the solutions of planar triangles via trigonometry, although with no proofs as befitted a practical work.[130] Armed with these new tools, he goes on to solve the usual surveying problems (heights of castles, etc.), but using plane trigo-nometry rather than the usual tools of practical geometry.

Text 1.5
Clavius on a Problem in Surveying
(from *Geometria practica*)

On the distance along the ground, whether it is accessible or inaccessible, by means of quadrant measurements at two stations in the same plane, when at its endpoint some perpendicular altitude is erected, even if [the base] is not seen at its lowest extreme. And here we determine the height.

Let the distance, or the sought length be *AB*, in plane *CB*, and erected at the endpoint *B* is some perpendicular altitude *BG*, although the endpoint *B* is not visible. Let the height of the measurer, from the eye to the feet, be *DA*. . . . Extend through *D* a parallel *EF* to *CB*, starting in the first station *D* and end-ing in the second station *E*, the furthest point; and line *DE*, the distance be-tween the stations, is known by an ordinary measurement. Then, guided by the side of the quadrant *HK* that has the sights, . . . set the sights so that the peak *G* may be seen, dropping perpendicular *HI*. And . . . angle *GDF* in min-utes, equal to arc *IL*, may be seen on the quadrant, clearly the complement of arc *IK*. For when thread *HI* is perpendicular to line *DF*, angle *GDF*, the com-plement of angle *DHI*, clearly will be equal to angle *IHL*, which is itself the complement of angle *DHI*. And we will call this angle *GDF* the angle of obser-vation. In the same way angle *GEF* is observed at the second station, by rays from the eye, through the quadrant's sights to the peak at *G*. Taking *EM* equal to *DN*, erect perpendiculars *M<H>* and *NO*. . . . Therefore, if we set *EM* and *DN* as the *sinus toti*, *MH* and *NO* will be tangents of the angles of observation at *E* and *D*. Also draw *DQ* parallel to *EG*, crossing *NO* at *P*. Angle *NDP* is equal to angle *E*. Therefore, the two angles *N* and *D* in triangle *NDP* are equal to two angles *M* and *E*, . . . and sides *DN* and *EM*, which are adjacent, are equal. Sides *NP* and *MH* will be equal, so *OP* will be the difference between the tangents of the angles of observation. Because of this, as *OP* is to *PN*, so is *GQ* to *QF*. And as *GQ* is to *QF*, so is *ED* to *DF*. . . . Hence if [the following] is done:

As OP, the difference between the tangents of the angles of observation is to PN (or HM), tangent of the smaller [angle], so is ED, the distance between the noted stations in a common measure to the other, that is, to DF,

---

[130] [Clavius 1604, 45–52].

[it] produces the sought distance, *DF* or *AB*, the same measure of the distance to the station; and if it is added to the distance *ED* between the stations, we will also learn the distance *EF*, or *CB*, to the furthest station.[131]

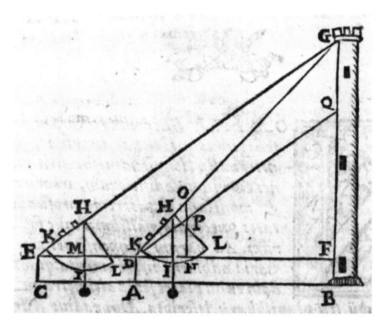

Figure 1.24
Finding the altitude of a tower if the base is inaccessible, from Clavius's 1604 *Geometria practica.*

Explanation: (See figure 1.24.) The goal is to determine the distance to a tower when its base *F* is inaccessible or hidden from view. Observers at two stations in a direct line from the tower, at *D* and *E*, measure the altitude of the pinnacle of the tower *G* (the "angles of observation" $\theta_1 = \angle GDF$ and $\theta_2 = \angle GEF$) with their quadrants. Slide $\triangle EMH$ to the right so that $\angle E$ is at *D*, defining *N* and *P*; extend *DP* to *Q*. Then $ON = DN \tan \theta_1$ and $MH = NP = DN \tan \theta_2$. So $OP = ON - NP = DN(\tan \theta_1 - \tan \theta_2)$, and hence $\dfrac{OP}{NP} = \dfrac{\tan \theta_1 - \tan \theta_2}{\tan \theta_1}$.

But $\dfrac{OP}{NP} = \dfrac{GQ}{QF} = \dfrac{ED}{DF}$; and *ED* is the measured distance while *DF* is the sought distance from the first station to the base of the tower. So

$$DF = ED \cdot \frac{\tan \theta_1}{\tan \theta_1 - \tan \theta_2}.$$

---

[131] [Clavius 1604, 54–55].

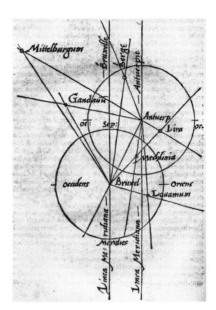

Figure 1.25
Gemma Frisius on surveying, from the
1540 edition of Apian's *Cosmographia*.

The merger of trigonometry with practical needs in geodesy and altimetry provided practitioners with much more powerful and precise mathematical tools. However, without corresponding improvements in the instruments used to measure distances and angles, the extra precision would be superfluous. The new methods were not adopted very widely at their outset. The power of geometry had been revealed to surveyors as early as 1533, with Gemma Frisius's introduction of the notion of triangulation on the surface of the earth (figure 1.25).[132] Although his techniques had required angle measurements, they had not employed trigonometry. Various instruments were invented for use in surveying through the sixteenth century, including a device called a "trigonometer," which formed with its arms a triangle similar to the triangle being measured on the ground. However, only the simple theodolite, measuring azimuths but not altitudes, seems to have gained much traction in practice. It would not be until the first half of the seventeenth century that the power of geometry in general and trigonometry in particular would become generally accepted in surveying practice.[133] This late adoption may have been aided at least in part by the wave of surveying applications in the trigonometry

---

[132] This appears first in Gemma Frisius's 1533 edition of Peter Apian's *Cosmographia* [Apian 1533a]. See analyses of Gemma Frisius's, Brahe's, and Snell's approaches to triangulation in [Haasbroeck 1968]. On triangulation in Gemma Frisius's work, see also [Taylor 1927] and [Pogo 1935]; the latter contains a facsimile edition.

[133] [Bennett 1991b], especially pp. 348–354.

Figure 1.26
Frontispiece of Aaron
Rathborne's 1616 *The Surveyor.*

textbooks but may have had more to do with logarithms, which we shall see
in chapter 2. A notable step forward was Aaron Rathborne's 1616 *The Surveyor* (figure 1.26), which introduces trigonometry in certain contexts and
even mentions Pitiscus and Napier in one of the earliest references to logarithms outside of mathematics and astronomy.[134] Rathborne was a member
of the peculiarly English trade of "mathematical practitioner." These men
earned their living, at least in part, through tutoring mathematics useful for
purposes such as engineering and gunnery rather than the higher pursuits of
natural philosophy.[135]

---

[134] [Rathborne 1616, 142].

[135] Much has been written about the culture of the English mathematical practitioners. For a start
on the literature, see [Taylor 1954] and [Taylor 1966]. A more recent account, arguing (in part)
that the upper classes were not entirely separate from the trade, is [Feingold 1984]. See also
[Bennett 1982], [Bennett 1991a], [Johnston 1994], [Neal 1999], [Hackmann 2000], and [Cor-
mack 2006], among others.

It is thus no surprise that England took the lead in the integration of trigonometry with navigation. This had not yet begun in the mid-sixteenth century, with European trigonometry still in its infancy and still firmly attached to astronomy; as Leonard Digges in his 1553 *Prognostication* had lamented, "but those who have tried [to introduce trigonometry] know how far this passes the capacity of the common man."[136] However, in 1581, surely before he had seen the flood of trigonometry textbooks that was just starting to appear, naval officer William Borough advocated using trigonometric tables to calculate the sun's azimuth, referring to Regiomontanus and the tables of Copernicus, Reinhold, and Rheticus, although apparently he had not seen Viète's *Canon mathematicus*.[137]

Trigonometry was circulating in England, but it did not really enter into English publications until its use in navigation became clearer near the end of the century as awareness of the practical value of the subject was growing.[138] An appendix to Thomas Blundeville's popular 1594 *Exercises* dedicated to astronomy, geography, and navigation,[139] larger than the rest of the book, contained the first trigonometric tables published in England (explicitly borrowed from Clavius). Blundeville illustrated the use of these tables to solve astronomical problems important for navigation and printed them in a compact size helpful for use at sea.[140]

Two major navigational books, both published in 1614, solidified the union of trigonometry with navigation. The first was a partial translation of Pitiscus's *Trigonometriae* by Ralph Handson, a friend of Aaron Rathborne and a student of Henry Briggs, of whom we shall say more in chapter 2. Handson added a section on navigation, "wherein is manifested, the disagreement betwixt the ordinarie sea-Chart, and the globe, and the agreement betwixt the globe, and a true sea-chart: made after Mercator's way, or Mr. Edw. Wright's projection: whereby the excellency of the art of triangles will be the more perspicuous."[141] (We shall discuss this projection shortly.) Among Handson's

---

[136] Quoted in [Taylor 1954, 52].

[137] Quoted in [Taylor 1957, 211]. Borough speaks of completing for himself the second half of Rheticus's *Canon doctrinae triangulorum* (Rheticus had calculated the table for arguments up to 45°), either unaware that one may simply read the columns backward to generate the entries for arguments greater than 45° (although Rheticus had provided arguments working backward on the right side of his table for this purpose) or hoping to provide tables with arguments up to 90° for easier use in "Navigation and Cosmographie."

[138] John Blagrave's *The Mathematical Jewel* [Blagrave 1585], a description of a new mathematical instrument, contains definitions of trigonometric functions; however, he solves triangles not with the functions but with his new instrument.

[139] [Blundeville 1594]; see also the facsimile edition [Blundeville 1971].

[140] [Waters 1958, 355–356].

[141] [Pitiscus (Handson) 1614, nautical section, 1].

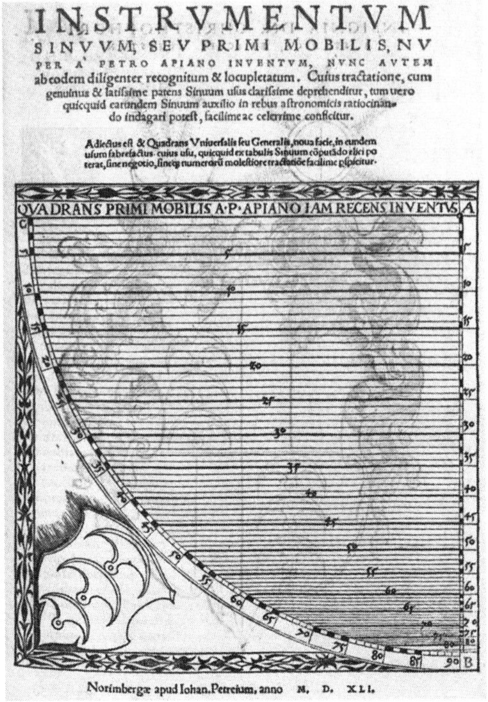

INSTRVMENTVM SINVVM, SEV PRIMI MOBILIS, NVPER A' PETRO APIANO INVENTVM, NVNC AVTEM ab eodem diligenter recognitum & locupletatum. Cuius tractatione, cum genuinus & latissime patens Sinuum usus clarissime deprehenditur, tum uero quicquid earundem Sinuum auxilio in rebus astronomicis ratiocinando indagari potest, facilime ac celerrime conficitur.

Adiectus est & Quadrans Vniuersalis seu Generalis, noua facie, in eundem usum fabrefactus. cuius usu, quicquid ex tabulis Sinuum cōputādo elici poterat, sine negotio, sineq; numerorū molestiore tractatiōe facilime cōspicitur.

QVADRANS PRIMI MOBILIS A·P·APIANO IAM RECENS INVENTVS

Norimbergæ apud Iohan. Petreium, anno M. D. XLI.

Figure 1.27
The title page of Peter Apian's 1541 *Instrumentum sinuum, seu primi mobilis.*

contributions was the "mid-latitude formula," which allowed sailors to determine, from the longitudes and latitudes of two places, their bearing and distance from one another. Ease of calculation, important for navigators, was important to Handson; he emphasizes the benefits of prosthaphairesis to convert multiplications to additions in the same year that his Scottish colleague John Napier was to render it obsolete.[142] Handson's book was aided into publication by his colleague John Tapp, who the same year published a new edition of Robert Norman's *The Newe Attractive* and William Borough's *Discourse on the Variation of the Cumpas*, to which he appended a set of navigational techniques for use with trigonometric tables.[143] Tapp's intent was to promote the "arithmetical sailing," that is, trigonometric methods with tables. The computational barriers to these methods, not inconsiderable in practice, were to become much more benign before the year was out.

However, in the meantime, the need to calculate—especially multiplication and division with trigonometric quantities—was a near-fatal disadvantage; while seamen might have been capable of the task, it was cumbersome

[142] For an account see [Waters 1958, 393–399].

[143] [Norman 1614]. See [Waters 1958, 559–562] for an account of Tapp's trigonometric navigation.

when required on a regular basis and, more seriously, prone to error. The alternative to calculation with trigonometric tables was the use of mathematical instruments, which worked much more quickly and easily, and the loss of precision caused by the use of a physical device was insignificant for navigation. Several such instruments had existed for centuries; see for instance the sine quadrant on the title page of Peter Apian's 1541 *Instrumentum sinuum seu primi mobilis* (figure 1.27). However, the needs of tradesmen and navigators in the context of the new practical mathematics seems to have brought instruments freshly into the discussion; in 1598 Thomas Hood and Galileo independently invented "sectors" with multiple uses that were predecessors to the slide rule.[144]

However, the sector that really made arithmetical navigation accessible was invented by Edmund Gunter around 1606. A young recent graduate of Oxford, Gunter would become associated with Henry Briggs and Edward Wright at Gresham College several years later. His fame rests on his instruments, especially the sector and a quadrant also named for him. Indeed, his connection with instruments and hence the class of mathematical practitioners seems at least once to have decreased his reputation. John Aubrey recounted his interview with Henry Savile for the first Savilian chair of geometry at Oxford:

> [Gunter] came and brought with him his sector and quadrant, and fell to resolving of triangles and doing a great many fine things. Said the grave knight, "Do you call this reading of Geometry? This is showing of tricks, man!" and so dismissed him with scorn, and sent for Briggs from Cambridge.[145]

It took Gunter until two years before his death to publish a book on his invention, the *De sector et radio* (1624), but his work had circulated widely in manuscript long before that.[146] Likely inspired by Hood's device, Gunter's sector is a simpler instrument honed for the purpose of calculation (figure 1.28).

---

[144] The origins of the sector are not entirely clear; see [Williams/Tomash 2003] for a survey and a description of the other lines on the instrument. On Hood and his sector, see [Johnston 1991] and [Taylor 2013]. On Galileo's sector see [Galileo 1978]. [Drake 1977] demonstrates that Hood and Galileo worked independently. For the contribution of Antwerp mathematician Michiel Coignet, see [Meskens 1997].

[145] [Aubrey 1982, 117]; see [Higton 2001] for a discussion of the context of the issue. Even today the attitude persists; the *Dictionary of Scientific Biography* entry says that "the tools he provided were of immense value long afterward," but that his contributions were "essentially of a practical nature," and that he was merely a "competent but unoriginal mathematician" [Pepper 1972, 593].

[146] [Gunter 1624]; see [Higton 2013] on the illustrations and diagrams in this work. On Gunter's sector and other contributions to navigation, see [Waters 1958, 358–392] and [Cotter 1981].

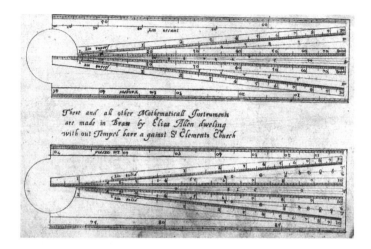

Figure 1.28
Gunter's sector, from his 1636 *Description and Use of the Sector,
Crosse-Staffe and Other Instruments.*

It has two arms fixed with a hinge at one end and various scales marked on
both sides of each arm. Scales for the sine, tangent, and secant allow the device to solve any triangle, plane or spherical. For instance, the sine scale is
marked so that the distance of any point from the hinge corresponds to the sine
of the angle indicated at that point. The arms open outward, and with a pair
of compasses the user is able to form similar triangles that correspond to
various ratios such as those that arise in the solutions of right-angled spherical
triangles.

Text 1.6
Gunter on Solving a Right-Angled Spherical Triangle with His Sector
(from *De Sectore et Radio*)

*In a rectangle triangle: To find a side by knowing the base, and the angle op-
posite to the required side.*
    As the Radius
      is to the sine of the base;
    So the sine of the opposite angle
      to the sine of the side required.
    As in the rectangle *ACB*, having the base *AB*, the place of the Sun 30° from
the equinoctial point, and the angle *BAC* of 23°30′ the greatest declination, if
it were required to find the side *BC* the declination of the Sun.
    Take either the lateral sine of 23°30′ and make it a parallel radius; so the
parallel sine of 30° taken and measured in the side of the Sector; shall give

the side required 11°30′. Or take the sine of 30° and make it a parallel radius; so the parallel sine of 23°30′ taken and measured in the lateral sines, shall be 11°30′ as before.[147]

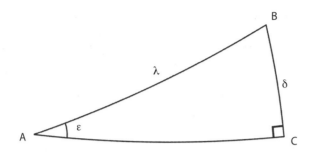

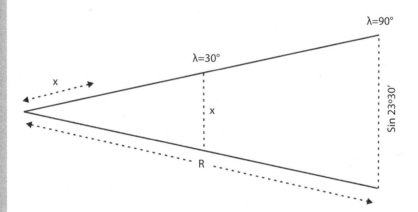

Figure 1.29
Finding the declination using Gunter's sector.

Explanation: (See figure 1.29) By "base," Gunter means the hypotenuse of the spherical triangle. Gunter's first example is a standard astronomical problem: find the sun's declination δ from its ecliptic longitude λ. The solution is $\sin \delta = \sin \lambda \sin \epsilon$ (where $\epsilon = 23°30′$ is the obliquity of the ecliptic), equivalent to the modern formula $\sin a = \sin A \sin c$ for a right triangle. But Gunter expresses it as $\dfrac{\mathrm{Sin}\,90°}{\mathrm{Sin}\,\epsilon} = \dfrac{\mathrm{Sin}\,\lambda}{\mathrm{Sin}\,\delta}$ for good reason.

The "line of sines" is the unequally marked scale near the middle in figure 1.28, ending at 90°, displayed on both arms of the sector. Set a compass

[147] [Gunter 1624, 76].

along the line of sines so that the distance between the two tips is equal to Sin 23°30′, that is, the distance along the line of sines from zero to 23°30′. Move the compass to the end of the sector, and spread the sector's arms so that their ends touch both ends of the compass. Without changing the angle of the pivot, move the compass inward (narrowing the gap between its tips) so that its two ends touch the two locations on the sector corresponding to 30°. We now have two similar triangles; from them, we have sin 90°/sin 23°30′ = sin 30°/x, where x is the new distance between the compass tips. From Gunter's ratio above, we know that x is equal to sin δ. Move the compass so that one of its tips is at the pivot. On the line of sines, the point corresponding to the other tip (11°30′) is δ.[148]

One of the scales on Gunter's instrument is entitled "meridional parts," and therein lies the final episode of this chapter. The shortest voyage between two ports is of course the great circle arc between them. However, traveling along this course is difficult because one's bearing changes continuously and so frequent course corrections are required. A simpler choice (although a slightly longer journey) is to travel along a path with a constant bearing, called a *loxodrome* or *rhumb line*. It would be helpful for a navigator to have in his possession a map with the property that a straight line on the map corresponds to a rhumb line on the ocean. A straight line drawn on the map, say, at a 45° angle upward and to the right, would follow a northeast bearing at every point. Pedro Nuñez had discovered the difference between great circles and rhumb lines in 1533. The first to construct a map with the desired property (in 1569) was none other than Gerard Mercator, a former pupil of Gemma Frisius.[149] For a map to achieve the required property, it turns out that the lines corresponding to latitude circles must be spaced not at equal intervals but at ever greater distances from each other as one moves from the equator to a pole (figure 1.30). Although this notion is at the heart of the Mercator projection,

---

[148] The reader may object that the arms cannot be spread far enough apart to fit R (the length of the sector) between the two 23°30′ indicators. However, elsewhere in the treatise Gunter explains how one may scale quantities up and down using linear scales (the "line of lines," marked from zero to ten printed on the other side of the sector. Using similar triangles as above, one may use compass distances of $R/10$ and sin 30°/10. On the "line of sines" side of the sector, separate the 23°30′ indicators by $R/10$. Then insert the compass points separated by sin 30°/10 at the appropriate place on the line of sines to find δ as before.

[149] [Mercator 1961]. The literature on Mercator is enormous; we point out only a few recent items. [Crane 2002] and [Taylor 2004] are two of the most recent biographies while [Monmonier 2004] is a social history of the projection, including the early history but also the modern debate with the alternative Peters projection. [D'Hollander 2005] deals with the projection itself. [Delevsky 1942] also considers the possible sources of Mercator's ideas. There also have been more than a few volumes of collected papers over the past two decades on Mercator and his historical context.

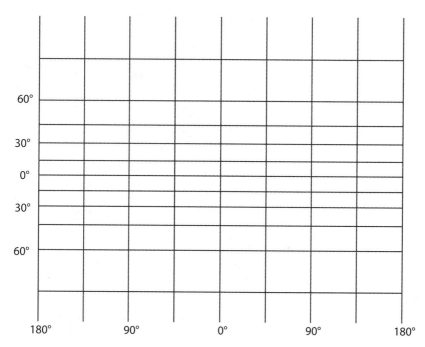

Figure 1.30
Longitude and latitude lines in Mercator's projection.

Mercator's latitude circles on his own map were not very accurately placed, and it is unclear what process he invoked to place them.

The first published solution to the problem of the spacing of the latitude lines is due to Edward Wright in his 1599 *Certaine Errors in Navigation* (figure 1.31).[150] Also known for his translation of Napier's *Mirifici logarithmorum canonis descriptio* in 1618 (published the year after Napier died), Wright also collaborated with Henry Briggs for many years; Wright, Briggs, and Edmund Gunter were together at Gresham College in 1615. Wright's interest in logarithms and mathematical instruments was practical—for the service of navi-

---

[150] [Wright 1599], although some of his table of meridional parts had previously appeared multiple times, first in [Blundeville 1594]. Indeed, the whole book nearly appeared under someone else's name before Wright was compelled to publish; clearly the demand for Wright's ideas was strong. In addition to various scholarly treatments (including within some of the books on Mercator we mentioned previously), the method has been described in several popular articles; see for instance [Rickey/Tuchinsky 1980], [Fernández Garcia/Jiménez Alcón/ Muñoz Prieto 2001], and [Maor 2002, 174–177]. On Wright and his work, see [Parsons/ Morris 1939] and [Waters 1958, especially pp. 219–229].

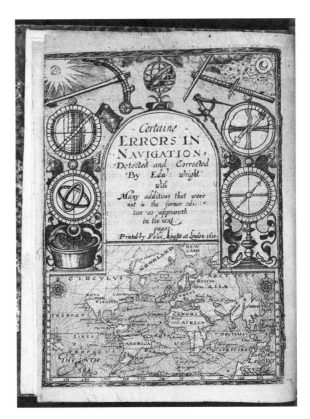

Figure 1.31
The frontispiece of the
second edition (1610) of
Edward Wright's
*Certaine Errors in
Navigation.*

gation. His *Certaine Errors* became a landmark, if one may put it that way,
for finding one's way at sea.

The idea behind Wright's solution to the problem of the spacing of the
latitude circles is straightforward. The latitude circles in figure 1.30 are all
drawn as if they have the same length, but on the globe their lengths vary in
proportion to cos ϕ, where ϕ is the latitude. Therefore, horizontal distances
(longitudes) have been stretched relative to the equator by a factor of the reciprocal of cos ϕ, that is, sec ϕ. To preserve bearings, the vertical distances
ΔA (where A is the northward distance on the map from the equator to the
latitude line corresponding to ϕ) must be stretched by the same ratio. So, at
latitude ϕ, ΔA should be proportional to sec $\phi \cdot \Delta A$ (at latitude 0°), but since
there is no stretching at the equator, at latitude 0° ΔA is equal to $\Delta \phi$. Hence
$\Delta A = k$ sec $\phi \cdot \Delta \phi$.

The modern calculus student will notice immediately that this is the same
as $A(\phi) = k \int_0^\phi \sec \varphi \; d\varphi$, but at Wright's time calculus was still many decades

away. So, to construct his table of meridional parts, Wright was forced into an onerous calculation (which he described as "an easy way laid open"):

> For . . . by perpetuall addition of the secantes answerable to the latitudes of each point or parallel unto the summe compounded of all the former secants, beginning with the secans of the first parallel's latitude, and thereto adding the secans of the second parallel's latitude, and to the summe of both of these adjoyning the secans of the third parallel's latitude, and so forth in all the rest, we may make a table which shall shew the sections and points of latitude in the meridians of the nautical planisphere: by which sections, the parallels are to be drawne.

Effectively, then, Wright uses a Riemann sum to compute $A(\phi)$. He chooses $\Delta\phi = 1'$ but helpfully refers readers to Rheticus's *Opus palatinum* should someone wish to take on the thankless task of improving the accuracy of the calculation by decreasing $\Delta\phi$ to $10''$.[151]

Wright was not the only English navigator working on the problem of meridional parts. John Dee (1527–1609), a friend of Mercator and Nuñez and a student of Gemma Frisius, had produced tables that predated Mercator's 1569 map, although the method he used to calculate them is unknown.[152] Later, Dee's colleague Thomas Harriot (1560–1621) (who himself served on an ocean-going expedition to Virginia with Sir Walter Raleigh in the 1580s) would also venture in this direction. Harriot's highly innovative work in mathematics and science never saw a printing press during his life. Today it exists only in manuscripts, notes, and modern scholarly editions. In mathematics he is known especially for his contribution to the theory of equations; closer to our interests here, he also was the first to state the area of a spherical triangle (although he did not prove his result). Harriot constructed tables of meridional parts in the 1580s or 1590s, not long after his voyage with Raleigh. He revisited the topic late in his life and in 1614 constructed a large table of meridional parts with the aid of finite difference interpolation.[153] We shall discuss this topic in chapter 2.

---

[151] [Wright 1599, chapter entitled "Faults in the common sea chart," from the 17th to the 19th page].

[152] The literature devoted to Dee is extensive, but not much attention has been paid to his interest in navigation; see [Taylor 1955], [Taylor 1957, 195–207], [Alexander 2005], and [Baldwin 2006]. For his table of meridional parts see [Taylor 1963, 415–433].

[153] The history of Harriot's contribution to the problem of meridional parts has been controversial. See [Taylor/Sadler 1953], [George 1956], [Lohne 1965/66], [Pepper 1967a], [Pepper 1967b], [George 1968], and especially [Pepper 1968] and [Pepper 1976]. [Taylor/Sadler 1953] and [Pepper 1967b, 23–25] reveal that Harriot somehow knew some formula for meridional parts.

2 ✵ Logarithms

Seldom in the history of mathematics has a major development come as suddenly or made such an immediate impact on society as the announcement of logarithms in 1614. Trigonometry had been gaining recognition since the 1580s for its potential to ally mathematics with efforts to analyze and control the physical world; it was becoming clear that the subject had the potential to extend well beyond the discipline of astronomy that had been its natural place since ancient Greece. In principle, the case for the use of mathematics in surveying, architecture, navigation and other practical pursuits was strong. However, practitioners did not rush to embrace the newcomer to their game. The reluctance was not merely conservatism. Application of trigonometry required skills not always available to the tradesman. Worse, often the necessary calculations were so tedious that the advantages were overweighed by the time commitment and vulnerability to error that were introduced by working with numerical tables. Some contemporary advances had the capacity to reduce the computational burden, especially the invention of sectors and the use of prosthaphaeresis. The latter—the application of the sine and cosine product-to-sum identities to replace multiplications with quicker and more reliable additions—was making a difference, but its benefits were being reaped only in astronomical circles. When logarithms were announced, they rendered prosthaphaeresis obsolete by moving beyond the products only of sines and cosines (although prosthaphaeresis was still practiced for some decades thereafter). Logarithms brought higher mathematics to the people: they removed the computational barriers that had prevented it from becoming a powerful tool in the sciences and the world of practitioners.

▪ Napier, Briggs, and the Birth of Logarithms

John Napier (1550–1617) was a Scottish landholder and served as baron of Merchiston, although he lived in the Merchiston castle only for the last nine years of his life. His passions were diverse, but the energies devoted to them were unremittingly high. His activism for the cause of Protestantism in opposition to Catholicism is highlighted by his 1593 *Plaine Discoverie of the Whole Revelation of Saint John*,[1] which uses the book of Revelation to advocate for a cleansing of the Scottish court of all but Protestants and to apply "justice" to the enemies of the church. Until near the end of his life, Napier's

---

[1] [Napier 1593]. For a survey of this work see [Havil 2014, chapter 2].

reputation rested on his combative theology. He developed agricultural technology as well and was responsible for several experiments and inventions that were applied in his lands.[2]

In the mathematical sciences, Napier cared most about efficiency in calculation. His *Rabdologiae*,[3] which appeared in the year of his death, describes Napier's rods (or bones), vertical rods engraved with numbers that simplified certain arithmetical calculations. Another of his works, *De arte logistica*, a manuscript perhaps intended to serve as a textbook, is wholly devoted to methods of computation.[4] The book was not published until over two centuries after his death by his descendant Mark Napier, who believed that John wrote it in the early 1590s.

However, without question his most lasting legacy has been the invention of logarithms. The idea likely began to occupy his attention already in the early 1590s, but logarithms were not to reach an audience until almost a quarter century later. His interest may have been spurred by a communication (via intermediary John Craig) with Tycho Brahe, where Napier became aware of the benefits of prosthaphaeresis; he may in turn have informed Brahe of his work on logarithms.[5] In any case, Napier continued his labors on the subject and his tables in silence for another two decades.

The silence ended in 1614 with the publication of a small book and set of tables entitled *Mirifici logarithmorum canonis descriptio*, today usually called the *Descriptio*.[6] Its full title, in Edward Wright's 1616 English translation, is as follows: "A description of the admirable table of logarithmes: with a declaration of the most plentiful, easy, and speedy use thereof in both kindes of trigonometry, as also in other calculations." A poem introducing the book puts it thus:

> The toylesome rules of due proportion
> Done here by addition and subtraction,
> By bipartition and tripartition,
> The square and cubicke roots extraction:
>> And so, all questions geometricall,
>> But with most ease triangles-sphericall.

---

[2] For general biographies of Napier's life and contributions, see [M. Napier 1834], [Gibson 1914], [P. Brown 1915], [Shennan 1989], [Gladstone-Millar 2006], and [Havil 2014].

[3] [Napier 1617]; translated to English in [Hawkins 1982b].

[4] [Napier 1839]; translated to English in [Hawkins 1982b]; see also [Steggall 1915].

[5] [M. Napier 1834, 361–363].

[6] [Napier 1614]; for English translations see [Napier (Wright) 1616] and [Napier (Filipowski) 1857]. [Naux 1966, vol. 1, 43–66] provides a useful survey of parts of the book.

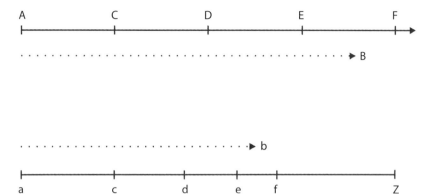

Figure 2.1
Napier's definition of a logarithm.

> The use is great in all true measuring
> of lands, plots, buildings, and fortification,
> So in astronomie and dialling,
> Geography and navigation.
>> In these and like, yong students soone may gaine,
>> The skilfull too, may save cost, time, & paine.[7]

The clearest sign that Napier invented logarithms with trigonometry specifically in mind is that after the initial 20 pages defining and describing logarithms, the remaining 37 pages are entirely devoted to trigonometric applications.

Napier's definition of a logarithm relies on the comparison of a magnitude increasing at an arithmetical rate with another magnitude increasing geometrically.[8] In figure 2.1, let point *B* move with constant velocity along the ray *ACDEF*, whose labeled points we consider to be one unit apart. At the same time, let point *b* move with same initial velocity along segment

[7] [Napier 1616, fourth and third pages before page 1].

[8] This passage has been described countless times; for just a few, see [Cajori 1913a], [Cairns 1928], [Sleight 1944], [Gridgeman 1973], [Phillips 1980], [Ayoub 1993], [Katz 1995], [Bruce 2000], and [Le Corre 2006]. One finds numerous statements that the base of Napier's logarithms is $1/e$, which is true but not relevant to how Napier himself was thinking. For an account of this definition in the context of the development of the notion of a function, see [Whiteside 1961c, 214–231]. On the role Napier's definition played in the gradually disappearing distinction between numbers and magnitudes, see [Neal 2002, chapter 5]. For a brief description of the scientific and computational context, see [Brioist 2004].

acdefZ, passing *c, d, e, f* as *B* passes *C, D, E, F* but in such a way that
$\dfrac{aZ}{cZ} = \dfrac{cZ}{dZ} = \dfrac{dZ}{eZ} = \dfrac{eZ}{FZ}$. Segment *aZ* is set equal to the radius of the trigonometric base circle; in Napier's case, $R = 10^7$. Then, as the length of *AB* in-
creases one unit at a time, *bZ* decreases as follows:

AB:	0	1	2	3	. . .
bZ:	R	$R\left(1-\dfrac{1}{R}\right)$	$R\left(1-\dfrac{1}{R}\right)$	$R\left(1-\dfrac{1}{R}\right)^3$	. . .

Napier defines the logarithm of the geometrically decreasing length *bZ* to be
the corresponding arithmetically increasing length, *AB*. Lengths on the scale
aZ are considered to represent sines.

Clearly Napier's logarithm, which we shall call "NLog," differed from
the modern function. As Napier goes on to prove, it does have some of the
desired properties. Of most interest to us (and to his audience of astronomers)
is his first proposition:

$$\text{If } \frac{a}{b} = \frac{c}{d}, \text{ then } \text{NLog}\,a - \text{NLog}\,b = \text{NLog}\,c - \text{NLog}\,d. \qquad (2.1)$$

We have already seen in chapter 1 that spherical trigonometric relations typically were presented as equalities of ratios, so this theorem would have caught
his readers' eyes immediately.

Text 2.1
Napier, Solving a Problem in Spherical Trigonometry with
His Logarithms
(from the *Descriptio*)

Let the *azimuth of the Sun rising* bee given *BS*, or *PZS* 70 degr. and the *elevation of the Pole*, 54 *degr.* which is *PB*, or the complement of *PZ*: and let the
difference ascensional be sought, that is, the complement of *SPB*. . . . And
because here likewise the extreme parts are set about the middle part,
therefore
Take the differentiall of the Suns azimuth, *BS*, or *BZS* 70 deg. which is
−1,010,683. Out of the Log. of the elevation of the pole *BP*, 54 deg. +211935.
And there will come forth +1,222,618 the differential of *SPB* 16 deg. 24′27″,
the arc of the *ascensional difference* sought for.[9]

---

[9] [Napier 1616, 53].

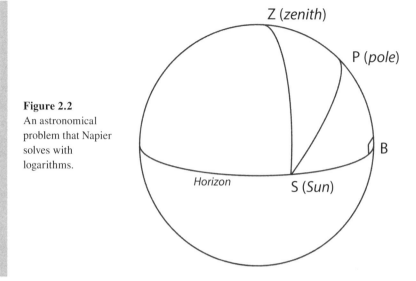

Figure 2.2
An astronomical
problem that Napier
solves with
logarithms.

Explanation: (See figure 2.2; this diagram is reconstructed from Napier's calculations.) The angle at B, the north point of the horizon, is 90°. Given $PB = 54°$ and $SB = 70°$, the goal is to find the complement of $\angle SPB$. Napier uses the following identity, expressed in ratios as usual:

$$\frac{\text{Tan } SB}{\text{Sin } PB} = \frac{\text{Tan } \angle SPB}{R}.$$

Since $N \log R = 0$, we may simply take the difference of the Napierian logarithms of the terms in the ratio on the left to get the Napier logarithm of the term on the right. (This, in fact, is why Napier chose NLog $R = 0$.) He then looks up the Napierian logarithm of sin 54° directly in his table, getting 211,935.[10]

Napier cleverly handles the logarithm of the tangent as follows. In figure 2.3 (a page from his table), the argument increases on the left from 20°0′ to 20°30′, and the complement of the argument decreases on the right from 69°30′ to 70°0′. Since $\dfrac{\text{Tan } \theta}{R} = \text{Sin } \theta / \text{Sin}(90° - \theta)$, in each row the difference between the two logarithms of sines is equal to the logarithm of the tangent of the argument. He tabulates this difference in the central column. Here Napier requires the logarithm of tan 70°, which is greater than R; it turns out to be equal to the negative of the logarithm of tan 20°, which is 1,010,683. Subtracting the logarithm of the sine of 54° (which is 211,935) gives −1,222,618, which according to our theorem is equal to the logarithm of the tangent of $\angle SPB$.

---

[10] Wright's edition uses $R = 10^6$ rather than the 10^7 in Napier's original tables, so the calculations shown here are given to one less decimal place than the table illustrated in figure 2.3.

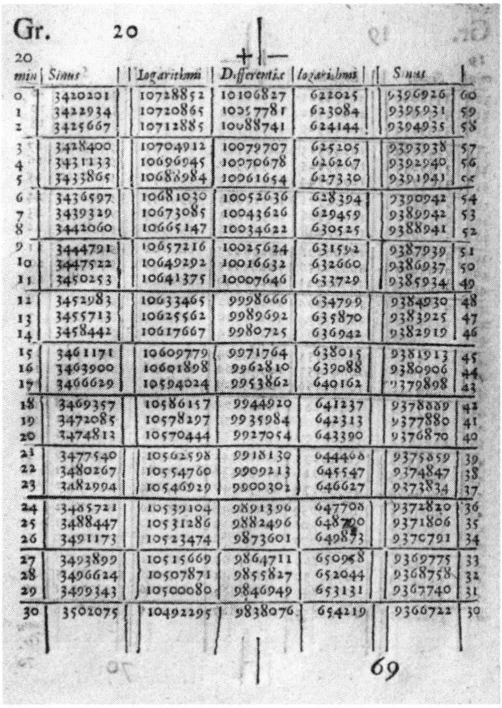
Gr. 20

20 min	Sinus	Logarithmi	Differentiæ	logarithmi	Sinus	
0	3420201	10728852	10106827	622025	9396926	60
1	3422934	10720865	10097781	623084	9395931	59
2	3425667	10712885	10088741	624144	9394935	58
3	3428400	10704912	10079707	625205	9393938	57
4	3431133	10696945	10070678	626267	9392940	56
5	3433865	10688984	10061654	627330	9391941	55
6	3436597	10681030	10052636	628394	9390942	54
7	3439329	10673085	10043626	629459	9389942	53
8	3442060	10665147	10034622	630525	9388941	52
9	3444791	10657216	10025624	631592	9387939	51
10	3447522	10649292	10016632	632660	9386937	50
11	3450253	10641375	10007646	633729	9385934	49
12	3452983	10633465	9998666	634799	9384930	48
13	3455713	10625562	9989692	635870	9383925	47
14	3458442	10617667	9980725	636942	9382919	46
15	3461171	10609779	9971764	638015	9381913	45
16	3463900	10601898	9962810	639088	9380906	44
17	3466629	10594024	9953862	640162	9379898	43
18	3469357	10586157	9944920	641237	9378889	42
19	3472085	10578297	9935984	642313	9377880	41
20	3474813	10570444	9927054	643390	9376870	40
21	3477540	10562598	9918130	644468	9375859	39
22	3480267	10554760	9909213	645547	9374847	38
23	3482994	10546929	9900302	646627	9373834	37
24	3485721	10539104	9891396	647708	9372820	36
25	3488447	10531286	9882496	648790	9371806	35
26	3491173	10523474	9873601	649873	9370791	34
27	3493899	10515669	9864711	650958	9369775	33
28	3496624	10507871	9855827	652044	9368758	32
29	3499343	10500080	9846949	653131	9367740	31
30	3502075	10492295	9838076	654219	9366722	30

69

Figure 2.3
A page from the table of logarithms in Napier's *Descriptio.*

Finally Napier uses the table backward to find the antilogarithm of +1,222,618, which (due to the sign switch) gives the complement of $\angle SPB$ as 16°24′27″.

One may wonder why we did not present an equivalent to the modern formula for products, $\log(ab) = \log a + \log b$: the reason is that Napier himself did not do so.[11] His presentation was geared to the theorems of spherical astronomy, so dealing with ratios was much more important. In fact, Napier's logarithms encounter an extra wrinkle when products are considered:

$$\mathrm{NLog}(ab) = \mathrm{NLog}\ a + \mathrm{NLog}\ b - \mathrm{NLog}\ 1. \tag{2.2}$$

The extra effort required to subtract NLog 1 every time we try to convert a multiplication to an addition almost doubles the computational workload. While this shortcoming was not particularly cumbersome for spherical astronomy, Napier was not satisfied. It did not take long for him to recognize

[11] One might also wonder if there is a method to find $\log(a+b)$ from the logarithms of a and b. In fact there is; it was discovered by Italian mathematician Bonaventura Cavalieri in 1639 and uses trigonometry. See [Govi 1875–76].

and propose a solution, eventually implemented by his colleague Henry Briggs after Napier's death.

Briggs (1561–1630) was the first professor of geometry at Gresham College, the center of the study of practical mathematical disciplines in early seventeenth-century England. He spent his early career interested in questions related to astronomy and navigation, the most immediate beneficiaries of the advances in computational efficiency that were swirling in the air. He spent his last decade serving as the first Savilian professor of geometry at Oxford (chosen over Edmund Gunter, who became professor of astronomy at Gresham College that same year). Without a doubt, his receipt of Napier's *Descriptio* was the defining moment of his career. He instantly recognized the power of the new idea, spent a month each in the summers of 1615 and 1616 in Scotland with Napier, and dedicated himself to elaborating and enhancing logarithms over much of the next decade. Among these efforts, Briggs helped to bring Edward Wright's translation of the *Descriptio* to press as early as 1616, shortly after Wright's passing and shortly before Napier's. His own first work on logarithms, a small privately circulated table called *Logarithmorum chilias prima*, was printed in 1617 and announces Napier's death.[12]

In a letter to Napier early in 1615, Briggs expressed his admiration for Napier's logarithms (indeed, in one account, at their first meeting they stared at each other for a quarter of an hour speechless with awe). He proposed one important revision of Napier's definition: namely, to keep $\log(R)=0$ but to make $\log(\mathrm{Sin}\ 5°44'21'')=\log(R/10)=10^{10}$. During his visit that summer, Napier agreed that a change should be made but counterproposed that $\log 1=0$ and $\log R=10^{10}$. Briggs readily agreed that this was better.[13] Napier's health did not permit him to proceed, so Briggs took on the task of constructing the table. The *Logarithmorum chilias prima* is a first attempt: it tabulates the function $10^{14}\log x$ for all whole numbers between 1 and 1,000, making it the first to compute logarithms in base 10. With $\log 1=0$ the new function was much easier to work with. There was a small cost to Briggs's table when used with trigonometry: to find the logarithm of the sine of an arc, the user needed to use a sine table first and then enter the sine value into the table of logarithms.

---

[12] For biographies of Briggs, see [Ward 1740, 120–129] and [Sonar 2002]. For accounts of Briggs's career and interactions with Napier, see [Hutton 1811b, 33–37] and [Sonar 2002]. The *Logarithmorum chilias prima* [Briggs 1617] is discussed in [Archibald 1955a] and [Jagger 2003, 54–59] and is recomputed in [Roegel 2010b].

[13] This is reported by Briggs in the preface to his *Arithmetica logarithmica* [Briggs 1624]. An online translation exists ([Bruce 2006]) but does not include the preface; the quotation is translated in [Gibson 1915, 126–127]. See [Gibson 1915] for a careful discussion.

▨   Interlude: Joost Bürgi's Surprising Method of Calculating a Sine Table

Meanwhile, German astronomical instrument maker Joost Bürgi (1552–1632) was working on logarithms independently and almost at the same time. Bürgi was also interested in the applications of trigonometry and aids to computation: he completed a sine table and was in possession of a proof of at least one of the formulas for prosthaphaeresis.[14] His work on logarithms may be found in his 1620 *Aritmetische und Geometrische Progress Tabulen*, which contains a table of antilogarithms.[15] Notwithstanding Bürgi's interest in matters of trigonometry, this manuscript contains no references to the subject. Neither did it impact the flood of logarithm tables that were about to appear.[16]

Until very recently, the method Bürgi used to compute his sine table was unknown. It had been known that something new was afoot from a reference to it by his friend Nicolaus Ursus in another manuscript:

> The calculation of the *Canon Sinuum* can be done . . . either in the usual way, by inscribing the sides of a regular polygon into a circle . . . , that is, geometrically. Or it can be done by a special way, by dividing a right angle into as many parts as one wants; and this is arithmetically.[17]

Ursus advertises that this new arithmetic method differs qualitatively from the traditional geometric methods, somehow bypassing the issues that had plagued trigonometric table makers since Ptolemy: namely, that values can be found only for angles of the form $3m/2^{no}$. But Ursus says no more about it, other than to credit Bürgi.

However, a recently discovered manuscript authored by Bürgi himself, the *Fundamentum astronomiae*, reveals an astonishing resolution to the mystery.[18] The method is indeed entirely new. Bürgi illustrates it with a sample

---

[14] [Thoren 1988, 38]. Kepler composed an introduction to Bürgi's sine tables, which was never published ([List/Bialas 1973]).

[15] See [Clark 2015] for an edition and commentary on the *Aritmetische und Geometrische Progress Tabulen*.

[16] It has been a matter of some debate for many decades whether Napier or Bürgi deserves priority for the discovery of logarithms; it seems that they both were working on the subject in the 1590s. Since Bürgi did not connect his logarithms with trigonometry directly, the question is not relevant here. For descriptions and a contrast of the approaches of both men as well as a critique of the nature of the debate, see [Clark/Montelle 2012].

[17] [Folkerts/Launert/Thom 2016, 138].

[18] The manuscript is announced, and the method described in [Folkerts/Launert/Thom 2016]. See also [Launert 2015] for an edition of the text with commentary. Finally, see [Wagner/

0°	0		0		0		0				0
		2,235,060		67,912		2,064			63		
10°	2,235,060		67,912		2,064		63				2
		2,167,148		65,848		2,001		61			
20°	4,402,208		133,760		4,065		124				4
		2,033,388		61,783		1,877		57			
30°	6,435,596		195,543		5,942		181				6
		1,837,845		55,841		1,696		51			
40°	8,273,441		251,284		7,638		232				7
		1,586,461		48,203		1,464		44			
50°	9,859,902		299,587		9,102		276				8
		1,286,874		39,101		1,188		36			
60°	11,146,776		338,688		10,290		312				9
		948,186		28,811		876		27			
70°	12,094,962		367,499		11,166		339				10
		580,687		17,645		537		17			
80°	12,675,649		385,144		11,703		356				11
		195,543		5,942		181		6			
90°	12,871,192		391,086		11,884		362				12

Figure 2.4
Bürgi's method of calculating sine tables. The arrows indicate the direction of calculation for the first of the four steps shown here.

calculation for a small sine table tabulated for every 10° of arc. In figure 2.4,[19] the arcs are on the left; we begin calculating at the rightmost column. The entries in this column are extremely rough approximations of sine values where the value at the bottom for the argument 90° is R, in this case 12 (although, in fact, *any* column of numbers whatever will do here). Divide the bottom entry by two and place the result in the next column to the left, displaced half a row upward. Then build the rest of this new column from bottom to top by adding to each entry the value above and to the right of that entry. Once this column has been completed, start another column to the left with a zero at the top. This time, generate new entries below it by adding to each new value the entry below it and to the right. This new column turns

Hunziker 2019] for an English translation as well as a speculation that Bürgi's method was transmitted from India.
[19] Bürgi's entries are in sexagesimal notation; they have been converted to decimal notation here.

out to be a better approximation of sine values (with a new value of R at the bottom, in this case 362) than the original column was. Repeat the process as many times as desired; in Bürgi's sample table it is done four times. The column on the left (with $R = 12,871,192$) already has remarkably accurate sine values, in error by no more than three in the last place. Whenever one chooses to stop, one can scale the resulting column to any value of R that one finds convenient.

How is Bürgi able to produce sine values seemingly by magic, without even referring to the geometric definition of the sine? The method has been proven to converge to the correct values using linear algebra: if we consider the sine columns as vectors, we can represent the operations that generate new columns in terms of matrices, and it turns out that the vector containing the correct sines is an eigenvector toward which the columns converge as the process is iterated. Of course Bürgi could not have thought like this. More likely, he recognized that successive nth order differences of a sine table produce tables that are alternately multiples of cosines and sines—a fact that had been used to good effect by table makers in India centuries before. Today, we see this phenomenon as an artifact of the pattern in the sequence of nth order derivatives of $\sin x$ ($\cos x$, $-\sin x$, $-\cos x$, $\sin x$, . . .). The "shifted" columns between the converging sine tables may be thought of as approximations to $\cos 5°$, $\cos 15°$, $\cos 25°$, . . . , $\cos 85°$ (with their own R values). The idea behind Bürgi setting the values for $\cos 85°$ equal to half the R value in the previous column may be that he knew that $\cos 75°$ (the cosine value in the table to be calculated after $\cos 85°$—obtained by adding $\sin 80°$, whose value is close to R) needed to be about three times $\cos 85°$. The rapid convergence of the columns of sine values might have been discovered simply by experimenting with nth order differences.[20] If this is indeed how Bürgi arrived at the method, its success would have held a mystery even for him.

In any case, Bürgi's approach unfortunately had no impact. It is possible that Henry Briggs was aware of it and perhaps of others.[21] But with no publication and little circulation, it was quickly forgotten.

▨ The Explosion of Tables of Logarithms

It became very clear, very quickly, that logarithms had the potential to revolutionize any activity that relied on computation. Over the next decades, dozens of tables of logarithms were computed and published, especially in England and in northern Europe. Figure 2.5 identifies some of these tables

---

[20] Investigations of nth order differences in general were being performed at the time, most famously by Thomas Harriot.

[21] [Folkerts/Launert/Thom 2016, 142–143].

Author	Work	Type	Logarithm of Sine	Shift	Arguments	Step size	Significant figures
Napier	*Descriptio* (1614)	Napierian	✓		0° – 90°	1'	7
Briggs	*Logarithmorum chilias prima* (1617)	Base 10	x		1 –1000	1	15
Speidell	*New logarithmes* (1619)	*See text*	✓	✓	0° – 90°	1'	6
Gunter	*Canon triangulorum* (1620)	Base 10	✓	✓	0° – 90°	1'	8
Gunter	*Canon triangulorum*, 2nd ed (1624)	Base 10	x		1 –10,000	1	8
Briggs	*Arithmetica logarithmica* (1624)	Base 10	x		1 – 20,000 90,000 – 100,000	1	15
Kepler	*Chilias logarithmorum* (1624)	*See text*	✓ (equally spaced)		1 – 10, 10 – 100, 100 – 100,000	1 10 100	8 8 8
Ursinus	*Trigonometria* (1625)	Napierian	✓		0° – 90°	10''	9
Wingate	*Arithmétique Logarithmétique* (1625)	Base 10 Base 10	x ✓	✓	1 – 1000 0° – 90°	1 1'	8 8
Vlacq	*Arithmetica logarithmica*, 2nd ed (1628)	Base 10 Base 10	x ✓	✓	1 – 100,000 0° – 90°	1 1'	11 11
Norwood	*Trigonometrie* (1631)	Base 10 Base 10	x ✓	✓	1 – 10,000 0° – 90°	1 1'	8 8
Vlacq	*Trigonometria artificialis* (1633)	Base 10 Base 10	x ✓	✓	1 – 20,000 0° – 90°	1 10'	11 11
Briggs	*Trigonometria britannica* (1633)	Base 10	✓	✓	0° – 90°	0.01°	15 (sin), 11 (tan)
Roe	*Tabulae logarithmicae* (1633)	Base 10 Base 10	x ✓	✓	1 – 100,000 0° – 90°	1 0.01°	8 11

Figure 2.5
Early tables of logarithms.

and their properties.[22] The first tables to appear after Briggs's *Logarithmorum chilias prima* were John Speidell's *New logarithmes*.[23] They were related to Napier's tables but not identical to them (which must have made them difficult to use, for they come with no instructions); the second edition is the first set of tables in base *e*. Speidell's work was so popular that it went through nine editions by 1627.[24] As figure 2.5 indicates, it was also the first set of

[22] For surveys of tables of logarithms (indeed, of other tables as well), see [Hutton 1811], [De Morgan 1868], [Glaisher et al. 1873], [Henderson 1926], and [Fletcher et al. 1962].

[23] Other than a reprinting of an excerpt of Napier's tables by Benjamin Ursinus. See [Speidell 1619].

[24] [Jagger 2003, 69, 76]. On Speidell's logarithms, see also [Carslaw 1916, 483–485] and especially [Jagger, n.d.].

tables that implemented an innovation that became standard for tables of logarithms through the twentieth century. To avoid some of his logarithms being negative and others positive (which left calculators prone to error), Speidell subtracted Napier's logarithms from 100,000,000, rendering all the values positive.[25] This process would have been familiar to Islamic astronomers on the other side of the Mediterranean; they had applied a similar process of displacement to their astronomical tables for centuries.

Our next set of tables, Edmund Gunter's *Canon triangulorum*,[26] exhibits both similarities with and differences from Speidell. Rather than following Napier, Gunter chose to adopt Briggs's definition of the logarithm as given in the *Logarithmorum chilias prima*, making Gunter's table the first formal publication of base 10 logarithms. No doubt he was influenced by Briggs to make this choice. But he also adopted Speidell's idea of displacement, adding a power of ten to each number. The *Canon triangulorum* tabulates logarithms of sines and tangents, although Gunter's 1624 edition also contains a table of ordinary "Briggsian" logarithms. Like Speidell, Gunter includes almost no explanations in the first edition. Given that this was the first table of its kind, one assumes that the audience would have been among the small circle who already understood logarithms. This book, incidentally, also contains the first known uses of the term "cosine" and "cotangent" (see the last few lines in figure 2.6).

Briggs must have been very busy in the intervening years, for in 1624 he published one of the most intensive computational projects of all time: his 1624 *Arithmetica logarithmica*.[27] This large book describes how to calculate logarithms in about 100 pages and how to use them in various contexts. Following this are 300 pages of base 10 logarithms (not of sines) to 15 significant figures for arguments between 1 and 20,000 and between 90,000 and 100,000. After such a vast effort, one may understand why he chose to publish at this point. Briggs intended to fill in the missing values and had gathered friends together to complete the task, but in this enterprise he was beaten in 1628 by Dutchman Adrian Vlacq.[28] The new book, published as a second edition of *Arithmetica logarithmica* (this was not the era of copyright), gives values only to 11 significant figures but does fulfill the promise of closing the gap in Briggs's table. Vlacq also provided tables of logarithms of the sine, tangent, and secant, using the displacement seen previously in Speidell and

---

[25] Modern tables usually add ten to the tabulated logarithms.

[26] [Gunter 1620].

[27] [Briggs 1624]; see [Briggs 2006] for an English translation.

[28] [Vlacq 1628]; see [Jagger 2003, 65–66]. There is strong evidence that Ezechiel De Decker actually computed these tables; see [Henderson 1926, 55–57], [van Haaften 1928], [Bruins 1980], and [van Poelje 2005].

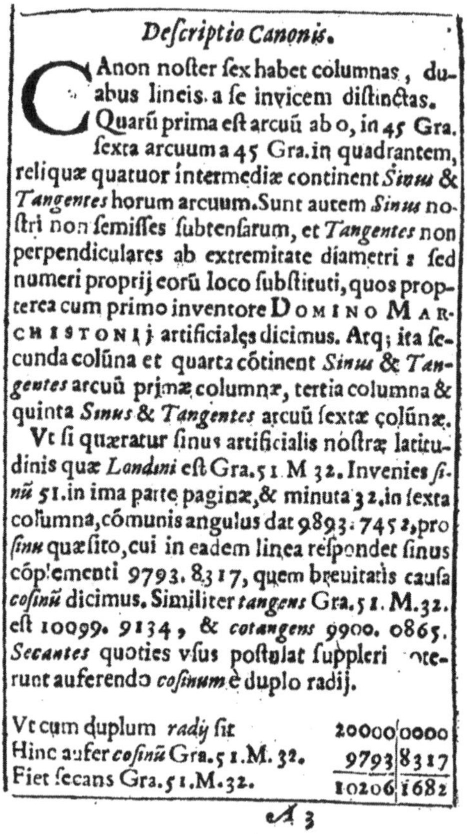

Deſcriptio Canonis.

CAnon noſter ſex habet columnas, duabus lineis. a ſe invicem diſtinctas. Quarũ prima eſt arcuũ ab o, in 45 Gra. ſexta arcuum a 45 Gra. in quadrantem, reliquæ quatuor intermediæ continent *Sinus* & *Tangentes* horum arcuum. Sunt autem *Sinus* noſtri non ſemiſſes ſubtenſarum, et *Tangentes* non perpendiculares ab extremitate diametri : ſed numeri proprij eorũ loco ſubſtituti, quos propterea cum primo inventore DOMINO MARCHISTONIJ artificiales dicimus. Atq; ita ſecunda colũna et quarta cõtinent *Sinus* & *Tangentes* arcuũ primæ columnæ, tertia columna & quinta *Sinus* & *Tangentes* arcuũ ſextæ colũnæ.

Vt ſi quæratur ſinus artificialis noſtræ latitudinis quæ *Londini* eſt Gra. 51 M 32. Invenies *ſinũ* 51. in ima parte paginæ, & minuta 32. in ſexta columna, cõmunis angulus dat 9893 . 7452, pro *ſinu* quæſito, cui in eadem linea reſpondet ſinus cõplementi 9793 . 8317, quem breuitatis cauſa *coſinũ* dicimus. Similiter *tangens* Gra. 51. M. 32. eſt 10099 . 9134, & *cotangens* 9900 . 0865. *Secantes* quoties vſus poſtulat ſuppleri poterunt auferendo *coſinum* è duplo radij.

Vt cum duplum *radij* ſit	20000	0000
Hinc aufer *coſinũ* Gra. 51. M. 32.	9793	8317
Fiet ſecans Gra. 51. M. 32.	10206	1682

A 3

Figure 2.6
The page of instructions for Gunter's *Canon triangulorum*, containing the first appearances of the words "cosine" and "cotangent."

Gunter's works. Vlacq's book seems to have upset Briggs and his friends, not because of any accusations of intellectual theft but because of omissions in the explanatory notes introducing the tables (including his work on interpolation by finite differences, below).[29]

Whatever hard feelings there had been, they could not have lasted long; Briggs's last major publication, the *Trigonometria Britannica*,[30] was helped through the press by Vlacq. The tables in this volume contain two innovations. Firstly, this was Briggs's first set of tables of logarithms of sines and tangents. Secondly and more notably, he introduced the division of a degree into decimal parts rather than sexagesimal: the increments in his table are 0.01° rather than the usual 1′. This initiated a small tradition of similar tables

[29] [Glaisher et al. 1873, 52]; see [Bruce 2002, 223] and the quotation from one of Briggs's letters in [Jagger 2003, 66].
[30] [Briggs 1633].

over the following decades (especially in the hands of John Newton[31]), the idea persisting either because the division is a logical one or because Briggs's 15-place sine tables were simply the best available (along with Pitiscus's table) and an obvious choice for copying. The decimal division of degrees did not last at the time, but with the invention of pocket calculators in recent years, it is now the dominant system.

Logarithms quickly found their way onto the continent through various reprints of existing tables. The first appearance was in Germany already in 1618, when Benjamin Ursinus published an extract of Napier's tables in his *Cursus mathematici practici*.[32] The real influx started in the mid-1620s, especially with Edmond Wingate's *Arithmétique Logarithmétique*[33] and Dénis Henrion's *Traicté des Logarithmes*,[34] both in France; Vlacq in the Netherlands (De Decker's own *Nieuw Telkonst* had been published in 1626 but is very rare); and Johann Faulhaber's *Zehntausend Logarithmi*[35] (a copy of Vlacq) in Germany. The appearances of these books in native languages rather than Latin is an indication that logarithms were reaching an audience well beyond scholars.

The widened audience had several effects. Logarithm tables proliferated, frequently giving fewer significant figures than the ones we have examined but nevertheless usually based on the entries in the original tables—the most likely sources being Briggs's two massive sets. The physical sizes of books containing logarithm tables decreased dramatically to adapt to their workplaces; for instance, in figure 2.7 we have a copy of the 1659 edition of Norwood's *Epitomie*[36] beside the first half of Rheticus's *Opus palatinum*.[37] It is clear which of these is more appropriate for a sea voyage. Logarithms of trigonometric functions persisted in the smaller editions, indicating that applications to navigation were far from a scholarly fantasy.

One of the first users of logarithms was none other than Johannes Kepler (1571–1630), who recognized the potential benefit to his heavily computational astronomical work by reading Ursinus's *Cursus mathematici practici*. Kepler published his own set of tables, the *Chilias logarithmorum*,[38] only a few years before the appearance of his decades-long life work, the *Rudolphine*

[31] See [Jagger 2003, 70].
[32] [Ursinus 1618].
[33] [Wingate 1625].
[34] [Henrion 1626].
[35] [Faulhaber 1631].
[36] The 1659 edition of [Norwood 1645]. These tables were called (following Napier's original word for logarithm in the *Constructio*) "artificial sines," "artificial tangents," etc., meaning logarithms of sines, tangents, etc.
[37] [Rheticus/Otho 1596].
[38] [Kepler (Hammer) 1960].

Figure 2.7
A copy of Norwood's 1659 *Epitomie* beside a
copy of half of Rheticus's *Opus palatinum*.

Tables. Although influenced by Napier, Kepler's definition of logarithms was
his own.[39] It is related to the so-called "logistic logarithms"[40] that were tabulated for centuries but have long since fallen into disuse. Kepler's tables are
logarithms of sines of arcs; however, unlike other such tables, it is the column
of sines that increases with a constant increment. He tabulates the logarithm
in a column to the right and works out the corresponding arc in a column to
the left.

Logarithms did not bring the production of tables of pure trigonometric
functions to a halt. However, the trigonometric tables by Pitiscus and Briggs
(in the *Trigonometria Britannica*) were so accurate that they became the basis
of many new sets of tables; hardly any substantial efforts to surpass them were
made until the early twentieth century.[41]

■   **Computing Tables Effectively: Logarithms**

The new logarithms provided users such a great computational saving that
they eventually provoked the famous remark attributed to Laplace: "an admirable artifice which, by reducing to a few days the labor of many months,
doubles the life of an astronomer, and spares him the errors and disgust in-
separable from long calculations."[42] This saving to the many came partly at
the cost of the few, especially Napier, Briggs, Vlacq, and de Decker, who spent
years either computing logarithms themselves or supervising others to do so.

---

[39] On Kepler's logarithms see [Belyj/Trifunovic 1972], [Gronau 1987], and [Gronau 2001]. [Kepler
(Peyroux) 1993] is a French translation of Kepler's writing on logarithms.

[40] "The logistic logarithm of any number of seconds is the difference between the logarithm of
3600″ and the logarithm of that number of seconds" [Hutton 1811b, 147].

[41] From Glaisher's report on mathematical tables: "Almost at once the logarithmic superseded
the natural canon; and since Pitiscus's time no really extensive table of pure trigonometrical
functions has appeared" [Glaisher et al. 1873, 42; see also p. 5].

[42] The earliest reference I have been able to find for this quotation is the article on astronomy in
the seventh edition of the *Encyclopaedia Britannica* (1842), vol. 3, 741.

We shall dwell on the methods of calculation only long enough to summarize Napier's method briefly to avoid wandering too far outside the bounds of trigonometry.

Napier himself was the first to divulge his methods for computing logarithms in *Mirifici logarithmorum canonis constructio*.[43] The book appeared in 1619, two years after his death, but was probably written even before the *Descriptio*.[44] It is easy to construct a table of a geometric sequence corresponding to an argument that increases linearly; simply multiply the previous entry by the same constant each time. It is much harder to do the reverse: to tabulate linked arithmetic and geometric sequences so that the column corresponding to the *geometric* sequence increases with a constant step size. To do what Napier wanted, namely, to have the step size of the geometric sequence increase according to the sines of increasing angles, was an even greater challenge.

Napier begins by constructing a collection of reference tables of geometric sequences, beginning with $R = aZ = 10^7$ (see figure 2.1) and working gradually to smaller and smaller values of the geometric sequences. The top entries in his first two tables are equal to R. Each successive entry in the first table repeatedly subtracts a 10,000,000th part of the entry above it, and in the second table a 100,000th part:[45]

10,000,000	×0.99999	10,000,000	×0.99999
9,999,999	×0.99999	**9,999,900**	×0.99999
9,999,998.0000001	⋮	9,999,800.001000	⋮
⋮		⋮	
9,999,900.0004950		9,995,001.222927	

Next Napier constructs a giant two-dimensional grid of numbers working downward in geometric sequences from R to ½R. Across the rows the multiplicative factor is 0.99; across the columns it is 0.9995.

10,000,000	**9,900,000**	9,801,000	. . .	5,048,858.8900
9,995,000	9,895,050	9,796,099.5000	. . .	5,046,334.4605
⋮	⋮	⋮		⋮
9,900,473.5780	9,801,468.8423	9,703,454.1539	. . .	4,998,609.4034

---

[43] [Napier 1619].

[44] The *Constructio* was translated to English in [Napier (Macdonald) 1889]. Accounts of Napier's method abound; see for instance [Naux 1966, vol. 1, 73–82], [Goldstine 1977, 6–10], [Bruce 2000], [Friedelmeyer 2006, 64–70], and [Havil 2014, 96–130].

[45] There is a small error in the value of the bottom right entry.

Napier already knows that NLog 10,000,000 = 0. He uses inequalities that he has demonstrated previously to obtain tight bounds for the logarithms of the four entries given in bold in the two tables above. Since the numbers in the rows and columns change geometrically, Napier knows that the logarithms of the other numbers in these tables must change arithmetically as one works downward or to the right, so this is enough information for him to find the remaining logarithms easily.

Now, Napier does not care so much about the logarithms he has just found; he wants the logarithms of *sines* of arguments. For instance: at one point he sets out to find NLog 7489557, which (he does not tell us) happens to be NLog(sin(48°30′)). Of course, it would be a miracle if this number were to be already in his tables; it is not. So, he searches through his large two-dimensional grid to find the number closest to 7489557. From here he applies one of his inequalities to get an estimate for the quantity he seeks. The inequalities work well in this case, because the entry he has located in the table is very close to 7489557. So, although Napier is estimating the logarithms of sines, there is nothing really trigonometric about his calculations; he could just as easily have used this method to find the logarithms of any sequence of values.

This was also the time of the rise of decimal fractional notation into public consciousness; various notations were being attempted, and their success in simplifying calculations was paying off. Although Napier did not invent the decimal point,[46] he uses it within the *Constructio*, and the popularity of his book has been credited with helping to spread the notation. He does not, however, take the additional step of converting his trigonometric tables to a unit circle, staying with his $R = 10^7$.

Briggs's methods for computing his logarithms do not involve trigonometry, so we shall pass over them here.[47]

▨ Computing Tables Effectively: Interpolation

As both trigonometric and logarithmic tables grew larger and larger, it became essential to find ways to generate entries using interpolation rather than directly calculating them all. Filling in tables in this way had been a healthy

---

[46] The decimal point is much older than previously thought. We shall soon publish an article on its appearance in the work of mid-fifteenth-century astronomer Giovanni Bianchini and will suggest a vector of transmission to the late sixteenth century.

[47] Interested readers may consult [Delambre 1821, vol. 1, 532–545], [Naux 1966, vol. 1, 99–114], [Goldstine 1977, 13–16], and [Hairault 2006]. See [Glaisher 1915] for an argument that Briggs's "radix method" was originally composed by William Oughtred.

tradition for many centuries; variants of linear interpolation had existed already in ancient Greece, and second-order interpolation had been practiced in both medieval India and medieval Islam.[48] However, these methods do not seem to have transmitted to Europe, and European table makers had to discover them for themselves.

The earliest of these pioneers was Thomas Harriot (1560–1621), who we mentioned in chapter 1 in the context of tables of meridional parts, which were required to space out the latitude lines on a Mercator map. Harriot spent part of his early life on a ship working for Walter Raleigh and surveying Virginia and studying the native culture there, providing a motivation for his interest in navigational problems. Within mathematics he is known most for his contributions to algebra. His colleagues knew of his work not through publications but through private circulation; his mathematical notes were not published during or even soon after his lifetime.[49]

So often, within these pages and elsewhere, we find that a European discovery of some mathematical concept has been preceded by some earlier figure in another culture. Surprisingly, this is not the case with interpolation. Although Greek, Indian, and Islamic astronomers all worked with either first- or second-order interpolation, we find no examples of direct implementation of third-order interpolation in these cultures, let alone a generalization to higher orders until Harriot. His work appears in notes with no explanatory text and must be decoded by scholars. The following, from his manuscript *Magisteria magna*, illustrates his approach.[50]

Suppose one has a table of entries, which in our illustration we shall assume exhibits constant third differences. (No trigonometric table will do this precisely, but with a small enough increment among arguments, the third differences are likely to be approximately equal.) The entries are in the rightmost column, and the first differences beside them to the left, then the second differences, and finally the constant third differences all the way to the left. Call the entries at the top of each column A, B, C, and D. If their values are known, then all the remaining differences and table entries may be determined from them working from the leftmost column to the right, as shown below.

[48] We have discussed these topics in [Van Brummelen 2009, 77, 111–113, 162–165].

[49] For general information on Harriot, begin with the biography [Shirley 1983]; the volumes of essays [Shirley 1974], [Fox 2000], and [Fox 2012]; and the source book [Shirley 1981].

[50] The manuscript has been reproduced and studied in [Beery/Stedall 2009]; the following is based on their commentary on pp. 10–11. See also [Lohne 1965/66, 31–34], [Goldstine 1977, 23–26], [Beery 2007], and [Beery 2008]. For Harriot's work on meridional parts using second differences, see [Pepper 1968, esp. 378–384].

$$D$$
$$C$$
$$B \qquad\qquad D+C$$
$$A \qquad\qquad C+B$$
$$B+A \qquad\qquad D+2C+B$$
$$A \qquad\qquad C+2B+A$$
$$B+2A \qquad\qquad D+3C+3B+A$$
$$C+3B+3A$$
$$D+4C+6B+4A$$

The binomial coefficients may be recognized in these expressions, and Harriot had expressions for them. Now imagine that one wishes to interpolate $n-1$ entries between each pair of entries in the rightmost column above. This requires that we replace the numbers A, B, C, and D at the top of the difference columns with a new set of values a, b, c, and d. A little algebra suffices to determine the following:

$$A = n^3 a$$
$$B = n^2 b + (n^3 - n^2)a$$
$$C = nc + \frac{n^2 - n}{2}b + \frac{n^3 - 3n^2 + 2n}{6}a$$
$$D = d. \tag{2.3}$$

Once these equations are rearranged to give a, b, c, d from A, B, C, D, it is a simple matter to start with a given table and interpolate $n-1$ entries between each pair of entries.

Harriot clearly intended this work (probably written about 1618 but based on methods composed around 1611) to be used to compute various tables, including trigonometric ones; he uses "pretend" sine values as examples in some of his calculations.[51] On Harriot's death his writings were bequeathed to Nathaniel Torporley for consideration for eventual publication (which never happened). There are some manuscripts written by Torporley on Harriot's work, including the first known use of superscripts for powers of quantities. Torporley knew Henry Briggs as well and called him a friend; the latter had given him a copy of *Logarithmorum chilias prima*. So it is possible that Harriot, through Torporley, was the source of Briggs's description of the use of

---

[51] [Beery/Stedall 2009, 13–15].

finite difference interpolation for table computations in the *Arithmetica logarithmica*.[52]

Text 2.2
Briggs, Completing a Table Using Finite Difference Interpolation
(from the *Arithmetica logarithmica*)

Given two consecutive integers and their logarithms: it is required to interpolate between them nine other equidistant numbers, and to find their logarithms.

If the second differences of the given logarithms are nearly equal, this will be an easy matter: but if the third differences cannot be neglected, this method will be found somewhat defective. Take two consecutive numbers *A*, and their logarithms *B*, together with their first differences *C*, and their second differences *D*. If the second differences are equal, multiply either of them into the numbers standing opposite the first ten natural numbers in the subjoined Table *E*; then, the last three figures having been cut off each of the products *F*, *G*, *H*, *I*, and *K*, the first five are to be added to the tenth part of the first difference of the two given logarithms, and the last five are to be subtracted from the same. The sums and the remainders will be the differences of the logarithms sought; and the successive addition of these differences to the smaller of the given logarithms, will give the logarithms required. For example, let the given numbers be 91235 and 91236, the first difference of their logarithms being 47601,4799.[53]

				E	
	47602,0016. C.			1. 45	products to be added
91235. A.	4,96016,14763,8639. B.	5217. B		2. 35	
	47601,4799. C.			3. 25	
91236. A.	4,96016,62365,3438. B.	5217. B		4. 15	
	47600,9582. C.			5. 5	
				6. 5	products to be subtracted
				7. 15	
				8. 25	
				9. 35	
				10. 45	

---

[52] On Briggs's interpolation methods with finite differences, see [Maurice/Sprague/Williams 1867], [Whiteside 1961a], [Whiteside 1961c, 233–236], [Naux 1966, vol. 1, 144–127], [Goldstine 1977, 26–32], and [Bruce 2002, 222–223].
[53] [Briggs 1624, 24–25]; translation by [Williams 1868, 73–74].

Numbers	Logarithms		Products		5217.Multiplicand	Multipliers
912350	4,96016,14763,8639		F--234	765	45	
	4760,1715	C+F	G--182	595	35	
912351	4,96016,19524,0354		H--130	425	25	
	4760,1662	C+G	I--78	255	15	
912352	4,96016,24284,2016		K--26	085	5	
	4760,1610	C+H	47601479	9 C		
912353	4,96016,29044,3626		47601714	7. C+F		
	4760,1558	C+I	47601662	5. C+G		
912354	4,96016,33804,5184		47601610	3. C+H		
	4760,1506	C+K	47601558	2. C+I		
912355	4,96016,38564,6690		47601506	0. C+K		
	4760,1454	C−K	47601453	8. C−K		
912356	4,96016,43324,8144		47601401	6. C−I		
	4760,1402	C−I	47601349	5. C−H		
912357	4,96016,48084,9546		47601297	3. C−G		
	4760,1350	C−H	47601245	1. C−F		
912358	4,96016,52845,0896					
	4760,1297	C−G				
912359	4,96016,57605,2193					
	4760,1245	C−F				
912360	4,96016,62365,3438					

Explanation: For convenience, we shall use modern decimal notation. Between the entries log $91235 = 4.96016147638639$ and log $91236 = 4.96016623653438$, Briggs wishes to insert the nine logarithms 91235.1, 91235.2, ..., 91235.9. The ten differences between these eleven logarithms must have an average of one-tenth the difference between the two extreme values, namely, 0.000000476014799. In this particular interval, Briggs assumes that the second differences remain constant at 0.00000000005217 (for one-unit increments, the number obtained from considerations outside of this interval); therefore, the second differences in his interpolated table must be one-tenth that amount. Table E provides multiplicative factors for determining the interpolated first differences (0.45, 0.35, 0.25, etc.) that result in this pattern while providing smooth transitions with similarly interpolated entries below 91235 and above 91236 (not shown here).

This method of presentation resembles medieval conceptions of how first and second differences must change from one entry to the next to provide a desired smoothly changing set of function values.[54] Unlike medieval authors, Briggs goes on to consider much more complicated situations involving differences as high as the fifth order.

Briggs used his interpolation methods to compute not just logarithmic but also trigonometric tables. Indeed, he would have needed to, in order to have any hope of completing the project he hinted at in his 1633 *Trigonometria Britannica*: to construct a table of sines with increments of one thousandth of a degree. In addition to improvements to Briggs's interpolation methods, the *Trigonometria Britannica* contains an extensive description of how he would approach the computation of such a massive sine table (although the table in the book has only 0.01° increments), a project that completed the proposals that Viète had made decades earlier.[55] He begins by showing how the chord of triple an arc depends on the chord of an arc in a way that brings to mind similar calculations by Jamshīd al-Kāshī in Samarqand two centuries earlier.[56] In the unit circle of figure 2.8, let $\overset{\frown}{BC} = \overset{\frown}{CD} = \overset{\frown}{DE}$ so that *BGFE* is the chord of the triple arc $\overset{\frown}{BCDE}$. Draw *CH* parallel to *DA*. Then triangles *BCG* and *GCH* are similar to *ABC*.[57] Let $BC = x$. Since $CG/BC = BC/AB$, $CG = x^2$; and since $GH/CG = CG/BC$, $GH = x^3$. Then, since *CDFH* is a parallelogram, $HF = CD = x$, so $EB = EF + BG + FH - GH = 3x - x^3$. So if one knows the original chord x, the chord *EB* of triple the arc is equal to $3x - x^3$; and if one wants to find the chord of one-third of a given arc *EB*, one needs to solve this cubic equation. By a related method Briggs reduces the problem of finding the chord of a quintisected arc to a quintic equation and similarly for divisions of arcs into other fractional parts.

Now that he is equipped with methods to trisect and quintisect chords, Briggs turns to the calculation of lengths of chords of various arcs, solving the polynomials that arise using numerical methods as Viète had proposed.

---

[54] See for instance [Van Brummelen 2009, 111–113; 164–165].

[55] See [Briggs/Gellibrand 1633, 1–49]. For a thorough modern account of the entire method, see [Bruce 2004].

[56] See [Rosenfeld/Hogendijk 2002–03].

[57] ∡*CBE* subtends the double arc $\overset{\frown}{CE}$, so ∡*CBE* = ∡*BAC*; and triangles *ABC* and *BCG* also share an angle at *C*, so they are similar. As for triangle *CGH*, since *CDFH* is a parallelogram, ∡*CHG* = ∡*CDF* = ∡*BCA*; and since triangles *BCG* and *CGH* also share an angle at *G*, they are similar.

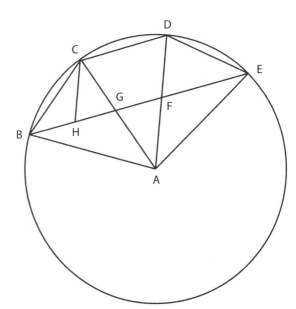

Figure 2.8
Briggs's trisection of a
given chord.

He begins by calculating chords in the sequence $60° \rightarrow 20° \rightarrow 100° \rightarrow 50° \rightarrow$ $25° \rightarrow 12°30' \rightarrow 6°15'$. From the latter value Briggs constructs a table of sines of multiples of 3.125° (since the sine of half the arc is half the chord of the full arc). Next, he quintisects this grid using his finite difference scheme, giving him a grid with a finer increment of 3.125°/5 = 0.625°. He quintisects the new grid again, ending up with a grid of 0.625°/5 = 0.125°; then again, leaving a grid of 0.125°/5 = 0.025°; and again, with 0.025°/5 = 0.005°; and finally once more, giving him his desired grid of 0.005°/5 = 0.001°. This arduous process made Briggs the first European to calculate a sine table using methods that extend beyond those suggested by Ptolemy's *Almagest* a millennium and a half earlier—or rather, we would have thought so had Bürgi's unpublished scheme remained undiscovered.

Napier on Spherical Trigonometry

Often forgotten in the research on the discovery of logarithms is the fact that Napier's *Descriptio* also contained several important developments in spherical trigonometry. Most notable is Napier's treatment of the ten identities for right-angled spherical triangles that we left in the hands of Viète in chapter 1: namely,

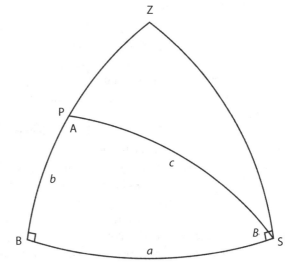

Figure 2.9
The triangle defining
Napier's circular parts.

$$\sin b = \tan a \cot A \qquad \sin a = \sin A \sin c$$
$$\cos c = \cot A \cot B \qquad \cos A = \sin B \cos a$$
$$\sin a = \cot B \tan b \qquad \cos B = \cos b \sin A$$
$$\cos A = \tan b \cot c \qquad \sin b = \sin c \sin B$$
$$\cos B = \cot c \tan a \qquad \cos c = \cos a \cos b,$$

where the right angle is at *C*. Viète seems not to have noticed the symmetries in these formulas, which we might intuit by noticing that every column of arguments of trigonometric functions in the above list follows the sequence *b*, *c*, *a*, *A*, *B* (looping back to the top of the column when one reaches the bottom). Napier located the symmetries, but he also found a remarkable geometric demonstration for them. His discussion stays in an astronomical context, but his method clearly may be generalized to arbitrary triangles.

We begin with the same pair of triangles that Napier worked with earlier. In figure 2.9, B is the north point on the horizon; S is the sun on the horizon; P is the North Pole; and Z is the zenith. Then angles B and S are right while $\widehat{BZ}$ and $\widehat{SZ}$ are 90° long. So ΔPBS is right while ΔPZS is *quadrantal* (has a side equal to 90°, namely $\widehat{SZ}$). We label the diagram in italics according to the now-conventional notation *a*, *b*, *c*, *A*, *B*. Napier now traverses around right-angled triangle BPS clockwise, starting at but not including the right angle, identifying the five elements of the triangle but taking the complements of the second through fourth. This produces the sequence of **circular parts** b, $\overline{A}$, $\overline{c}$, $\overline{B}$, a (where the bar refers to the complement of the quantity). He does the same for the quadrantal triangle proceeding counterclockwise, starting at but not including the 90° side, arriving at the same sequence of circular parts a, b, $\overline{A}$, $\overline{c}$, $\overline{B}$.[58] (The beginning and end of the sequence are immaterial; the sequence is to be thought of as a loop.)

Next, Napier constructs figure 2.10 (of which figure 2.9 is a part), which Gauss eventually would name the *pentagramma mirificum*. Three of the solid arcs forming the pentagram are extensions of the original triangle PBS; the other two are the equator CZQE corresponding to the pole of the sun S and the celestial equator FOQD. Indeed, every solid arc in the pentagram has astronomical significance, as indicated in the diagram. The pentagram has a number of properties: the five dotted arcs forming the inner pentagram are all 90°; each arc of the inner pentagon is the equator of the opposing vertex; the five triangles comprising the "petals" around the outside each have a right angle at the tip; and most importantly, the five triangles contain the same sequence of circular parts b, $\overline{A}$, $\overline{c}$, $\overline{B}$, a (or their complements), starting at different places in the sequence.

Now, if one takes any three of these circular parts, the three will form a sequence of three terms in a row (say, $\overline{B}$, a, b), or one term will be isolated and the other two will be adjacent to each other (say, a, $\overline{A}$, $\overline{c}$). Napier gives the following two results:

- In the case where three terms are in sequence, the sine of the middle part is equal to the product of the tangents of the other two parts (say, $\sin a = \tan \overline{B} \tan b$);
- In the case where one term is isolated, the sine of that circular part is equal to the product of the cosines[59] of the other two parts (say, $\sin a = \cos \overline{A} \cos \overline{c}$).

These two rules, applied to all possible sets of three circular parts, generate all ten of the identities for right-angled triangles.[60] These statements, known as "Napier's rules of circular parts," would become a staple device to help students to memorize the ten identities.[61] For some while into the nineteenth

---

[58] This requires some simple arguments; for instance, since $\widehat{ZB}$ and $\widehat{ZS}$ are 90°, the angle at Z is equal to $\widehat{BS}$, which is a.

[59] The term "cosine" did not yet exist; it was expressed as the "sine of the complement."

[60] Although the rules that would be preserved in textbooks deal with right-angled triangles, Napier also shows how to use circular parts to handle the quadrantal triangle.

[61] On Napier's treatment of the circular parts, see [von Braunmühl 1900/1903, vol. 2, 12–14], [Sommerville 1915] (with some mathematical extensions), [Hawkins 1982b, vol. 2, 97–108],

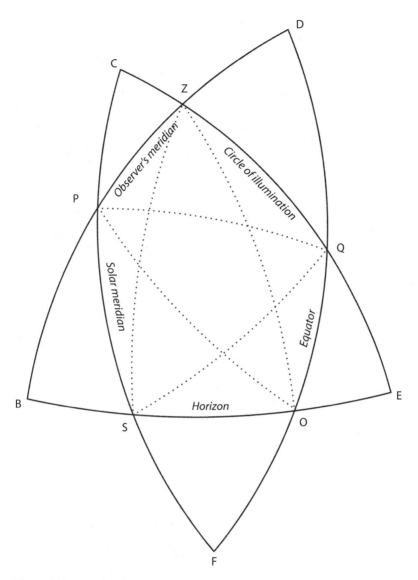

Figure 2.10
Napier's rendition of the *pentagramma mirificum*.

century the mathematics underlying the mnemonics was forgotten by most; usually only the mnemonic remained.[62] The patterns of the arcs and angles in Napier's rules were revisited from time to time, including by the late eighteenth-century analyst Johann Heinrich Lambert with the aid of stereographic projection,[63] and were even analyzed in the late nineteenth century using permutation groups.[64]

In the *Descriptio* Napier introduces these rules not directly as products but rather in logarithmic form as sums. Indeed, the structure of the identities, involving as they do only products of trigonometric quantities, made it easier to solve oblique spherical triangles by separating them into pairs of right-angled triangles. Today's standard identities for oblique triangles, the Laws of Cosines for sides and angles, combine sums and products in a way that is not amenable to the use of logarithms. Indeed, it was not until electronic computation became prevalent and logarithmic calculation became obsolete that the Laws of Cosines became the preferred tools.

Even so, Napier also went on to treat certain cases of oblique triangles without breaking them into pairs of right triangles, and he discovered new theorems that avoid combinations of sums and products. For the case where all three sides are known and the angles are to be determined, Napier gives one of the *half-angle formulas*:

$$\sin^2 \frac{A}{2} = \frac{\sin(s-b)\sin(s-c)}{\sin b \sin c} \tag{2.4}$$

(expressed here in modern terminology) where s is the triangle's semi-perimeter,[65] relying on Regiomontanus's statement of the spherical Law of Cosines and the prosthaphaeretic formulas for its demonstration. Napier goes on to state an equivalent of the cosine half-angle formula,

$$\cos^2 \frac{A}{2} = \frac{\sin s \, \sin(s-a)}{\sin b \, \sin c}, \tag{2.5}$$

[Silverberg 2008], and [Dietrich/Girstmair 2009, 221–224]. Augustus De Morgan showed that Napier had been anticipated here by Nathaniel Torporley in his famously obscure *Diclides Coelometricae*, but it seems that no one noticed [De Morgan 1843]. One finds the *pentagramma* also, for instance, in [Oughtred 1657, 22–25].

[62] The *pentagramma mirificum* was restored to memory through a series of articles in the late nineteenth and early twentieth centuries; see [Lovett 1898], [Moritz 1915], and [Ransom 1938].

[63] [Lambert 1792]; see also [von Braunmühl 1900/1903, vol. 2, 130–132].

[64] [Pund 1897].

[65] [Napier 1614, 47–48]; see also [von Braunmühl 1900/1903, vol. 2, 14–15] and [Dietrich/Girstmair 2009, 224–225].

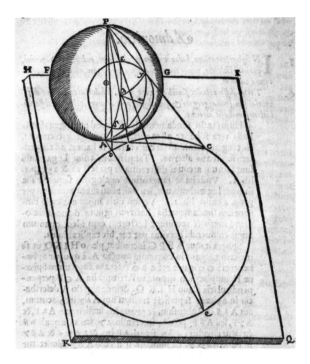

Figure 2.11
Napier's use of a
spherical projection to
demonstrate a theorem
about spherical
triangles.

but does not give the tangent half-angle formula, although it is a trivial combination of (2.4) and (2.5). Here we witness an illustration of how a new, essentially external discovery (logarithms) can trigger a reorganization of priorities and choices within another mathematical theory (spherical trigonometry): the Laws of Cosines became much less important while the half-angle formulas were discovered in the same book that announced logarithms and quickly rose to a prominent place.

One last contribution from the *Descriptio* is a very early use of a spherical projection to solve spherical triangles for the case where the three sides are known. In figure 2.11, the triangle to be solved is $A\lambda\gamma$. Drop a perpendicular from λ to $A\gamma$, defining μ. Suppose that A is a pole, and map the circle with center λ and radius $\lambda\gamma$ onto the plane tangent to the sphere at A by drawing lines from the opposite pole P through the various points and onto the plane. This forms the circle $dbce$. In the configuration on the plane, by Euclid's *Elements* III.36, $Ab : Ae = Ad : Ac$. But the various terms in this ratio may be replaced by trigonometric equivalents, and Napier has

$$\frac{\text{Tan}\frac{A\mu-\mu\gamma}{2}}{\text{Tan}\frac{A\lambda+\lambda\gamma}{2}} = \frac{\text{Tan}\frac{A\lambda-\lambda\gamma}{2}}{\text{Tan}\frac{A\lambda}{2}}. \tag{2.6}$$

Since the three sides of $A\lambda\gamma$ are known, the only quantity in this formula that is unknown is $A\mu - \mu\gamma$, so we may use the formula to find it. But one of the known sides is $A\gamma = A\mu + \mu\gamma$, so we can solve for $A\mu$ and $\mu\gamma$. We have now split the given triangle into two right triangles, both of which have two known sides, so we may apply Napier's rules of circular parts to find the various angles. Whether or not Napier was the original inspiration for this unique method, we shall see that the idea of projecting a triangle onto a sphere to solve it would reappear more than once and would have currency as far into the future as the early nineteenth century.

One more of Napier's contributions to spherical triangles is found not in the *Descriptio* but in the *Constructio*. Although published five years after the *Descriptio*, the *Constructio* appears to be an earlier work (for instance, although it contains the ten identities for right-angled spherical triangles, it shows no awareness of the symmetries that lead to Napier's rules). The bulk of the *Constructio* does what it claims to do—namely, to explain how to compute his table of logarithms. However, a section at the end of the book lists various methods for solving spherical triangles, each of them designed to work well with logarithmic computations. One of those methods, designed to handle the case when a side b and the two angles adjacent to it are known (A and C), translates to the following formulas in modern notation:

$$\log\sin\frac{A+C}{2} + \log\sin(A-C) + \log\tan\frac{b}{2} - \log\sin(A+C)$$
$$- \log\sin\frac{A-C}{2} = \log\tan\frac{c+a}{2} \qquad (2.7)$$

and

$$\log\sin\frac{A-C}{2} + \log\tan\frac{b}{2} - \log\sin\frac{A+C}{2} = \log\tan\frac{a-c}{2}. \qquad (2.8)$$

The variables $\dfrac{a+c}{2}$ and $\dfrac{a-c}{2}$ may be determined from the known quantities using these formulas, and from them (by adding and subtracting these two quantities), the values of a and c can be found.

Even after removing the logarithms, (2.7) is more complicated than it needs to be. With some algebra, it simplifies to

$$\frac{\tan\frac{a+c}{2}}{\tan\frac{b}{2}} = \frac{\cos\frac{A-C}{2}}{\cos\frac{A+C}{2}}, \qquad (2.9)$$

one of four formulas today called ***Napier's Analogies***.[66] Similarly, (2.8) simplifies to

$$\frac{\tan\frac{a-c}{2}}{\tan\frac{b}{2}} = \frac{\sin\frac{A-C}{2}}{\sin\frac{A+C}{2}}, \tag{2.10}$$

another of Napier's Analogies.[67] These formulas were to have a life as long as spherical trigonometry itself (that is, until the middle of the twentieth century).

Following this section of the *Constructio*, Henry Briggs added a series of notes on the preceding trigonometric propositions. Within his commentary he solves two spherical triangles using these methods; and within the second solution he uses the remaining two of Napier's Analogies in logarithmic form. In modern notation, they are

$$\frac{\tan\frac{A+C}{2}}{\cot\frac{B}{2}} = \frac{\cos\frac{a-c}{2}}{\cos\frac{a+c}{2}} \tag{2.11}$$

and

$$\frac{\tan\frac{A-C}{2}}{\cot\frac{B}{2}} = \frac{\sin\frac{a-c}{2}}{\sin\frac{a+b}{2}}.^{68} \tag{2.12}$$

Briggs provides no explanation for them, although they may be obtained by applying the polar triangle to Napier's two analogies (2.9) and (2.10). In none of these cases do Napier or Briggs ever provide any proofs of these results (even in the *Trigonometria Britannica*, where all four analogies appear). Proofs were to be provided, finally, in Oughtred's 1657 *Trigonometria*, where Oughtred uses an analemma construction to prove the first two.[69]

▧ Further Theoretical Developments

The period after the invention of logarithms and before calculus was not especially active within trigonometry itself. Astronomers were satisfied with adapting their existing astronomical work to the new trigonometric methods. As we have seen, practitioners of the other mathematical sciences were

---

[66] "Analogy" is an archaic word expressing an equality of ratios.

[67] [Napier 1619, 56]; for a translation see [Napier (Macdonald) 1889, 73]. For discussions see [von Braunmühl 1900/1903, vol. 2, 16–17] and [Dietrich/Girstmair 2009, 226–227].

[68] [Napier 1619, 61]; for a translation see [Napier (Macdonald) 1889, 80].

[69] [Oughtred 1657, 34–35, 28–29, 38–39]. (The pages are in sequence; the pagination is out of order.)

coming to terms with the dramatic new tools that were being discovered. Nevertheless, there are some developments to report. One of the earliest is the discovery of the formula for the area of a spherical triangle and the related concept of the spherical excess. One may wonder why such an obvious question as the determination of the area of a spherical triangle took so long to arise. The reason is straightforward: simply, there was no need for it. The notion of area does not arise as a meaningful concept in spherical astronomy. In new applications of trigonometry such as surveying, it would take many years before the difference in area between a plane and a spherical triangle would become practically significant.

The first person to discover the area formula was Thomas Harriot in 1603; he states that the area of the spherical triangle is proportional to the amount by which the sum of the triangle's angles exceeds 180°. In modern terms, the area of a triangle on a unit sphere is

$$\frac{\pi}{180}(A + B + C - 180°). \tag{2.13}$$

The quantity in parentheses, the **spherical excess**, appears here for the first time; it would go on to make several further appearances in various identities. Harriot presents the formula in the context of his work on stereographic projection, which happens to include the earliest assertion that angles between great circles are preserved under this mapping.[70]

However, Harriot did not publish this (or most of his other) work; the first record we have of the formula in print with a supporting argument is by French mathematician Albert Girard (1595–1632). A member of the Reformed Church, Girard spent most of his working life in the Netherlands. There he likely worked with Snell, from whom he may have learned of the polar triangle. In 1626 Girard wrote a work on trigonometry that included a set of tables,[71] but curiously his announcement of the area formula is not to be found there; perhaps he had yet to discover it. Rather, it appears as the centerpiece of a 15-page article entitled "De la mésure de la superfice des triangles et polygones sphericques, nouvellement inventée" added to the end of his most well-known book, *Invention Nouvelle en l'Algèbre*.[72] Girard seems not quite convinced by his own argument, referring to it only as a "probable conclusion." The *Invention Nouvelle* itself is known today not for spherical geometry but rather for an early form of the Fundamental Theorem

---

[70] [Lohne 1965/66, 27–31] contains an account of Harriot's argument, which is essentially the standard one found in textbooks; see also [Pepper 1968, 366–367].

[71] For a survey of this work, see [Bosmans 1926].

[72] [Girard 1629]. Girard's text was reprinted in [Girard 1884], including the treatise on areas; the algebraic part of the treatise was translated to English in [Schmidt/Black 1986].

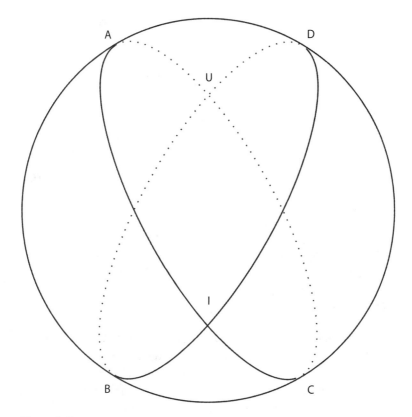

Figure 2.12
Cavalieri's demonstration of the area of a spherical triangle.

of Algebra and various innovative uses of algebraic notation (including the
first appearance of $\sqrt[3]{\ }$).

The first complete proof of the area formula to appear in print came
three years later in Italian mathematician Bonaventura Cavalieri's 1632 *Directorium generale uranometricum*[73] (just three years before his first book
on his theory of indivisibles, a predecessor to the integral calculus, on
which his reputation relies today). Cavalieri's proof is essentially the same
as that found in modern textbooks. In figure 2.12, *AID* is the triangle in
question; the sides of the triangle are extended to great circles. Since $\overparen{DAB}$
and $\overparen{ABC}$ are semicircles, $\overparen{DA} = \overparen{BC}$; by similar arguments the other sides of
triangles *AID* and *CUB* are also equal, so the two triangles are the same.

[73] [Cavalieri 1632].

Next Cavalieri considers the lune enclosed by semicircles $\overparen{DAB}$ and $\overparen{DIB}$, concluding that

$$\frac{area\ of\ sphere}{area(DABI)} = \frac{360°}{\angle ADI}\ .$$

Similarly,

$$\frac{area\ of\ sphere}{area(ADCI)} = \frac{360°}{\angle DAI}\ \text{and}\ \frac{area\ of\ sphere}{area(IBUC)} = \frac{360°}{\angle AID}.$$

Combining the three lunes gives the front hemisphere plus triangles AID and CUB, which have the same area; therefore,

$$\frac{area\ of\ sphere}{hemisphere + 2\Delta AID} = \frac{360°}{\angle ADI + \angle DAI + \angle AID};$$

a little algebraic manipulation of this expression gives

$$\frac{area\ of\ sphere}{\Delta AID} = \frac{360°}{\frac{1}{2}(spherical\ excess)}.$$

A few pages after the proof and after some corollaries, Cavalieri finds the area on the earth's surface in the triangle between his home town of Bologna, Alexandria, and the North Pole—the first time that the area of a portion of the earth's surface was computed taking into account the sphericity of the earth.[74] Although calculating areas such as this were mere curiosities and would remain so for a long time, eventually during the great surveys of the nineteenth century the topic would become much more important. Each of the three scholars working on the area problem (Harriot, Girard, and Cavalieri) seems to have been unaware of the others' work; indeed, two decades later, Gilles de Roberval published the result and later had to be informed by John Pell of Harriot's earlier discovery.[75]

Pell (1611–1685) did not do much work in trigonometry himself. However, his treatise *Controversiae de vera circuli mensura*,[76] a work of cyclometry that demonstrated that Christian Longomontanus had been wrong in asserting that $\pi = \frac{78}{43}\sqrt{3}$, was based on a new fundamental result, the tangent double-angle formula:

---

[74] The demonstration is in [Cavalieri 1632, 316–317]; the calculation of the area of the triangle is in [Cavalieri 1632, 320].
[75] [Malcolm 2002, 161].
[76] [Pell 1647].

$$\frac{R^2 - \text{Tan}^2\,\theta}{2R^2} = \frac{\text{Tan}\,\theta}{\text{Tan}\,2\theta} \tag{2.14}$$

$\left(\text{in modern terms, } \tan 2\theta = \dfrac{2\tan\theta}{1 - \tan^2\theta}\right).$[77] We are interested here not so much in the theorem itself but in its later use as an example of the new algebraic methods that were entering into mathematics through the analytic geometry of René Descartes (1596–1650). Descartes's *Géometrie*[78] opens with the sentence "any problem in geometry can easily be reduced to such terms that a knowledge of the lengths of certain straight lines is sufficient for its construction."[79] In other words, by labeling certain line segments with letter names, one may manipulate algebraically the relations among these quantities to solve geometric problems. This led eventually to the Cartesian coordinates and analytic methods in geometry that are now the staple of high school mathematics.

Descartes's *Géometrie* was written in French; it became more widely known through its translation into Latin by Dutch scholar Frans van Schooten (1615–1660). An algebraist by training, van Schooten met Descartes as a young scholar and became one of Descartes's most active promoters. The success of the first edition of the translation led to a much larger second edition containing much more commentary by van Schooten.[80] The second volume also contains various contributions by van Schooten's students and colleagues, including Christiaan Huygens and Jan de Witt;[81] here we are interested especially in "Tractatus de concinnandis demonstrationibus geometricis ex calculo algebraïco."[82] In this passage, van Schooten uses Pell's tangent double-angle theorem as an example, illustrating how Descartes's methods may be used to prove a geometric theorem.

In figure 2.13 (which shows the entire derivation), $\angle BAE = 2\angle BAD = 2\theta$. Let $a = AB$, $x = BF = \text{Tan}\,\theta$, $y = BG = \text{Tan}\,2\theta$, and $z = AG$. Then by the Internal Angle Bisector Theorem (*Elements* VI.3), $FG/BF = AG/AB$, so $BF \cdot AG = FG \cdot AB$. Converting these expressions to their representations in terms of a, x, y, z, we get

$$xz = (y - x)a, \text{ so } z = \frac{ay - ax}{x}.$$

[77] On the controversy see [Malcolm/Stedall 2005, especially 119–121], [Sørenson/Kragh 2007] and [van Maanen 1986]; the latter includes a detailed description of the correspondence especially among Pell, Charles Cavendish, and Merin Mersenne on the construction of the proof.

[78] [Descartes 1637].

[79] [Descartes 1637, 297]; the translation is from [Descartes (Smith/Latham) 1925, 2].

[80] [Descartes 1649], [Descartes 1659], and [Descartes 1661].

[81] See [Bissell 1987, 41–43] for a survey of the second edition of van Schooten's edition of the *Géometrie*.

[82] [Descartes 1661, vol. 2, 341–421]; the relevant passage is on pp. 366–368.

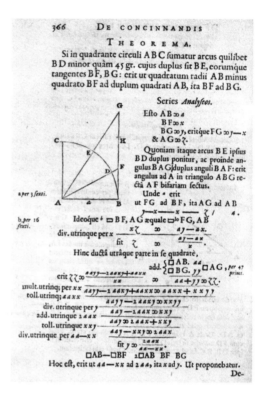

Figure 2.13

Van Schooten's proof of the tangent double angle formula, using Cartesian methods.

Squaring both sides and noting that $z^2 = a^2 + y^2$, we have

$$a^2y^2 - 2a^2xy + a^2x^2 = a^2x^2 + x^2y^2,$$

which simplifies to

$$y = \frac{2a^2x}{a^2 - x^2}.$$

Rearranging, we have

$$\frac{a^2 - x^2}{2a^2} = \frac{x}{y},$$

and if we substitute in the meanings of these terms, we have the sought result. The difference in character between this proof and those that came before is as stark as can be. The use of an equation with fourth-power terms such as a^2y^2, for instance, indicates that van Schooten has left Euclidean geometry far behind. The power of the new geometry revealed itself clearly; and while traditional geometrical methods did not disappear overnight, analysis was the path of the future.

▧ Developments in Notation

The names of the primitive functions, "sine," "tangent," and "secant," had been established already by late sixteenth-century textbook authors in continental Europe such as Fincke and Magini. However, the abbreviations of these and other trigonometric quantities in mathematical discourse were not standardized for some time. Symbolic algebra was gradually gaining a foothold throughout mathematics as a common language, and this would lead eventually to the algebra familiar to us today. But even by the end of the seventeenth century, the process had not yet been completed.[83]

We have seen in the previous chapter Fincke's introduction of the abbreviations "sin," "tan"/"tang" and "sec" for the sine, tangent, and secant; for the complementary functions he would append "comp" or "compl" at the end of the expression. Some other authors picked this up, but there were alternatives. A number of writers preferred single-letter abbreviations; for instance, Adrianus Romanus adopted "S" for sine, "P" (*prosinus*) for tangent, and "T" (*transsinuosa*) for secant, following Viète's nomenclature.[84] It took some while for the complementary functions to be considered on their own rather than as variants of the primitive functions. The various notations generally added a symbol or phrase to the notation for sine, tangent, or secant to indicate that (for instance) the sine of the complementary arc was intended. Indeed, even the terms "cosine" and "cotangent," introduced by Edmund Gunter as early as 1620, took a long time to catch on.

In fact, on the continent through the mid-seventeenth century, very little attention was paid to symbolic language in trigonometry; most texts continued to express identities and other theorems rhetorically.[85] It was in England, which ironically had received its trigonometry through Pitiscus's mostly symbol-free *Trigonometria*, that authors began to experiment with new notational systems. The earliest work of interest in this respect is an unsigned appendix to the 1618 edition of Wright's translation of Napier's *Descriptio* which has been attributed to William Oughtred.[86] In figure 2.14 we find the treatment of formulas for right-angled spherical triangles (with the right angle at *A*). The appearance is startlingly modern, with the first printed appearance

---

[83] The thorough [Cajori 1928–1929, vol. 2, 142–179] is still the fundamental survey on the history of notations in trigonometry. See also [Tropfke 1923, vol. 5, 37–47].

[84] [Romanus 1609, 48].

[85] One exception to this is Albert Girard, whose *Tables des Sinus, Tangentes & Secantes* [Girard 1629] uses both the "tan"/"sec" abbreviations and the letters "H," "P," and "B" to represent hypotenuse, perpendicular, and base (using the nomenclature of Rheticus and Viète).

[86] [Napier 1618]. See [Glaisher 1915], which also contains the text and an analysis of this appendix.

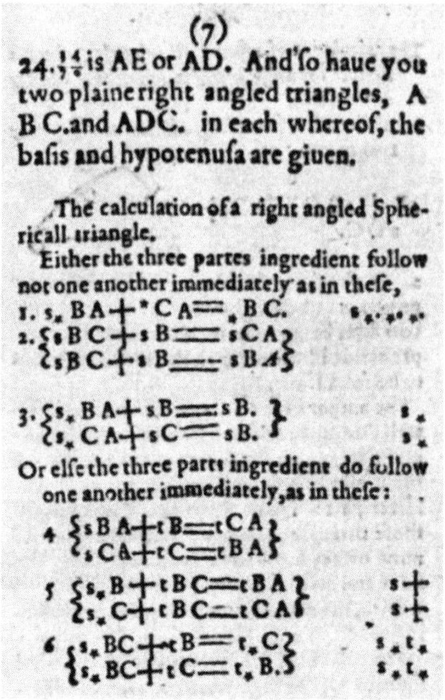

(7)

24.¦¼ is AE or AD. And so haue you two plaine right angled triangles, A B C.and ADC. in each whereof, the basis and hypotenusa are giuen.

The calculation of a right angled Sphericall triangle.

Either the three partes ingredient follow not one another immediately as in these,

1. s* BA+*CA=*BC.

2. {s BC+s B=s CA / s BC+s B=s BA}

3. {s* BA+s B=s B. / s* CA+s C=s B.}

Or else the three parts ingredient do follow one another immediately, as in these:

4 {s BA+t B=t CA / s CA+t C=t BA}

5 {s* B+t BC=t BA / s* C+t BC=t CA}

6 {s* BC+t B=t* C / s* BC+t C=t* B.}

Figure 2.14
Identities for the right-angled spherical triangle from an anonymous appendix to Wright's translation of Napier's *Descriptio.*

of the equal sign after Recorde's 1557 invention, and (elsewhere in the appendix) the first use of "×" to represent multiplication.[87] The letters "s" and "t" stand for "sine" and "tangent" while a subscripted "*" stands for "complement." So, the first expression in line 6,

$$s_*BC + tB = t_*C,$$

corresponds to

$$\cos BC \tan B = \cot C.$$

The use of the "+" sign indicates that the author assumes the use of logarithms throughout, thereby converting products to sums.[88] The equal sign was inadvertently promoted by the introduction of logarithms; before them, most theorems (especially in spherical trigonometry) had been represented as ratios: "as *a* is to *b*, so is *c* to *d*."[89]

[87] [Cajori 1928–1929, vol. 2, 149].

[88] [Napier 1618, appendix, 7]; see also [Glaisher 1915, 151–152]. The first formula is missing two *s*s in front of the isolated asterisks.

[89] For discussions of the role of ratios and the gradual shift to fractions and equations in the seventeenth century, see [Sylla 1984] and [Grosholz 1987].

The most influential English texts in the following decades continued the notational experimentation. Richard Norwood's *Trigonometrie*[90] uses "s" for sine and "t" for tangent, adding a "c" afterward for complement and "sec" for secant.[91] He uses the same notation without feeling the need to define it in his later *Epitomie*,[92] a work on the application of trigonometry to nautical problems. We do find modern-looking equations in Norwood's books similar to those from the anonymous 1618 work above, but they are not used consistently throughout the text. Oughtred's notation in his influential *Trigonometria*[93] is similar; we also begin to see an intermixing of the ratio equality notation ($a : b :: c : d$) with modern equalities. John Newton's *Institutio mathematica* and *Trigonometria Britannica*[94] also use the single letter "s" and "t" for sine and tangent respectively but add a "c" for complement in front of these letters rather than behind. The words "cosine" and "cotangent" make a reappearance in the latter work, perhaps as a result of his choice of notation, and the word "cosecant" appears perhaps for the first time.[95] The same notation is found in John Seller's *Practical Navigation*.[96] The lack of commentary on these notations suggests that the authors believed their readers to be familiar with them or could learn them from context. A consensus on notation was building but not toward the abbreviations we use today.

Practical and Scientific Applications

Proposals to apply trigonometry to practical and scientific concerns, already begun in the late sixteenth century, continued through the seventeenth. However, take-up by practitioners accelerated dramatically with the increased computational power brought to bear by the invention of logarithms. Existing applications became more compelling to potential users; new applications found a smoother path of entry. It is well beyond our scope to write a history of the numerous disciplines that began to use trigonometry at this time; several examples will suffice to give the flavor of what was happening.

Astronomy had been the original motivator for logarithms, so it is no surprise that astronomers were the first to bring them into common practice. One of the enthusiastic early adopters was Johannes Kepler; we have

---

[90] [Norwood 1631].

[91] [Norwood 1631, 20]; see also [Miura 1989, 21]. We also find in the *Trigonometrie* the same equation notation we have just seen in the text attributed to Oughtred, with "c" replacing the "*" for the complement, as well as the use of Napier's circular parts.

[92] [Norwood 1645].

[93] [Oughtred 1657].

[94] [J. Newton 1654] and [J. Newton 1658] respectively.

[95] [J. Newton 1658, 24]. This seems not to have been noticed before.

[96] [Seller 1669].

already described his variant on Napier's definition of the logarithm. In his 1627 *Rudolphine Tables*, Kepler introduced logarithms—that "outstanding invention"—immediately, describing variations by Briggs and Bürgi, illustrating their application, and using them extensively in the first eleven chapters and elsewhere.[97] A sure indication that logarithms had found a permanent place in astronomical calculations by mid-century is their appearance in John Newton's textbook 1657 *Astronomia Britannica*,[98] a year before his companion volume *Trigonometria Britannica*. The first half of the book deals with spherical astronomy; its methods are illustrated and demonstrated exclusively with logarithms.

Text 2.3

John Newton, Determining the Declination of an Arc of the Ecliptic with Logarithms

(from the *Astronomia Britannica*)

The Sun's greatest declination being given, to find his Declination in any point of the Ecliptic.

Let *DFHG* represent the Solsticial Colure, *DBAG* the Equator, *FAH* the Ecliptic, *I* the pole of the Ecliptic, *E* the pole of the Equator, *ECB* a meridian line passing from *E*, through the Sun at *C*, and falling upon the Equator *DAG* at right angles at the point *B*. Therefore, in the Rectangle spherical Triangle *ABC* we have known: 1. The Hypotenuse *AC* the Sun's distance from the next Equinoctial Point, whether Aries or Libra and may be supposed to be in 10 degrees of Cancer, and that being nearer unto Libra than Aries, I take his distance from Libra which is two signs and 20 degrees, or 80 degrees. 2. We have known the angle *BAC* the Sun's greatest declination, which by the accurate observation of *Tycho* is found to be 23 deg. 31 minutes and 30 seconds. And in Decimal numbers 23.525. Hence to find the present declination the proportion is,

> As the Radius is to the sine of the Sun's greatest declination, so is the sine of the planet's distance from the next equinoctial point, To the sine of the declination required.

And by this proportion, together with the help of the Canon of artificial sines and tangents, I find the declination of the Sun at the time proposed thus. First, I seek for the sine of Radius or 90 degrees, the measure of the right angle at *B*, and I find the sine thereof 10,000000, next I seek the sine of 80 degrees, and likewise the sine of 23.5250, these two I add together and from their ag-

[97] [Kepler 1627, 9–26]; the quotation is on p. 10.
[98] [J. Newton 1657].

gregate, I subtract the Radius and remaineth is the sine of the declination sought, as in the following work you may perceive.

As Radius

To the Sine of *BAC*	23.525	9,6011352
So sine of *AC*	70.	9,9729858
To the sine of *BC*	22.02910	9,5741210[99]

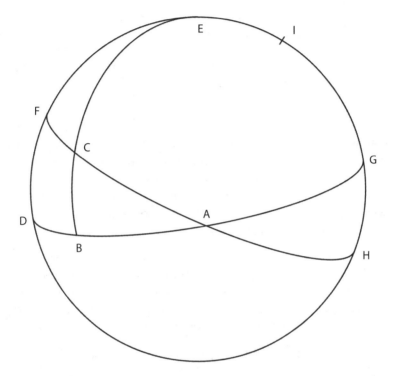

Figure 2.15
From John Newton's *Astronomia Britannica*, the diagram for the determination of the declination of a given arc of the ecliptic.

Explanation: (See figure 2.15.) This is the standard problem of finding the declination of a given arc of the ecliptic λ measured from the vernal equinox, which in this diagram is the point antipodal to A. In this case λ is given to be 100° (10° into Cancer), but since the declination is the same regardless of whether it is measured forward from the vernal equinox (Aries) or backward from the autumnal equinox (Libra), we may take $\lambda = 80°$. Notice, however, that when Newton performs his calculations, he mistakenly sets $\lambda = 70°$. The value of the "greatest declination," equal to the obliquity of the ecliptic, is taken to

---

[99] [J. Newton 1657, 9–10].

be Tycho Brahe's $\epsilon = \angle BAC = DF = 23.525°$. (Newton, a proponent of decimal numeration, converts away from degrees, minutes and seconds.)

The formula is given as an equality of ratios:

$$R : \text{Sin } \epsilon : : \text{Sin } \lambda : \text{Sin } \delta;$$

this corresponds to the modern $\sin \delta = \sin \lambda \sin \epsilon$ (with Newton adopting $R = 1$). Next, Newton turns to his table of logarithms of sines, which according to standard practice of the time he calls "artificial sines." His table of logarithms (in the *Trigonometria Britannica*, not the *Astronomia Britannica*) incorporates the standard displacement of ten, so the tabulated function is $f(x) = 10 + \log \sin x$. Then $\delta = f^{-1}(f(80°) + f(23.525°) - f(90°)) = f^{-1}(9.9729858 + 9.6011352 - 10) = f^{-1}(9.5741210) = 22.02910°$.[100]

Navigators adopted logarithms just as quickly; especially in England, logarithms were incorporated among the more knowledgeable seamen already by the early 1630s.[101] In other aspects of navigation, however, practice often trailed well behind the introduction of new methods. For instance, we find in Richard Norwood's popular 1645 *Epitomie*[102] that the plain sea chart had not been replaced in practice by the Mercator chart, even though several decades had passed since Wright's *Certaine Errors in Navigation*:

> Although the ground of the Projection of the ordinary Sea-Chart being false, (as supposing the Earth and Sea to be a plain Superficies [surface]) and so the conclusions thence derived must also for the most part be erroneous; yet because it is most easy, and much used, and the errors in small distances not so evident, we will not wholly neglect it.[103]

Indeed, he spends as much space on plain sailing as he does on Mercator sailing.[104] The *Epitomie*, a popular selection of Norwood's writings for navigational purposes published in various editions throughout the second half of the seventeenth century, took its sections on plain and Mercator sailing from his *Trigonometrie*. Among the various problems on plain sailing, we find the following:

> A Marchant-man, being in the Latitude of 43 degrees, falls into the hands of Pyrats; who amongst other things take away his Sea-

---

[100] Function notation is of course anachronistic to this time period; it is provided to aid the modern reader through the calculations.

[101] For an account of the introduction of logarithms to navigational practice, see [Waters 1958, 402–456.]

[102] [Norwood 1645].

[103] [Norwood 1645, 1].

[104] For a short survey of Norwood's methods see [Miura 1989], especially pp. 24–26.

compasse. But when he is gotten cleare, hee sayles away as directly as he can, and after two dayes meets with a man of War; who also had bin the day before in the Latitude of 43 deg. and had sayled thence SE by S, 37 leagues. He desirous to find those Pyrats, the Marchant-man tells him, he left them lying too and fro where they tooke him, and hee had sayled since at least 64 leagues, betweene the south and west: what course shall the man of Warre shape to find these Pyrats?[105]

Norwood solves this problem of plane trigonometry using the Law of Sines, applying logarithms in the same manner that we have just seen in the astronomical problem above. He does not report whether the pirates were still there when the man of war arrived.

The introduction to the section on Mercator sailing in the *Epitomie* contains an important addition that had not been found in his original *Trigonometrie*:

> Now that which hee [Edward Wright] hath shewed to performe by the Chart it selfe [the table of meridional parts], we will here shew to worke by the Doctrine of plaine Triangles, using the helpe of the Table of *Logarithme Tangents*, beginning at 45 degrees 00 minutes, and so increasing upwards, accounting every 30 min. to be one Degree of the Meridian line, as the Tangent of 45 deg. 30 min. to be one degree of the Meridian line, and the Tangent of 46 deg. 00 min. to be two Degrees 00 min. of the Meridian line, and so forwards; so that every Minute is two minutes of the Meridian line, and although that these be not the same Meridionall parts that are in the Doctrine of Triangles, yet they proceed in the same proportion as the *Secants* added together doe, and shall produce the same solution to every Probleme of Sayling by *Mercators*-Chart, as ye other Tables doe.[106]

In chapter 1, we reported that Edward Wright's 1599 *Certaine Errors in Navigation* had asserted that, to determine the distance A above the equator of a line of a given latitude in the Mercator map projection, one needs to calculate an equivalent of

$$A(\phi) = k \int_0^\phi \sec \varphi \, d\varphi.$$

[105] [Norwood 1645, 7–8].
[106] [Norwood 1645, 14].

In the absence of calculus, Wright had computed these **meridional parts** effectively by performing a Riemann sum (as the quotation above says, "the *Secants* added together"). In an unpublished work, probably around the year 1614 Thomas Harriot had derived by a complicated argument a method for computing meridional parts that is equivalent to the modern formula

$$A(\phi) = k \ln \tan\left(\frac{\pi}{4} + \frac{\phi}{2} \right),$$

in a sense finding the integral of the secant (but not by any methods related to the calculus). However, the result remained buried in Harriot's papers.[107] The appearance of the statement in Norwood's *Epitomie* quoted above appears to be the result of a chance discovery by teacher Henry Bond that a table of meridional parts bears a nearly perfect resemblance to a table of logarithms of tangents.[108] The story of the attempts to deduce the relation is long and complicated (and beyond the scope of this chapter), involving the offering of a prize by Nicholas Mercator (no relation to Gerard) and some of the most important figures in the development of the calculus.

Harriot had also anticipated another incursion of trigonometry into the sciences, namely, Snell's law in optics, which states that when a light ray crosses a boundary from one medium to another, the sines of the angles of incidence are in the same ratio to each other as their refractive indexes.[109] This principle of geometric optics was stated by several scholars, including Snell himself in his manuscript, released around 1620; René Descartes in his appendix on optics to the 1637 *Discourse on Method*; and Pierre Herigone in his *Cursus mathematicus* of the same year.[110]

Starting with the edition of 1659, Norwood's *Epitomie* also contains a short section demonstrating how trigonometry and logarithms can be used in geography: in particular, determining the distance between two points on the earth's surface. This problem, essentially the same as the medieval Islamic problem of determining the *qibla* (the direction of Mecca), is solved as follows. In figure 2.16, D is London, with latitude $\phi_D = 51°30'$; O is Jerusalem, with $\phi_O = 31°40'$ and with a difference in longitude of $\Delta\lambda = \angle DBO = 46°$; B is the North Pole; and $\widehat{FM}$ is the equator. Norwood begins by dropping a per-

---

[107] [Pepper 1968] contains an account and reconstruction of Harriot's work on this topic.

[108] See [Halley 1695, 202] and [Rickey/Tuchinsky 1980, 164].

[109] See [Fishman 2000]; see also [Goulding 2012] on correspondence between Harriot and Kepler relating to questions of refraction, within which Harriot withheld some information from Kepler to retain his priority.

[110] Kepler also came close to a correct statement [Hallyn 1994]. On Snell's writings on the subject see [Vollgraff 1936]; on Descartes, see [Smith 1987] and [Schuster 2000]. See [Sabra (1967) 1981], [Joyce/Joyce 1976], and [F. J. Dijksterhuis 2004] for descriptions of the evolution from geometrical to physical optics in the works of Descartes, Newton, and others. The sine law had also appeared in the medieval period in the work of Ibn Sahl; see [Rashed 1990].

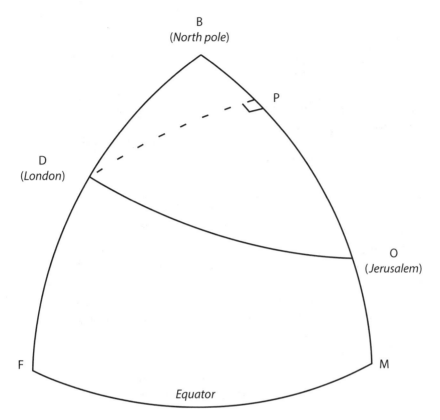

Figure 2.16
Norwood's calculation of the distance from London to Jerusalem.

pendicular from D onto $\widehat{BO}$ at P, splitting oblique triangle DBO into two right triangles. Then

$$\frac{R}{\text{Cos}\,B} = \frac{\text{Tan}\,\widehat{DB}}{\text{Tan}\,\widehat{BP}},^{111} \qquad (2.15)$$

where $\widehat{DB} = 90° - \widehat{DF} = 38°30'$. Norwood finds $\widehat{BP}$ using logarithms:

$$\log \text{Tan}\,\widehat{BP} = \log \text{Cos}\,B + \log \text{Tan}\,\widehat{DB} - \log R, \qquad (2.16)$$

and arrives at $\widehat{BP} = 28°55'$; so $\widehat{OP} = \widehat{BO} - \widehat{BP} = 90° - \phi_O - \widehat{BP} = 29°25'$. Then

$$\frac{\cos \widehat{BP}}{\cos \widehat{OP}} = \frac{\cos \widehat{DB}}{\cos \widehat{DO}},^{112} \qquad (2.17)$$

---

[111] This is equivalent to the modern expression $\cos B = \cot \widehat{DB} \tan \widehat{BP}$.
[112] This theorem is demonstrated in Norwood's *Trigonometrie* [Norwood 1631, vol. 2, 42].

which he applies (again using logarithms) to solve for $\widehat{DO} = 38°51'$. Accounting for the size of the earth, Norwood finds the distance to be 2,331 miles. Norwood's procedure, applying (2.15) and (2.17) rather than applying the Law of Cosines or some other identity, is ideal for use with logarithms since the identities require only ratios, not additions or subtractions. Techniques like this eventually would become standard in navigational practice when dealing with oblique triangles.

We conclude this chapter tarrying one last time with Richard Norwood, turning to his 1639 *Fortification, or Architecture Military*.[113] Norwood was not the first to apply trigonometric methods to military architecture; Bartholomew Pitiscus had already added a chapter on the subject in the 1612 edition of his influential *Trigonometriae*.[114] Samuel Marolois's 1627 *Fortification*,[115] with corrections and commentary by Albert Girard, had become a standard text and quickly spread through northern Europe with editions in Dutch, German, and French by 1629 and an English edition a few years later.[116] The first half of Marolois's book treats the dimensions of forts mathematically (although with no apparent use of trigonometry) and discusses practical matters of fort construction afterward. Girard's somewhat caustic commentary, restricted mostly to the mathematical section, includes the following remark:

> The Authour hath here above so disposed his calculations, that instead of explaining them briefly, he confounds them teadiously, as if heretofore there had bene noe certaine rule sett downe in writing, to calculate lines and angles as is ordinarily done by the Trigonometrie of plaine Triangles, although there have bene a number of Authours, which have treated of them, th'one after one manner, th'other after another; and the most part of them commixed with long discourses, which moved me not longe since, to putt into light some tables of Sines in a portable volume . . . the manner & order thereof [of Girard's book] being much more facile, & easier to conceive, than the reading of our Authour in his former editions, being obscure, troublesome, and hard to be attayned unto.[117]

Girard's complaint, more than the vexation of an ignored author, was addressed in a chapter added to the second edition of Edmund Gunter's *De-*

---

[113] [Norwood 1639].

[114] [Miura 1986, 68–69].

[115] [Marolois 1627].

[116] [Marolois (Girard, Hexham) 1638].

[117] [Marolois (Girard, Hexham) 1638, 6]. Girard also inserts advertisements for trigonometry in Marolois's *Geometria theoretica ac practica* [Marolois 1647, 20–24, 27–28].

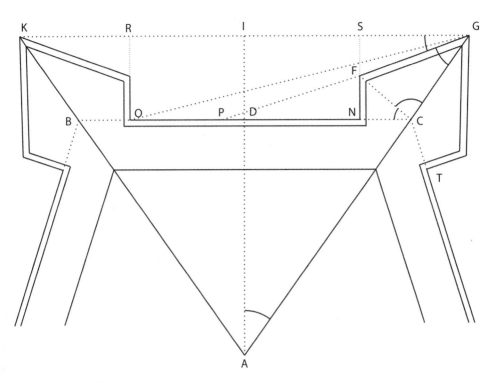

Figure 2.17
A pentagonal fort from Norwood's 1639 *Fortification*.

scription and Use of the Sector, Cross-Staff and Other Instruments, published
a decade after Gunter's death.[118] Norwood's more extensive *Fortification*,
likely inspired by said chapter, sets out the calculations of fortifications in a
similar manner, "chiefly aimed to show the application of the doctrine of triangles, according to that late invention of Logarithmes."[119] Norwood, like
Marolois and Gunter, begins with forts built in the shapes of regular polygons. Figure 2.17 shows a regular pentagonal fort, with bulwarks around the
vertices. The book begins just as Euclid's *Elements* does, with definitions of
the basic terms followed by axioms. For instance, *ON* is the **curtain**, *FN* is

---

[118] [Gunter 1636, 49–75]. The authorship of this chapter is unclear; possibilities include Gunter
himself, Samuel Foster, and John Twysden.

[119] [Norwood 1639, second page of dedicatory epistle]. Even so, the second half of Norwood's
book concerns itself with practical matters of constructing forts. Twice in those pages
(pp. 125, 134) Norwood says that his original intent was to show the benefits of trigonometry
with logarithms, but the treatise expanded to other topics as he wrote.

the *flank*, *NT* is the *gorge* of the bulwark, and *NC* is the *gorge line*. Some of the axioms are as follows:

1. A Fort is made to the intent that a few men might be able to defend themselves and the place, against a greater number. . . .
3. And because the sides thus enclosing a Fort, are not apt for the defence of themselves, especially when an enemy is nearest, and so the defence most necessary, therefore the sides of the Fort have . . . flankes to defend them, which flankes are also themselves flanked by the Curtains or sides; these flankes in [figure 2.17] are represented by [*FN*, for instance]. . . .
8. And forasmuch as the front of a bulwarke needes the more defence for that it lyes farthest from the flanke defending it, &c. therefore it is so to be drawn that it may be defended by shot from as great a part of the Curtaine as conveniently may be, which part of the Curtaine is called the second flanke; thus in [figure 2.17] the second flanke is represented by [*OP*].[120]

The next two chapters determine the angles between walls in various parts, working from a square fort up to a regular dodecagon. The trigonometric work begins after that, where Norwood poses various problems about lengths of certain walls. He begins with the regular pentagonal fort of figure 2.17. The first problem poses the following:

The length of the Curtaine, and of the Front of the Bulwarke given, to finde what the other sides and lines should be.[121]

The key is that *PG* must not exceed 720 feet; otherwise the diamond point of the bulwark *G* will be out of range of musket shot. In axiom 17 Norwood had asserted that, with this range, the curtain must be about 420 feet, the front of the bulwark *FG* must be 280 feet, and ∡*FCN* forming the flank must be about 40°. So in this problem he begins with these values.

We begin with △*SGF*. Since it is a right-angled triangle, we have

$$\frac{Radius}{FG} = \frac{Sin \angle SGF}{SF},^{122}$$

and since Norwood already knows that ∠*SGF* = 19°30′, *SF* = 93.47 feet.[123] In the same triangle, replacing the sine with a cosine, he finds *SG* = 263.94 feet;

[120] [Norwood 1639, 5–7].
[121] [Norwood 1639, 23].
[122] We would normally call the radius of the trigonometric base circle *R*, but have avoided that here since there is a point *R* in the diagram.
[123] Norwood uses logarithms to do the calculations throughout, as we saw in the example of the calculation of the distance from London to Jerusalem.

therefore $IG = \frac{1}{2}ON + SG = 473.94$ feet, and the distance between the two diamonds is twice that, $KG = 947.88$ feet.

Norwood turns next to ΔIAG. Since the fort is a regular pentagon, $\angle IAG$ is 36°, and working with this triangle as he just did with ΔSGF, he finds $AG = 806.31$ feet and $AI = 652.32$ feet. Next he examines ΔFCG. The angles of this triangle were all given in the previous chapter ($\angle FCG = 86°$, $\angle FGC = 34°30'$, and $\angle GFC = 59°30'$) and FG is known, so the Law of Sines gives $FC = 158.98$ feet and $CG = 241.84$ feet. So $AC = AG - CG = 564.47$ feet. From ΔFCN, applying the sine we find $FN = 102.19$ feet, so $ID = SN = SF + FN = 195.66$ feet; and applying the cosine we find $NC = 121.78$ feet, so $BC = 2(NC + DN) = 663.56$ feet. Next, Norwood uses ΔFPN to find $PN = 288.58$ feet (but he uses both sine and cosine rather than the tangent), so OP, the *second flank*, is equal to $ON - PN = 131.42$ feet. Finally, in ΔROG, we know that $RG = SG + ON = 683.94$ feet, so $\angle ROG = 74°02'$ (this time using the tangent function); and using the sine we arrive at a value for OG, the *longest line of defense*, of 711.4 feet—just within the 720-foot range of musket shot.[124]

The theory of fortification became obsolete with advances in artillery in the second half of the seventeenth century. Nevertheless, this example and the others in this section illustrate clearly that trigonometry's interaction with the physical world had progressed far beyond Maurice Bressieu's hesitant introduction of a surveying problem in 1581. By the middle of the seventeenth century, whatever barriers had restricted trigonometry to geometry and astronomy had long since vanished; the subject was so embedded in the sciences and the practical arts that its application to new contexts no longer raised an eyebrow.

---

[124] [Norwood 1639, 23–26]. It is worth noting that Norwood feels the need to conclude these calculations with a description of the meaning of the decimal point and fractional decimal notation. Even at this time, decimal numeration was not yet universally understood.

3 ⚛ Calculus (From the Mid-Seventeenth Century to the Mid-Eighteenth Century)

Although trigonometry began to be applied to scientific and practical disciplines in the late sixteenth century, the subject itself continued to grow actively mostly on its own and in conjunction with astronomy until the middle of the seventeenth century. This internal life did not really come to a halt until the early twentieth century (if even then); however, with the advent of calculus and analysis, advances or alterations in the theory itself were increasingly shaped by interactions with other movements, especially in mathematics and physics. At the same time, trigonometry continued its gradual transition toward its current role as part of the toolbox of mathematical functions used by science and industry. These transformations at the foundation of trigonometry were part of a larger movement of mathematics away from geometric conceptions and toward analysis, caused by the emergence of calculus. This process took some while, but by the middle of the eighteenth century trigonometry would have looked utterly foreign to a practitioner trained just a century earlier.

▨ Quadratures in Trigonometry Before Newton and Leibniz

The ideas at the heart of calculus—differentiation and integration—have had precursors in various limited contexts since Archimedes, whose works had become well known by the end of the sixteenth century. As symbolic algebra gradually became a universal mathematical language in the seventeenth century and coordinate geometry presented a framework for analyzing curves with the assistance of algebra, interest in infinitesimal methods to solve problems involving areas and volumes gradually increased. Eventually this would culminate in what we now call calculus, but the path to its realization in the works of Newton and Leibniz was long and complicated.

Among the various way stations in this journey was an episode within astronomer Johannes Kepler's so-called "war on Mars," in his *Astronomia nova*.[1] This transformative book, in which two of Kepler's three laws of planetary motion appear for the first time, found Kepler dealing with the intractable problem of Mars's orbit, more than once needing to calculate the incalculable. Just as Edward Wright, working out the mathematics of Merca-

---

[1] [Kepler 1609]. Editions of the *Astronomia nova* may be found in [Kepler 1858–1870, vol. 3] and [Kepler (Hammer/von Dyck/Caspar) 1937–present, vol. 3]; [Kepler (Donahue) 1993b] is a translation of the work into English.

tor projection, had needed to approximate what we would call the integral of the secant function, Kepler found himself needing to calculate an integral in the absence of calculus. One of these moments came about in his study of attractive and repulsive forces of planetary magnetism. He finds that the "sine of the equated anomaly is the measure of the strength of the planet's approach towards the sun in this place." But then, "the measure of the distance of the reciprocation traversed by these continuous increments of power is quite another thing." Effectively, Kepler needs to integrate the strength to calculate this distance. To do this he uses the following statement, related to a theorem of Archimedes,[2] given here in modern notation:

$$\frac{\Sigma_{\theta=1°}^{x} \operatorname{Sin}\theta}{\Sigma_{\theta=1°}^{90°} \operatorname{Sin}\theta} \approx \frac{\operatorname{Vers} x}{\operatorname{Vers} 90°}. \tag{3.1}$$

To understand how this equation might be seen as a step toward calculus, think of the summation term in the denominator of the left side as a conversion factor to radian measure. The rest of the equation is essentially a Riemann sum approximation to the formula $\int_0^x \sin\theta\, d\theta = \operatorname{vers} x = 1 - \cos x$. Kepler seems uneasy about the reader's reaction to this result, which he calls a "geometrical *faux pas* and fallacious principle." To set minds at rest, he demonstrates that the approximation is a good one by setting $x = 15°$, $30°$, and $60°$, finding that "the effects of the two procedures differ imperceptibly." Of course, we would interpret the result oppositely, that is, the ratio of the sums is an approximation of the versed sine rather than vice versa. So, whether we can ascribe to Kepler an awareness of a sort of limit process remains open to interpretation.[3]

Integration was not the only intractable mathematical problem that Kepler faced in the *Astronomia nova*. In chapter 60, he determines the true position, or **anomaly**, of a planet as a displacement from the aphelion. In figure 3.1 A is the aphelion, S is the sun at a focus of the elliptical orbit with eccentricity e, and P is the planet; the anomaly, then, is $a = \angle ASP$. The circle drawn on diameter AB is called the **auxiliary circle**, and P' is the extension onto this circle of a perpendicular from AB through P. The true anomaly may be found via a simple trigonometric calculation from the value of the ***mean***

---

[2] Archimedes's theorem is in *On the Sphere and Cylinder*; see [Boyer 1947, 267–268]. Earlier in the *Astronomia nova*, Kepler quotes a similar statement from Girolamo Cardano's *De subtilitate* ([Cardano 1550, 303]); for a translation see [Cardano (Forrester) 2013, vol. 2, 774].

[3] [Kepler 1858–1870, vol. 3, 390–391], [Kepler (Hammer/von Dyck/Caspar) 1937–present, 353–354], and [Kepler (Donahue) 1993b, 556–558]. For commentaries on Kepler and the pseudo-integration of trigonometric quantities see [Günther 1888], [Eneström 1889; 1912–13], [Boyer 1947, 268], and [Baron 1969, 109].

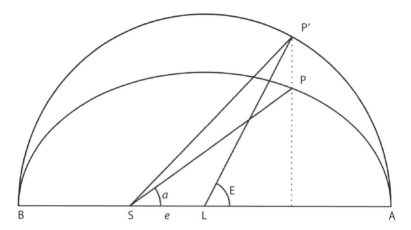

Figure 3.1
A diagram from Kepler's *Astronomia nova* (simplified).

anomaly, a quantity proportional to $M = \text{area}(ASP')$, and the ***eccentric anomaly***, a quantity proportional to $E = \text{area}(ALP')$. The key relation in modern terms is

$$M = \text{area}(ASP') = \text{area}(ALP') + \text{area}(SLP') = E + e \sin E. \qquad (3.2)$$

This allows M to be found from E; but in most cases one is much more interested in finding E from M since M is a linear function of time and therefore easy to calculate in any case. Unfortunately, (3.2) cannot be solved for E. In his later *Epitome astronomiae Copernicae*[4] he returns to the problem, resorting to iteration to approximate a solution. He begins with an estimate E_1 and calculates the right side of (3.2), then adds to E_1 the amount by which the resulting quantity differs from M. This produces a new estimate E_2, to which the same procedure may be applied.[5] Kepler's process is equivalent to fixed-point iteration where $E = f(E) = M - e \sin E$, and one iterates $E_{n+1} = f(E_n)$ in the hope that the iterations converge to the solution. Here Kepler was (presumably unknowingly) following a long-standing tradition of fixed-point iteration in astronomy that had begun in early India and continued in medieval Islam, which had included its use with the Kepler equation itself.[6]

It was not long after Kepler that predecessors to integration began to come more clearly into focus, especially in the work of two figures. The first, Bo-

[4] [Kepler 1617–1621].
[5] See [Swerdlow 2000], from which this account is derived. See [Colwell 1993], [Goldstine 1977, 47, 64], and [Badolati 1985] on the history of the Kepler equation.
[6] See [Van Brummelen 2009, 129–133, 147–148, 159–162].

naventura Cavalieri, we have met before. The second, Gilles Personne de Roberval (1602–1675), was active in both mathematics and physics but did not publish much during his lifetime: he released only two books, one on mechanics and another on Aristarchus's astronomy.[7] However, early in his career he came in contact with the circle of mathematicians connected with the famous mathematical correspondent, Marin Mersenne. Although Roberval likely did not receive as much credit for his discoveries as he might have if he had been more forthcoming with his work, the Mersenne communication network ensured that his name and some of his work was known at least to his colleagues.

During the 1620s and 1630s, both Cavalieri and Roberval worked on problems related to integration: that is, finding the areas and volumes of curved figures by dividing them into infinitely many parts. Cavalieri likely developed his theory before Roberval and called it the theory of "indivisibles." Cavalieri may have been inspired by Archimedes; indeed, he believed that the ancients must have anticipated his work. His central idea was that an area may be considered to be composed of infinitely many adjacent lines and that solids are composed of infinitely many adjacent slices of area. If two solids have the same areas in every horizontal cross-section, Cavalieri asserts that they must have the same volume.

The title of Roberval's *Traité des Indivisibles*,[8] finally published long after his death in 1693, suggests that he followed Cavalieri; indeed, Evangelista Torricelli once accused Roberval of plagiarizing Cavalieri. (Later, Roberval returned the favor by accusing Torricelli of plagiarism of his own work.) However, the title is misleading. Mersenne had posed to Roberval two problems with respect to the "roulette" or cycloid: to find the area under this curve (i.e., its "quadrature") and to construct a tangent line to it. Roberval solved them both by 1634; here we examine his quadrature. Roberval defines the cycloid kinematically: in figure 3.2, circle $AEFGB$ slides to the right along segment AC, whose length is equal to the semicircle so that the vertical diameter AB ends at CD. At the same time, A moves counterclockwise toward the top of the moving circle at the same speed as the circle so that it reaches D precisely when B does. The path traced out by A (through $E_2, F_2, \ldots, D$) is half the arch of a cycloid. To define a curve mechanically in this way, similarly to Napier's description of a pair of moving points in his definition of a logarithm, was becoming tolerated but was not yet universally approved.

---

[7] In Roberval's case this reluctance to publish has been attributed to the protection of his university professorship.

[8] [Académie Royale des Sciences 1693, 190–245].

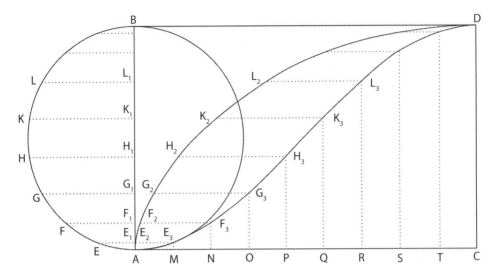

Figure 3.2
Roberval's quadrature of the cycloid.

Descartes, for example, considered such curves of no geometrical interest (but he went on to attack the problem in any case).[9]

To determine the area under the cycloid, Roberval begins by dividing the left semicircle into equally spaced arcs AE, EF, FG, and so on. He draws horizontal lines from each arc's endpoint, forming E_1, F_1, G_1, ... The lines are extended to the cycloid at E_2, F_2, G_2, ... Finally, copies of EE_1, FF_1, ... are placed with their left endpoints at E_2, F_2, G_2, ... Roberval calls the curve $AE_3F_3G_3$ ... D formed by the right endpoints of these shifted segments the "companion curve."[10] Consider the area between the cycloid and the companion curve. Its horizontal cross sections E_2E_3, F_2F_3, G_2G_3, ... were constructed to have the same widths as the cross sections EE_1, FF_1, GG_1, ... within the left semicircle. Also, rectangle $ABDC$ (which one may verify easily that it is four times the area of the semicircle) is cut in half by the companion curve. Therefore, the area $AE_3F_3G_3$ ... DC is twice the area of the semicircle, so the area $AE_2F_2G_2$ ... DC under the cycloid is three times the area of the semicircle.[11]

[9] [Jesseph 2007] is an account of this controversy.

[10] On the role of companion curves in Roberval's work, see [Hughes 2002].

[11] [Académie Royale des Sciences 1693, 191–193]. See also the free translation/interpretation of the treatise in [Walker 1932], especially pp. 174–176. Also, on the quadrature of the cycloid see [Whitman 1943] and [Vita 1973]. [Jesseph 2007] takes up the debate that ensued between Descartes and Pascal on the nature and importance of the curve, and [Hara 1969], [Hara

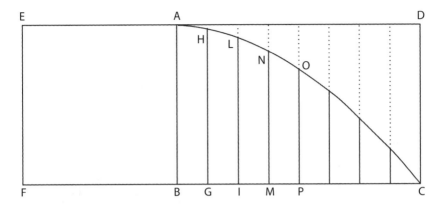

Figure 3.3
Roberval's sine curve.

As far as the history of calculus is concerned, the key to the difference between Roberval and Cavalieri is their handling of the cross sections. Roberval thinks of EE_1, E_2E_3, and so on as small strips of areas, not just as line segments. On the other hand, Cavalieri was satisfied with dividing the areas into infinitely many lines. But as far as the history of trigonometry is concerned, more interesting than their differing views of infinitesimal areas is the fact that $AE_3F_3G_3 \ldots D$ turns out to be the curve of a cosine function— the first time that a sinusoidal arc was drawn.

In this case the cosine curve is the end result of a geometric construction, so it is not really the graph of a quantity that changes with respect to an argument. However, Roberval leaves no room for doubt that he is able to think almost in this way just a couple of pages later in the same treatise. In figure 3.3, let AB be the radius of a circle, and draw BC perpendicular to AB with length equal to one-quarter the circumference of the circle. Starting from B, cut off equal line segments BG, GI, . . . corresponding to arcs on this one-quarter circumference, and draw perpendiculars GH, IL, . . . according to the sines of these arcs. This produces the sine curve $AHLNOC$. Roberval proves that the area under the curve is equal to square $ABFE$, a result that corresponds to the modern definite integral $\int_0^{\frac{\pi}{2}} \sin x \; dx = 1$. This isn't Roberval's only integral; much of the remainder of his book involves other quadratures that correspond to various definite integrals of trigonometric quantities.

1971a], and [Hara 1971b] deal with the cycloid in the works of Pascal and Wallis. For a biography of Roberval's life and work, see [Auger 1962]; the cycloid is discussed on pp. 39–49.

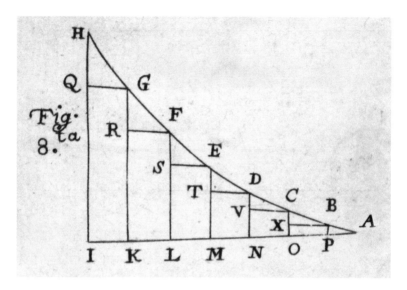

Figure 3.4
The tangent curve in Gregory's *Exercitationes geometriae.*

Roberval did not give his curve a name. In fact, he didn't even publish this book; it did not appear in print until more than half a century later in 1693. Nevertheless, presumably due to his connection with Mersenne's network of correspondents, the curve quickly was named after him and remained known by his name into the eighteenth century. The quadrature of curves quickly became a favorite topic. French polymath Honoré Fabri turned his attention to these questions in two works, his 1659 *Opusculum geometricum de linea sinuum et cycloide* and 1669 *Synopsis geometrica*,[12] where he has no problem admitting the new curves to the purview of geometric investigation (especially quadrature, including finding not just areas but also volumes of revolution and centers of gravity) and refers explicitly to the "sine curve."[13] Curves representing the other two major trigonometric functions were not far behind. The tangent curve may be found in James Gregory's 1668 *Exercitationes geometriae*[14] (figure 3.4), where he also states clearly that the integral of the tangent is the logarithm of the secant.[15] The secant curve is in John

---

[12] [Fabri 1659] and [Fabri 1669] respectively; the latter work contains the former.
[13] On Fabri's mathematics see [Fellmann 1959].
[14] [Gregory 1668a].
[15] [Gregory 1668a, 25]; the tangent curve is in the flyout page at the back of the book. See also [von Braunmühl 1900/1903, vol. 2, 41]. The curve is not drawn for the entire first quadrant, so the asymptote does not appear.

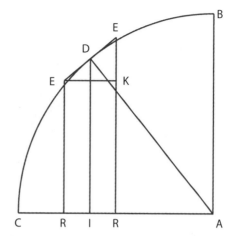

Figure 3.5
A tangent line to a circle in Pascal's
Traité des sinus du quart de cercle.

Wallis's *Tractatus de motu*.[16] In all cases the curves are drawn for what we would call just the first quadrant of the argument. This is a signal that, even though the distinction between geometrically and mechanically generated curves was fading, the geometric origins of these curves were still fundamental to the authors' conceptions.

In the same year that Fabri published his quadratures relating to the sine curve, Blaise Pascal (1623–1662) took another step forward while yet remaining in the past. His interest in mathematics started early; he became involved in the Mersenne circle in his teens and was a friend of Roberval. Prompted by Mersenne, the cycloid had become a standard proving ground for the methods of indivisibles, and priority disputes for the various methods of solution were in the air. In 1658 Pascal turned mathematicians' eyes back to the cycloid by establishing a contest that posed six problems related to its quadrature. Via this contest, he wrote his own account of the events concerning the cycloid. He published his contribution in 1659 under the pseudonym A. Dettonville[17] in what was to be his last mathematical work. He spent the last few years of his life in religious contemplation, away from scientific pursuits.

Pascal's mathematical style was in some ways rather old-fashioned; he did not show much interest in the latest developments in symbolic algebra or analytic geometry. Nor does he introduce or consider the sine curve, although in general his predisposition was to allow geometry to include mechanically generated curves like the cycloid.[18] However, he did not need them here. The

---

[16] The curve and the relevant passage appear in the *Tractatus de motu* in Wallis's *Opera mathematica* [Wallis 1695, 926] but not in the original publication [Wallis 1670, 555].
[17] [Pascal 1659].
[18] See [Jesseph 2007], especially pp. 425–430.

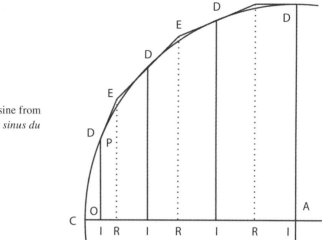

Figure 3.6
Integration of the sine from
Pascal's *Traité des sinus du
quart de cercle.*

following passage is taken from one of the essays penned under the Dettonville name, entitled *Traité des Sinus du Quart de Cercle*. Pascal begins with
the quarter circle in figure 3.5. A tangent to the circle is drawn at D with endpoints both called E, and perpendiculars are dropped to AC, defining the two
points R. He shows that $DI \cdot EE = AB \cdot RR$[19] and then continues as follows.

Text 3.1
Pascal, Finding the Integral of the Sine
(from *Traité des Sinus du Quart de Cercle*; see Figure 3.6)

Proposition I. The sum of the sines of any arc of a quadrant is equal to the
portion of the base between the extreme sines, multiplied by the radius. . . .

Proof of Proposition I. I say that the sum of the sines DI (each of them multiplied of course by one of the equal small arcs DD) is equal to the segment
AO multiplied by the radius AB.

Indeed, let us draw at all the points D the tangents DE, each of which intersects its neighbor at the points E; if we drop the perpendiculars ER it is clear
that each sine DI multiplied by the tangent EE is equal to each distance RR
multiplied by the radius AB. Therefore, all the quadrilaterals formed by the
sines DI and their tangents EE (which are all equal to each other) are equal to
all the quadrilaterals formed by all the portions RR with the radius AB; that is
(since one of the tangents EE multiplies each of the sines and since the radius
AB multiplies each of the distances), the sum of the sines DI, each of them
multiplied by one of the tangents EE, is equal to the sum of the distances RR,

[19] This may be shown by dropping a perpendicular from the left-hand E onto RE (the right-hand
E), defining K, and considering similar triangles.

each multiplied by *AB*. But each tangent *EE* is equal to each one of the equal arcs *DD*. Therefore, the sum of the sines multiplied by one of the equal small arcs is equal to the distance *AO* multiplied by the radius.

Note. It should not cause surprise when I say that all the distances *RR* are equal to *AO* and likewise that each tangent *EE* is equal to each of the small arcs *DD* since it is well known that, even though this equality is not true when the number of sines is finite, the equality is nevertheless true when the number is infinite because then the sum of all the equal tangents *EE* differs from the entire arc *BD*, or from the sum of all the equal arcs *DD*, by less than any given quantity: similarly, the sum of the *RR* from the entire *AO*.[20]

Explanation: Consider the given arc *BP* (where *P* is any point between *C* and *D*, and *O* is the point below *P* on *CA*) to be divided into infinitely many arcs among successive points labeled *D*. In the statement of the proposition, the phrase "sum of the sines" assumes implicitly that each sine *DI*[21] is multiplied by the infinitesimal arc from the given point *D* to the next point *D*. So the theorem claims that

$$\Sigma DI \cdot DD = AO \cdot AB. \tag{3.3}$$

In modern notation, this is equivalent to $\int_p^b \sin\theta \; d\theta = \cos p - \cos b$, where *p* and *b* refer to the arcs from *C* to *P*, and *C* to *B*, respectively. Now, for each point *D* we know that $DI \cdot EE = AB \cdot RR$. Therefore $\Sigma DI \cdot EE = \Sigma AB \cdot RR$. But $\Sigma RR = AO$, and each *EE* is equal to a corresponding *DD* (since the two quantities are infinitesimal). This proves the result.

(Pascal goes on to deal with the equivalents of integrals of *n*th powers of the sines, although he does not provide closed-form expressions for their solutions.) Since the definite integral of the sine between two bounds is found to be the difference between the cosines of the bounds, Pascal's theorem is a more general statement than we have seen before and is a clear step toward an expression of the indefinite integral of the sine.[22] But Pascal does not take what might seem to us to be an obvious final step. Nor is the connection to the problem of finding the tangent to a curve, and hence the Fundamental Theorem of Calculus, to be found—even though figure 3.5 suggests it. In fact, Leibniz would later be inspired by this figure to his theory of the calculus

---

[20] [Pascal 1659, *Traité des sinus du quart de cercle* 1–3]; translation by Dirk Struik in [Struik 1969, 239–241].

[21] Recall that sines were still considered to be lengths within circles so that $DI = \sin \angle CAD = R \sin \angle CAD$, where $R = AB = AD = AC$ is the radius of the circle.

[22] For commentaries on and analyses of Pascal's quadratures, see [Bosmans 1923], [Russo 1962], [Vekerdi 1963], [de Lorenzo 1985], [Loeffel 1986], and especially [Merker 1995] and [Merker 2001, 101–110].

and reflected that "Pascal seemed to have had his eyes obscured by some evil fate."[23]

▣ Tangents in Trigonometry Before Newton and Leibniz

Calculus, of course, begins with the two problems of finding areas between curves and of finding tangent lines to curves. Its great discovery is the fact that the solutions of these problems are inverse processes of each other— integration and differentiation respectively. In the decades prior to the discovery of this fundamental theorem, both problems occupied many mathematicians, but they were not yet linked. These problems were thorny because before one began, one needed to take a position on what sorts of curves are valid objects of geometrical study and what sorts of methods are to be permitted. Descartes, for instance, distinguished between algebraic and mechanical curves, the latter involving composition of motions. He insisted upon algebraic solutions for the former but allowed solutions based on motion for the latter.[24]

Roberval's cycloid was a classical example of a curve generated from the composition of motions. Now, Roberval does seem to have had an analytic method of finding tangents, but it is lost, and what is left to us in his publications is a mechanical method. In figure 3.7, as before, the cycloid is generated by a pair of motions: point A travels counterclockwise around the generating circle at the same velocity as the circle itself slides from left to right. At some arbitrary moment, the generating circle has moved to the right (represented in dashes) while A has traveled upward on the circle to X where we wish to draw the tangent line to the cycloid. Draw a segment XY of any length rightward from X, representing the circle's horizontal motion. Next, draw a tangent XZ to the generating circle at X with the same length as XY.[25] Finally, complete the parallelogram $XYWZ$. Since A moves on the circle at the same speed as the circle moves to the right, XZ and XY in effect represent velocity vectors of the two motions. Therefore the diagonal XW will be the sought tangent line.[26]

---

[23] [Leibniz 1920, 17]. On Pascal's influence on Leibniz, see also [Costabel 1962].

[24] See [Baron 1969, 163–166] on this topic, including Descartes's approach to finding the tangent to the cycloid.

[25] The tangent line to the circle is simply the perpendicular to the radius that meets the circle at X.

[26] [Académie Royale des Sciences 1693, 192–193]; for an English paraphrase see [Walker 1932, 176–177]. The fundamental paper on Roberval's method of tangents is [Pedersen 1968]. For a discussion of this method as it transmitted to Newton, see [Wolfson 2001]. The priority dispute over the dates of discovery is considered in [Walker 1932, 149–157]. This method is valid only if the composed motions are physically and mathematically independent; see [Duhamel 1838],

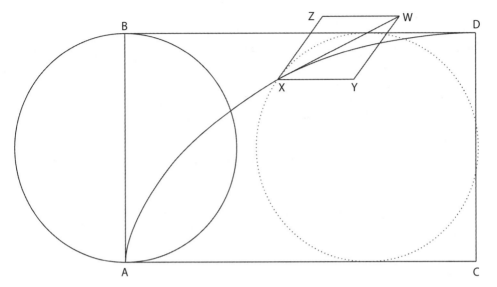

Figure 3.7
Roberval's construction of a tangent to the cycloid.

Roberval deals with tangents to a variety of curves in another work, *Observations sur la Composition des Mouvemens*,[27] in which he repeats his argument for the tangent of the cycloid. But in this later work he goes on to find the tangent to his "companion" curve to the cycloid using similar methods (figure 3.8).[28] This is the first time that a tangent to the sine curve was constructed.

Methods of finding tangents to curves continued to develop through the efforts of such luminaries as Torricelli, Descartes, Fermat, Wallis, and Gregory. Since this story is itself tangential to the history of trigonometry, we shall pass over most of it. One episode, however, is of interest: English mathematician Isaac Barrow's (1630–1677) method of finding the derivative of the tangent curve. Barrow, Isaac Newton's mentor and patron, came as close to the calculus as one can get without quite arriving at it. He had a geometric version of the Fundamental Theorem of Calculus (the statement that quadratures and tangents are inverse problems of each other), which had a major impact on both Newton's and Leibniz's work.[29] His work is a curious mixture

[Auger 1962, 63], [Baron 1969, 174–175], and [Wolfson 2001, 208–210]. For a defense of Roberval see [Pedersen 1968, 175–176].

[27] [Académie Royale des Sciences 1693, 69–111].

[28] [Académie Royale des Sciences 1693, 109–110].

[29] A good place to start on Barrow's life and mathematics is [Feingold 1990]. On Barrow's influence on both Newton and Leibniz, see especially [Feingold 1993]. For a discussion of Barrow's influence on Leibniz with respect to their presentations of the Fundamental Theorem of Calcu-

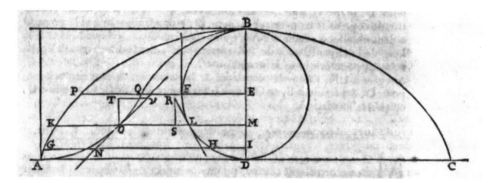

Figure 3.8
Roberval's construction of a tangent to the sine curve.

of the old geometric and the new analytic methods; Barrow uses symbolic algebra in a recognizable form, yet the argument still retains substantial geometric elements.[30] His analytical method of finding tangents to curves, related to an earlier method by Fermat, is powerful and may be applied to many curves yet falls slightly short of a "calculus."[31] Nevertheless, the case has been made that Barrow deserves at least some of the credit for the invention of the calculus.[32]

Text 3.2
Barrow, Finding the Derivative of the Tangent
(from *Lectiones geometricae*)

Let *DEB* be a quadrant of a circle, and *BX* a tangent to it; and let the line [i.e., curve] *AMO* be such, that taking *AP* any how in *AV*, equal to the arc *BE*, and erecting *PM* perpendicular to *AV*; let *PM* be equal to the tangent *BG* of the arc *BE*.

Assume the arc *BF* = *AQ*, and drawing *CFH*, as also the perpendiculars *EK*, *FL* to *CB*, call *CB*, *r*; *CK*, *f*; *KE*, *g*. Then because it is *CE* : *EK* :: arc *EF* : *LK*,

---

lus, see [Nauenberg 2014]. [Malet 1997] discusses Barrow's role in the seventeenth-century replacement of indivisibles by infinitesimals.

[30] Discussions of old and new methods in Barrow's work on tangents and in his mathematics in general may be found in [Sasaki 1985], [Mahoney 1990], [Maierù/Toth 1994], [Hill 1996], [Hill 1997], and [Panza 2008].

[31] On Barrow's method of tangents, see [Child 1916], [Boyer 1949, 182–186], [Whiteside 1961c, 358–363], and [Baron 1969, 239–252]. On the method applied to the trigonometric tangent function, see [Barrow 1916, 121–123], [Coolidge 1951, 458–459], and [Roy 2011, 110–111].

[32] [Barrow 1916, vii]. [Feingold 1993] analyzes the scholarship for and (mostly) against this claim, declaiming the "predilection of historians to concentrate on (and aggrandize) the work and significance of the truly great" (p. 312).

or $CE : EK :: QP : LK$,[33] that is, $r : g :: e : = LK$; it shall be $CL = f + \dfrac{eg}{r}$. And

$$LF = \sqrt{rr - ff - \frac{2fge}{r}} = \sqrt{gg - \frac{2fge}{r}}.$$ But it is $CL : LF :: (CB : BH ::) CB : QN.$

That is, $+ \dfrac{eg}{r} : \sqrt{gg - \dfrac{2fge}{r}} :: r : m - a$. Or (by squaring) $+ \dfrac{2fge}{r} : gg - \dfrac{2fge}{r}$

$:: rr : mm - 2ma$. Whence (rejecting as necessary) we obtain the equation

$rfma = grre + gmme$. Therefore (by substitution) $rfmm = grrt + gmmt$. Or

$\dfrac{rfmm}{grr + gmm} = t$. Or (since m is $= \dfrac{rg}{f}$) it shall be $t = \dfrac{rr}{rr + mm} \quad m = \dfrac{CBq}{CGq} BG$

$= \dfrac{CKq}{CEq} BG$.[34]

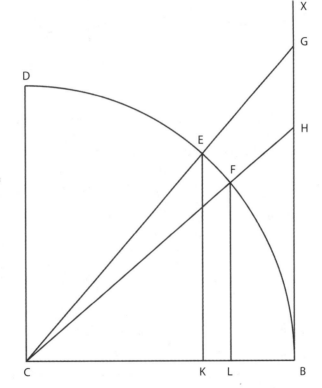

Figure 3.9
Finding the
derivative of the
tangent from
Barrow's
*Lectiones
geometricae.*

---

[33] The text has QF rather than QP.

[34] [Barrow 1735, 179–180]. There is another English translation in [Barrow 1916, 122–123], but since notation and language are important to the historical interpretation here, we have chosen to reproduce a translation much closer to Barrow's time. The original Latin is in [Barrow 1670, 84].

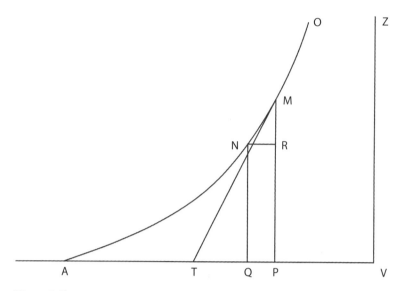

Figure 3.10

The tangent line to the tangent curve from Barrow's *Lectiones geometricae*.

Explanation: Figure 3.9 gives the situation in geometric terms: let $x = \angle BCG$; then $y = \text{Tan } x = BG$. Figure 3.10 gives the situation in coordinate terms: it displays the curve $y = \text{Tan } x$ almost in a modern way (although Barrow feels the need to explain the construction), with the origin at A, $x = AP$, and $y = PM$. The goal is to construct a tangent line MT to the tangent curve at M. Barrow begins by considering an x infinitesimally smaller than the one defined by $\angle BCG$, namely, $\angle BCH$ (figure 3.9). This corresponds to $x = AQ$ and $y = QN$ in figure 3.10, producing the ***differential triangle*** NRM—which, since it is infinitesimal, is similar to TPM. Since $m = PM$ (figure 3.10) $= BG$ (figure 3.9) by construction, to draw the tangent line it remains only to determine the length of the ***subtangent*** $t = PT$. Barrow operates by three rules, which we shall illustrate as they arise.

With $r = CB$, $f = CK$, and $g = KE$ (figure 3.9), we define two infinitesimal quantities $a = MR$, $e = NR$ (figure 3.10). Then, since $\widehat{EF}$ is infinitesimal it may be equated to a straight line, so $\dfrac{CE}{EK} = \dfrac{\widehat{EF}}{LK} = \dfrac{QP}{LK}$, which gives $CL = f + LK = f + \dfrac{ge}{r}$. Then, using the Pythagorean Theorem on ΔCFL, we have $LF = \sqrt{r^2 - f^2 - \dfrac{2fge}{r} - \dfrac{g^2 e^2}{r^2}}$. Barrow now applies his rule 1, which grants him permission to ignore any power or products of infinitesimals "for these terms will be equal to nothing." Therefore he drops the $\dfrac{g^2 e^2}{r^2}$ term (since e is an infinitesimal quantity), and the resulting expression for LF then simplifies to $\sqrt{g^2 - \dfrac{2fge}{r}}$. Barrow then squares the ratio equality $\dfrac{CL}{LF} = \dfrac{CB}{QN}$ and applies his first rule again, giving

$$\frac{f^2 + \dfrac{2fge}{r}}{g^2 - \dfrac{2fge}{r}} = \frac{r^2}{m^2 - 2ma}.$$

Cross-multiplying and multiplying through by r, we have

$$f^2 m^2 r + 2f\,gem^2 - 2maf^2 r - 4fgema = r^3 g^2 - 2fger^2;$$

rule 1 allows Barrow to ignore $4fgema$ since both e and a are infinitesimal. His next step applies the clever rule 2. Both sides of the remaining equality contain infinitesimal terms. Since we know the two sides are equal, we know that the *non-infinitesimal* terms must be equal. So $f^2 m^2 r$ and $r^3 g^2$ may be canceled. After a bit of simplification, Barrow has

$$rfma = ger^2 + gem^2.$$

Now that Barrow's equation contains only first-order infinitesimals, he invokes rule 3, which allows him to treat ΔPMT as similar to the infinitesimal ΔRMN. Effectively, he multiplies the left side by m/a and the right side by the t/e, and all the infinitesimal quantities vanish. Solving for t produces Barrow's sought result,

$$t = \frac{rfm^2}{gr^2 + gm^2}.$$

Barrow's final step interprets the algebraic quantities back into geometric terms (the q is to be interpreted as "squared").

Barrow's result, converted to modern notation, is as follows:

$$\frac{d}{dx}\tan x = \frac{m}{t} = \frac{m}{\dfrac{rfm^2}{gr^2 + gm^2}} = \frac{g(r^2 + m^2)}{rfm} = \frac{KE \cdot CG^2}{CB \cdot BG}$$

$$= \frac{BG \cdot CG^2}{CB \cdot CB \cdot BG} = \frac{CG^2}{CB^2} = \sec^2 x.[35]$$

---

[35] Barrow's three rules may be found in [Barrow 1735, 173–174]. J. M. Child asserts: "although Barrow does not mention the fact, he must have known (for it is so self-evident) that the same two diagrams can be used for any of the trigonometrical ratios. Therefore, *Barrow must be credited with the differentiation of the circular functions*" [Barrow 1916, 123]. This seems overly charitable to Barrow; it measures self-evidence by modern standards and does not consider that the trigonometric curves were new; not all of them had even appeared in print by this time.

The strikingly modern appearance of Barrow's argument extends also to his visual expression of the tangent function as a curve in Cartesian coordinates. Also, with the benefit of hindsight, one can see that Barrow would be able to extend this pattern of calculation to a number of other fluent quantities, bringing him to the verge of something that might be entitled to be called a calculus.

▨ Infinite Sequences and Series in Trigonometry

Meanwhile, interest continued in another subject that for centuries had involved potentially infinite processes, namely, cyclometry. Ludolph van Ceulen's calculation of π to 35 decimal places in the first decade of the seventeenth century was not to be improved on substantially until 1699, but that does not mean that interest in the calculation waned. Up to van Ceulen's time, the standard methods for calculating π were geometric in the style of Archimedes, approximating the circumference of the circle with the perimeter of regular polygons with an increasing number of sides and computing the ratio of their circumferences to their diameters. However, the analysis of π may also be approached as a problem of the quadrature of the circle, which was how James Gregory (1638–1675) thought of it. Gregory, a young Scottish scholar interested in optics and astronomy, spent some of his formative years in Padua, Italy, studying with mathematicians in the Cavalierian school. This seems to have provoked his interest in questions of quadrature, which he approached using indivisibles and infinitesimals.[36] He published two books related to his stay in Italy, *Vera circuli et hyperbolae quadratura* and *Geometriae pars universalis*.[37] In the former, Gregory attempts to prove that the precise quadrature of the circle is impossible; the failure of this project, pointed out by Christiaan Huygens, prompted a controversy between the two.[38]

As the title of *Vera circuli et hyperbolae quadratura* suggests, Gregory uses a single method to find the areas of sectors of the circle, ellipse, and hyperbola simultaneously. In figure 3.11, $\overset{\frown}{ACB}$ is the conic section to be considered and M is its center; the goal is to find the area $MACB$. Tangents AT and BT to the curve are drawn, and T is joined with M, defining C. Points A, B, C are joined. Then $i_0 = area(MAB)$ is a lower bound for the sought area, and $I_0 = area(MATB)$ is an upper bound. Draw the tangent to the curve at C, defining U and V, and join MU and MV. Configurations $MAUC$ and $MCVB$ now

---

[36] For surveys of Gregory's approach to infinitesimal arguments, see [Scriba 1957] and [Crippa 2019, 35–92].

[37] [Gregory 1667] and [Gregory 1668b] respectively.

[38] See [Dijksterhuis 1939] and [Scriba 1983].

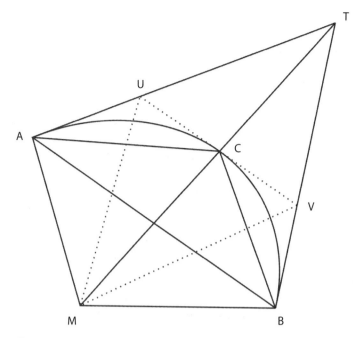

Figure 3.11
Gregory on areas in the circle and hyperbola.

share the same properties as the original diagram, and so $i_1 = area(MACB)$
and $I_1 = area(MAUVB)$ are improved lower and upper bounds of the sought
area respectively. The process may be repeated on both halves of the new diagram, generating i_2, I_2, and so on. From the properties of conic sections,
Gregory derives the following relations:

$$i_{n+1} = \sqrt{i_n I_n} \text{ and } I_{n+1} = \frac{2i_{n+1}I_n}{i_{n+1} + I_n}. \tag{3.4}$$

That these sequences converge to the area A of the sector is obvious from the
geometry, but Gregory demonstrates the convergence arithmetically by showing that the difference between i_n and I_n decreases by more than half at each
iteration. Gregory thought so much of this convergence that he described it
as a new sixth primitive arithmetical operation, after addition, subtraction,
multiplication, division, and the extraction of roots. More relevant here is the
fact that the area of the sector of the circle is

$$A = I_0 \sqrt{\frac{i_0}{I_0 - i_0}} \arctan \sqrt{\frac{I_0 - i_0}{i_0}}. \tag{3.5}$$

Therefore, the iteration (3.4) brings the arc tangent function into the realm of arithmetic, provided that one allows Gregory to call the convergence of a sequence an arithmetic operation. Even more interesting, if one allows I_0 to be less than i_0, the curve transforms from a circle to a hyperbola, and the circular arc tangent in (3.5) becomes a hyperbolic arc tangent. This link between the circle and the hyperbola, which Gregory uses here to approximate both π and ln (10), is also the earliest hint that trigonometric and exponential/logarithmic functions have some sort of familial relationship. It would take some decades before the fruits of this relationship would become clear.[39]

Meanwhile, not far away, the young and almost famous Isaac Newton (1642–1727) had already experienced his "annus mirabilis" in 1666, the unprecedented year of progress and discovery that would establish him eventually as one of the world's intellectual giants.[40] His discoveries in this year included linked contributions to quadrature and to finding tangents to curves, with methods so general that he may be called the discoverer of the calculus. However, just as Barrow and Harriot before him, Newton chose not to publish his findings immediately. He was provoked into a sort of publication of his work on quadrature when he came upon Nicolaus Mercator's 1668 *Logarithmotechnia*.[41] Mercator, a German mathematician who spent most of his later life in London and was well connected with the mathematics community there, approached the problem of the quadrature of the hyperbola by converting it to an infinite series:

$$\ln(1+x) = \int_0^x \frac{1}{1+t}\, dt = x - \frac{x^2}{2} + \frac{x^3}{3} - \frac{x^4}{4} + \cdots.^{42} \tag{3.6}$$

In terms of quadrature methods, this was squarely within Newton's territory. With rumors swirling that Mercator was about to publish a similar book on the circular functions, Newton rushed to circulate his findings on paper, if not in print. The result, his *De analysi per aequationes numero terminorum infinitas*, was shown to John Collins of the Royal Society in 1669 and

---

[39] On the *Vera circuli et hyperbolae quadratura* and this argument, see [Heinrich 1901], [Dehn/Hellinger 1939], [Dehn/Hellinger 1943], and [Scriba 1983]. This account is taken from the two articles by Dehn and Hellinger.

[40] A summary of the literature on Isaac Newton's life and works would fill a book in itself. We confine ourselves here to noting four recent scientific biographies, [Westfall 1980], [Christianson 1984], [Hall 1992], and [Panza 2003]; the handbook [Gjertsen 1986]; and the collection [Cohen/Smith 2002]. See [Hofmann 1951] and [Fierz 1972] for surveys of Newton's work as a mathematician.

[41] [Mercator 1668].

[42] See [Naux 1971, vol. 2, 56–59] and [Edwards 1979, 161–164] for an account of Mercator's derivation.

distributed among friends and colleagues; it was eventually published in 1711.[43] It seems to have fended off the threatened scoop; Mercator's work on the circular functions never appeared.

Inspired by John Wallis's *Arithmetica infinitorum*,[44] Newton had been solving quadrature problems by representing curves using infinite series. The key to his algebra was his use of the binomial theorem, generalized to apply not just to whole-numbered powers but to fractional and negative powers as well.[45] However, in the *De analysi* Newton does not appeal to the binomial theorem, relying instead on algebraic generalizations of arithmetic procedures that allow him to divide and find square roots of polynomials. Once he has derived a series expression for a quantity, Newton effects his quadrature using the theorem that opens the *De analysi*: effectively, the reverse of the power rule $\int ax^{m/n} \, dx = \frac{an}{m+n} x^{\frac{m+n}{n}}$. Newton's ability to move easily back and forth between geometric and algebraic expressions and his facility and boldness with infinite expressions were an audacious step forward. As geometric curves, trigonometric functions were subject to Newton's methods; mechanical curves were just as amenable to his approach.[46]

Text 3.3
Newton, Finding a Series for the Arc Sine
(from *De analysi*)

Let *ADLE* be a Circle, the Length of whose Arch *AD* is to be investigated. Draw the Tangent *DHT*, and having compleated the indefinitely small Rectangle *HGBK* and put $AE = 1 = 2AC$, it shall be as *BK* or *GH* the moment of the base *AB* (x) to *HD* the Moment of the Arch *AD* :: *BT* : *DT* :: $BD \left(\sqrt{x-xx}\right) : DC\left(\frac{1}{2}\right) :: 1 \, (BK) : 1/2\sqrt{x-xx} \, (DH)$. And so $1/2\sqrt{x-xx}$ or $\sqrt{x-xx}/(2x-2xx)$ is the Moment of the Arch *AD*. Which being reduced makes $\frac{1}{2}x^{-1/2} + \frac{1}{4}x^{1/2} + \frac{3}{16}x^{3/2} + \frac{5}{32}x^{5/2} + \frac{35}{256}x^{7/2} + \frac{63}{512}x^{9/2}$, &c. Wherefore by Rule the second the Length of the Arch is $x^{1/2} + \frac{1}{6}x^{3/2} + \frac{3}{40}x^{5/2}$

---

[43] It appears as the first of several writings in [I. Newton 1711]. For an English translation, see [I. Newton (Whiteside) 1967–1981, vol. 2, 206–247]. See also [Roy 2011, 140–149].

[44] [Wallis 1656]. For an English translation see [Wallis (Stedall) 2004]. On Newton's debt to Wallis especially with respect to interpolation and the binomial theorem, see [Guicciardini 2009, 139–158].

[45] See [Whiteside 1961b] and [Hilliker 1974]. The result had been anticipated by Henry Briggs almost half a century earlier [Whiteside 1961a], but apparently Newton remained unaware of this. On Newton's general approach to series, see also [Ferraro 2008, 53–78].

[46] "Let these remarks suffice for geometric curves. But, indeed, if the curve is mechanical it yet by no means spurns our method" [I. Newton (Whiteside) 1967–1981, vol. 2, 239].

$+\frac{5}{112}x^{7/2} + \frac{35}{1152}x^{9/2} + \frac{63}{2816}x^{11/2}$,   &c.   or $x^{1/2}$ into $1 + \frac{1}{6}x + \frac{3}{40}x^2 + \frac{5}{112}x^3$

$+\frac{35}{1152}x^4 + \frac{63}{2816}x^5$, &c.[47]

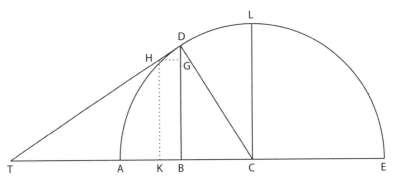

Figure 3.12
Newton's derivation of a series for the arc sine.

Explanation: The circle in figure 3.12 has radius ½; let $x = AB$. The goal is to determine the length of $\overset{\frown}{AD}$ in terms of x. DGH is an infinitesimally small differential triangle; in modern notation we are interested in

$$\frac{DH}{GH} = \frac{d(\overset{\frown}{AD})}{dx}.[48]$$

Considering $\overset{\frown}{DH}$ to be straight, employing similar triangles (ΔBDC and ΔGHD), and using Pythagoras (ΔBDC), Newton gets $DH = 1/2\sqrt{x - x^2}$ $= \sqrt{x - x^2}/2(x - x^2)$ as the moment of the arc length when GH, the moment of x, is one; in other words, this expression for DH is the derivative of the arc length on the circle with respect to x. Newton converts this latter expression into an infinite series by first performing the algebraic analog of the square root algorithm (which is itself based on the binomial theorem) on $\sqrt{x - x^2}$ and then using long division to divide the result by $2(x - x^2)$. This gives

$$\frac{1}{2}x^{-1/2} + \frac{1}{4}x^{1/2} + \frac{3}{16}x^{3/2} + \frac{5}{32}x^{5/2} + \frac{35}{256}x^{7/2} + \frac{63}{512}x^{9/2} + \cdots.$$

This is the derivative of the arc length, so to find the arc length itself, Newton must integrate. He applies rules 1 and 2 (the inverse power rule, and the sum rule for integrals) to arrive at

$$\overset{\frown}{AD} = x^{1/2} + \frac{1}{6}x^{3/2} + \frac{3}{40}x^{5/2} + \frac{5}{112}x^{7/2} + \frac{35}{1152}x^{9/2} + \frac{63}{2816}x^{11/2} + \cdots.$$

[47] [Newton 1964–1967, vol. 1, 18].
[48] Newton refers to the numerator and denominator of this expression as "moments" of the changing quantities.

To see how this corresponds to the modern arc sine series, join D to A and E, forming a right triangle inscribed in the semicircle. The Pythagorean Theorem on $\triangle ABD$ gives $AD = \sqrt{x}$. Then $\sin \angle AED = AD / AE = \sqrt{x}$; but $\angle AED = \frac{1}{2}\angle ACD = \frac{1}{2} \cdot 2\widehat{AD}$, so $\widehat{AD} = \arcsin \sqrt{x}$.

Therefore

$$\arcsin \sqrt{x} = x^{1/2} + \frac{1}{6}x^{3/2} + \frac{3}{40}x^{5/2} + \frac{5}{112}x^{7/2} + \frac{35}{1152}x^{9/2} + \frac{63}{2816}x^{11/2} \cdots,$$

or

$$\arcsin x = x + \frac{1}{6}x^3 + \frac{3}{40}x^5 + \frac{5}{112}x^7 + \frac{35}{1152}x^9 + \frac{63}{2816}x^{11} + \cdots.$$

Immediately after this, Newton describes an algebraic process that allows one to invert a given series in order (for instance) to work backward from a given area under the curve to the x value that gives that area. Two of his examples of this process produce

$$\sin x = x - \frac{1}{6}x^3 + \frac{1}{120}x^5 - \frac{1}{5040}x^7 + \frac{1}{362880}x^9 \cdots \tag{3.7}$$

and

$$\cos x = 1 - \frac{1}{2}x^2 + \frac{1}{24}x^4 - \frac{1}{720}x^6 + \frac{1}{40320}x^8 - \frac{1}{3628800}x^{10} \cdots; \tag{3.8}$$

this marks the first time the Maclaurin series for the sine and cosine appeared in Europe.[49]

Several years later, in a letter to Henry Oldenburg dated June 13, 1676, Newton enclosed a list of infinite series related to various curves. One of these states an expression for the chord of arc $n\theta$ in terms of the chord of θ. Converted to sines, it becomes

$$\sin n\theta = n\sin\theta - \frac{n(n^2 - 1^2)}{3!}\sin^3\theta + \frac{n(n^2 - 1)(n^2 - 3^2)}{5!}\sin^5\theta - \cdots. \tag{3.9}$$

---

[49] [I. Newton (Whiteside) 1967–1981, vol. 2, 237]; see also [Edwards 1979, 205–209], as well as pp. 204–205 for an account of Newton's method of inversion of series. We have seen these series before in medieval India, although they arose in an entirely different way; see [Van Brummelen 2009, 113–121]. The question of a possible transmission from India to Europe has led to spirited arguments on both sides, of which [Joseph 2009] is one of a number of examples. Here, we shall reflect the magnitude of primary evidence available on the question of transmission by saying nothing. Series of inverse trigonometric functions would reappear in eighteenth- and nineteenth-century China, described in chapter 4.

For odd n this series is finite; for even n it is infinite. Newton does not give the equivalent cosine series, although one presumes that he must have known it. Perhaps he was inspired by Viète's earlier work on multiple-angle formulas; he had read Viète's writing on the subject. The series would eventually be derived by Abraham de Moivre in 1698 and later by James Stirling.[50] Formulas of this sort were soon pursued by Johann and Jacob Bernoulli, Jacob Hermann, and especially the computational mathematician Thomas Fantet de Lagny (1660–1734). In 1705 the latter also wrote a paper establishing formulas for $\tan n\theta$ and $\sec n\theta$ in terms of $\tan\theta$ and $\sec\theta$ respectively. For $\tan 5\theta$, for instance, we have

$$\tan 5\theta = \frac{5\tan\theta - 10\tan^3\theta + \tan^5\theta}{1 - 10\tan^2\theta + 5\tan^4\theta}. \tag{3.10}$$

The coefficients in this and de Lagny's other equations arise from alternating terms in the binomial expansion, and the formula for arbitrary n may be proved via induction.

Collins wrote to Gregory in March 1670 reporting one of Newton's series (for finding the area of a part of a circle), but without describing where it came from.[51] In a response to Collins dated almost a year later Gregory gives seven series, including the following:

$$\text{arc Tan}\,x = x - \frac{x^3}{3R^2} + \frac{x^5}{5R^4} - \frac{x^7}{7R^6} + \frac{x^9}{9R^8},^{52} \tag{3.11}$$

$$\text{Tan}\,x = x + \frac{x^3}{3R^2} + \frac{2x^5}{15R^4} + \frac{17x^7}{315R^6} + \frac{3233x^9}{181440R^8},^{53} \tag{3.12}$$

and

$$\text{Sec}\,x = R + \frac{x^2}{2R} + \frac{5x^4}{24R^4} + \frac{61x^6}{720R^5} + \frac{277x^8}{8064R^7}.^{54} \tag{3.13}$$

[50] Newton's letter is given with English translation in [I. Newton (Turnbull/Scott/Hall/Tilling) 1960–1977, vol. 2, 20–47]; see also [Roy 2011, 148–152]. De Moivre's proof is in [de Moivre 1698].

[51] The letter is published in [Turnbull 1939, 88–89].

[52] Setting $x=1$ in this equation gives the series for π named after him, $\frac{\pi}{4} = 1 - \frac{1}{3} + \frac{1}{5} - \frac{1}{7} + \cdots$.

[53] The coefficient of the last term of this series is in error; it should be 62/2835.

[54] [Turnbull 1939, 170]. In modern notation, the other four series are for the functions $\log\sec x$, $\log\tan\left(\frac{x}{2} + \frac{\pi}{4}\right)$, arc $\sec\left(\sqrt{2}e^x\right)$, and $2\,\text{arc}\,\tanh\left(\tan\frac{x}{2}\right)$.

Each of these series is equivalent to the Maclaurin series for the given function.[55] Gregory gives no clue, in this or any of his other letters or publications, how he arrived at them or the many other series found in his correspondence. It is possible that he had some knowledge of the general Taylor series formula, but he may have found the series by some other means.[56]

A couple of years later, another young mathematician became interested in the quadratures of the circle and of the conic sections. Gottfried Wilhelm Leibniz (1646–1716) began his career interested in various aspects of philosophy but on his arrival in Paris in 1672 studied mathematics with Christiaan Huygens. It did not take long for the mathematical neophyte to make impressive discoveries; the snippet that we shall describe below dates from 1673 and was eventually published in the *Acta Eruditorum* in 1682. He derived a powerful new method for the quadrature of curves using his "transmutation theorem," which converts the original curve into something more tractable. The construction relies on the tangents to the original curve and in modern terms is equivalent to the formula for integration by parts, although Leibniz would not have thought of it in this way. Applying his technique to the circle and inspired by Mercator's use of polynomial division to generate (3.6), Leibniz arrived at the following formula for the area of a sector of a circle:

$$\text{Area} = Rz - \frac{z^3}{3R} + \frac{z^5}{5R^3} - \frac{z^7}{7R^5} + \cdots, \tag{3.14}$$

where R is the radius of the circle and z is the distance from one of the endpoints of the arc to the intersection of the two tangent lines taken from the two endpoints. Applying this to a quadrant of a unit circle, Leibniz has

$$\frac{\pi}{4} = 1 - \frac{1}{3} + \frac{1}{5} - \frac{1}{7} + \cdots. \tag{3.15}$$

This seems to have been new in Paris to Huygens and Leibniz, although of course not to Newton and Gregory. The series is utterly impractical for actually computing digits of π; in the letter to Henry Oldenburg now called the *epistola posterior*, Newton remarked that it would take over five billion terms

---

[55] Our continuing use of the word "function" may be anachronistic, but as Malet points out, one is at a loss to find another word to describe the collection of entities for which series were calculated [Malet 1993, 121]. The series are terminated as shown here; of course, they might have been continued for as long as patience allowed.

[56] One such reconstruction, [Malet 1993], relies on Gregory's method for finding tangents to curves, which itself is related to (but goes beyond) Barrow's method described earlier in this chapter.

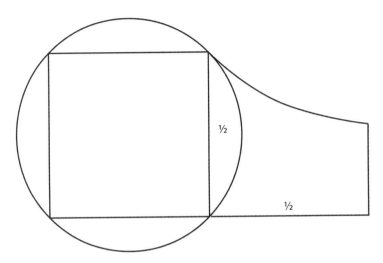

Figure 3.13
Leibniz on quadratures of the circle and hyperbola.

and a millennium of time to use it to achieve a value accurate to 20 decimal places.[57]

Leibniz was certainly aware of this; in fact, he went on to manipulate the series by combining pairs of terms. This results in

$$\frac{\pi}{8} = \frac{1}{1 \cdot 3} + \frac{1}{5 \cdot 7} + \frac{1}{9 \cdot 11} + \cdots. \tag{3.16}$$

This series bears a startling resemblance to what one gets if one substitutes $x = 1$ into Mercator's series (3.6) for the natural logarithm and divides by four:

$$\frac{1}{4}\ln 2 = \frac{1}{2 \cdot 4} + \frac{1}{6 \cdot 8} + \frac{1}{10 \cdot 12} + \cdots. \tag{3.17}$$

Leibniz recorded this in figure 3.13 where the area of the circle inscribed about a square of side length ½ is $\pi/8$, and the area of the hyperbola of diameter $\sqrt{2}$ is $\frac{1}{4}\ln 2$. We have here another anticipation of the relation between trigonometric and logarithmic functions, this time in the context of quadrature.[58]

---

[57] [I. Newton (Turnbull/Scott/Hall/Tilling) 1960–1977, vol. 2, 120, 139].

[58] On Leibniz, the transmutation technique, and the quadrature of the circle, see [Leibniz (Child) 1920, 42–45], [Hofmann 1974, 54–61], [Horváth 1983], and [Brown 2012, 87–92]. On Leibniz on series in general, see [Costabel 1978] and [Ferraro 2008, 25–54], among others.

▧ Transforming the Construction of Trigonometric Tables with Series

It took no time at all for Newton to recognize that these infinite series could transform the age-old practice of generating sine tables, "on which the whole subject of trigonometry depends."[59] In the same *epistola posterior* to Oldenburg just a few pages after his discussion of calculating π, Newton provides his thoughts on how series might be used. Given angles φ and θ, and the sine and cosine of θ, he produces a geometric argument that, given the sine and cosine of $\phi + n\theta$ (for some n), determines the sines and cosines of $\phi + (n \pm 1) \theta$. Thus, for instance, one can fill in a table of the sines and cosines of multiples of 5° if one happens to know the sine and cosine of 5°. But this is nothing new: it is just the same problem with which all builders of sine tables had had to contend. The nub of the problem in this case is to come up with the sine and cosine of 5° in the first place. However, Newton points out that the series (3.8) may be used to solve the "knotty problem" of finding the cosine of 5° (or any geometrically inaccessible angle). Once one has series for the sine and cosine, any sine or cosine value may be determined, no matter how knotty, and the theoretical problem of computing a trigonometric table is solved.[60]

In practice, however, trigonometric tables had become large enough that computing more than a tiny fraction of their entries using infinite series—or even using the old-fashioned geometric methods—would be impractical. In several works, Newton turned instead to finite-difference interpolation, beginning with tables of square, cube, and fourth roots and also applying his methods to trigonometric and logarithmic tables. Apparently unaware of both Harriot's unpublished and Briggs's published works on the topic, Newton started his work with interpolation from scratch and was able to generate a number of the most important formulas in the field. His interest in the subject began with a 1675 inquiry from accountant John Smith, who wanted to compose a table of square, cube, and fourth roots of the numbers from 1 to 10,000. Newton's reply advises Smith to compute every 100th entry directly,

---

[59] [I. Newton (Turnbull/Scott/Hall/Tilling) 1960–1977, vol. 2, 125, 143]. Earlier in Newton's career, around 1665, he had written a few pages on the construction of a table of sines, apparently following the method proposed six decades earlier by Viète [I. Newton (Whiteside) 1967–1981, vol. 1, 485–488].

[60] [I. Newton (Turnbull/Scott/Hall/Tilling) 1960–1977, vol. 2, 125–126]; see also [von Braunmühl 1900/1903, vol. 2, 65–66]. Of course, Newton did not the issue of approximation in the calculation of sines since you cannot come to the end of an infinite series. However, series make it easy to approximate the sine of any angle that might be needed; and in any case, regardless of what method one uses, as soon as a square root enters any direct calculation, the resulting sine values are necessarily approximate in any case.

then every tenth entry using constant third differences, and finally the remaining entries using constant second differences.[61]

By the time he wrote the *epistola posterior* in late 1676, Newton's thinking had progressed considerably. In particular, he had come to realize that if one has a sequence of values of the function to be tabulated for an arithmetically increasing sequence of arguments $a, a+1, a+2, \ldots$, then under the assumption that the nth order differences are constant, it is possible to find a polynomial of degree n (Newton called it a "parabola," regardless of the value of n) that approximates the curve throughout this range of arguments.[62] To evaluate a function value for some x, we have the "Gregory-Newton" formula

$$f(a+h) = f(a) + h\Delta(a) + \frac{h(h-1)}{2!}\Delta^2(a) + \frac{h(h-1)(h-2)}{3!}\Delta^3(a) + \cdots, \quad (3.18)$$

where Δ^i are the ith order forward differences of f.[63] Although this formula had already been discovered by several others (Harriot, Briggs, Gregory, Mercator, and Leibniz among them), Newton went considerably further in two later treatises, the unfinished *Regula differentiarum* (1676) and the *Methodus differentialis* (eventually published in 1711;[64] see figure 3.14). One of the main advances in this later work is the ***divided difference*** formulas, which allow the arguments of the known function values to be separated by different amounts. Clearly, this is more useful in scientific and other observational contexts than it is for building a trigonometric table.[65]

Now, although Newton was renowned for the pleasure he took in extensive computations, he did not compose trigonometric tables himself, either in the old geometric style or with series. We find the new methods of table construction, complete with calculations, laid out in a passage by Abraham

---

[61] The letter to Smith is available in [I. Newton (Whiteside) 1967–1981, vol. 4, 14–21]. Newton's interest in interpolation also stemmed from John Wallis's study of the area of a quadrant of a circle; this led to Newton's discovery of the binomial theorem. See for instance [Whiteside 1961c, 236–246] and [Goldstine 1977, 78–80].

[62] Newton's clearest expression of this topic in the *epistola posterior* actually occurs in a section that he eventually excluded from the letter; it is available in [I. Newton (Whiteside) 1967–1981, vol. 4, 22–35].

[63] That is, $\Delta(a) = f(a+1) - f(a)$, $\Delta^2(a) = \Delta(a+1) - \Delta(a)$, $\Delta^3(a) = \Delta^2(a+1) - \Delta^2(a)$, and so on.

[64] [I. Newton 1711, 93–101].

[65] On Newton's work with finite difference interpolation, see [Fraser 1918] (which includes a facsimile and translation of the *Methodus differentialis,* editions of the letter to Smith and a lemma in Newton's *Principia Mathematica* that refers to finite difference interpolation), [Fraser 1919] (which contains a translation of part of the *epistola posterior*), [Fraser 1920], [Fraser 1927] (which announces the discovery of, and provides an edition and translation of, the *Regula differentiarum*), [Whiteside 1961c, 232–251], [I. Newton (Whiteside) 1964–1967, vol. 2, 165–173] (edition and translation of the *Methodus differentialis*), [I. Newton (Whiteside) 1967–1981, vol. 4, 36–41] (edition and translation of the *Regula differentiarum*), [Goldstine 1977, 68–84], and [Roy 2011, 158–175].

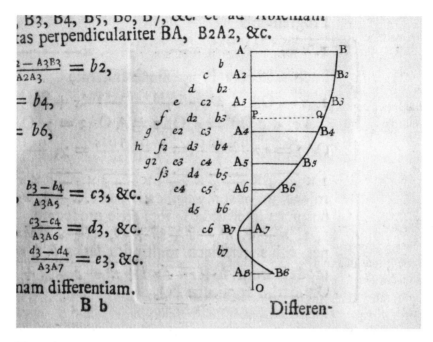

Figure 3.14
nth order differences from Newton's 1711 *Methodus differentialis*.

Sharp (1651–1742). The one-time assistant to John Flamsteed at the Royal Observatory in Greenwich included this work in a collection of mathematical tables compiled by Henry Sherwin in 1705.[66] This transformative text, so different in character from the dozens on tables that had gone before, contains only two geometrical diagrams. The first simply gives the definitions of the six standard trigonometric functions and the versed sine, and the second is used to prove an identity near the end of the work. On the second page, where Sharp begins the computations, we find Newton's series for all six functions given up to powers between 15 and 19. Calculating up to the fifth power term of the sine series, Sharp gets $\sin 5' = 0.001454440530542\mathit{02321343}$ (where the italicized digits are all in error). Perhaps he should have included a few more

---

[66] On Abraham Sharp's life and work see [Cudworth 1889] and [Connor 1942]. His only mathematical work was the 1717 *Geometry Improved*, which works with polyhedra and contains extensive computations similar in precision to his trigonometry. The tables, which were reissued a number of times throughout the eighteenth century, are in [Sherwin 1705]; the passage containing Sharp's description of table computation is on pp. 44–52. See [Cudworth 1889, 171–175] for an account of the book's compilation and [von Braunmühl 1900/1903, vol. 2, 85] for some of the mathematics.

terms. The other two sines he computes at the outset ($\sin 29°55'$ and $\sin 60°5'$) are both accurate to 21 places.

Sharp shows some practical awareness of issues of convergence. He instructs the reader to use the sine series to compute sines of arcs less than $30°$ and to use the cosine series for sines of arcs greater than $60°$; in these cases, the series converge quickly. For arcs between $30°$ and $60°$, instead of series, Sharp uses the identities

$$\sin(30° + x) = \sin(90° - x) - \sin(30° - x) \tag{3.19}$$

and

$$\sin(60° - x) = \sin(60° + x) - \sin x. \tag{3.20}$$

These identities allow Sharp to use the series only for arguments that lead to quick convergence. Nevertheless, in an act of computational bravado he uses a series to calculate $\sin 44°37'$, "which being so very near $45°0'$ must necessarily be as troublesome and laborious as any that need be proposed."[67] Working up to the 19th power he arrives at the value 0.7023601432666095521385, correct to all but the last two places. As for the crucial entry $\sin 1'$ that had been sought by so many of his predecessors, Sharp says:

> By a continued bisection [of an arc whose sine has been found previously] the sine of an arc a little less than $0°1'$ may be found, and from that by proportion the sine of $0°1'$.[68] But the sine of $0°1'$ may be obtained from the length of its arc by the series in the other method with incomparably less labour and greater accuracy.[69]

Indeed, in nine short pages Sharp's passage wipes away the geometric approach that had been standard since Claudius Ptolemy, replacing it with the power and simplicity of the new analysis.

Sharp's obsession with exceedingly precise calculations was met with the sort of puzzled admiration one reserves for those who complete important but impossibly tedious tasks. And indeed, the mathematical accomplishment that Sharp is best known for today is his calculation of π to 72 places[70]— important for his trigonometric calculations because the series assume that the argument is in what we call radian measure today. Thus, to use a series one first needs to convert the argument from degrees to radians by multiply-

---

[67] [Sherwin 1705, 48 (incorrectly numbered 41)].

[68] Here Sharp is referring to the often-used fact that the ratio of two sine values is approximately equal to the ratio of the two given arcs (provided that they are small). This fact had been in common use to find the sine of the smallest arc in a table.

[69] [Sherwin 1705, 52].

[70] On Sharp's calculation of π see [Cudworth 1889, 169–170] and [Beckmann 1971, 144].

ing by $\pi/180$. Sharp's efforts nearly doubled the known number of decimal places of π. He was not the first to convert to series; that honor belongs to Isaac Newton, who in 1665 had computed π to 16 places using his arc sine formula and the relation $\frac{\pi}{6} = \arcsin\frac{1}{2}$. Sharp instead used Gregory's arc tangent series (3.11) in 1699 with $x = \sqrt{\frac{1}{3}}$, giving

$$\frac{\pi}{6} = \frac{1}{\sqrt{3}}\left(1 - \frac{1}{3\cdot3} + \frac{1}{3^2\cdot5} - \frac{1}{3^3\cdot7} + \cdots\right). \tag{3.21}$$

Sharp's record calculation was not to stand for long. In 1706, deep within the pages of William Jones's *Synopsis Palmariorum Matheseos*, we find the following:

> There are various other ways of finding the Lengths, or Areas of particular Curve Lines, or Planes, which very much facilitate the Practice; as for instance, in the Circle, the Diameter is to the Circumference as 1 to
>
> $$\overline{\frac{16}{5} - \frac{4}{239}} - \frac{1}{3}\overline{\frac{16}{5^3} - \frac{4}{239^3}} + \frac{1}{5}\overline{\frac{16}{5^5} - \frac{4}{239^5}}, \& c.$$
>
> $$= 3.14159, \& c. = \pi.^{71} \tag{3.22}$$

> This Series (among others for the same purpose and drawn from the same Principle) I receiv'd from the Excellent Analyst, and my much Esteem'd Friend Mr. John Machin; and by means thereof, Van Ceulen's Number . . . may be Examin'd with all desirable Ease and Dispatch.[72]

This passage, notable as the first appearance of the symbol π, refers to the result of a 100-digit calculation given 20 pages earlier, praised by Jones:

> As Computed by the Accurate and Ready Pen of the Truly Ingenious Mr. John Machin: Purely as an Instance of the Vast advantage Arithmetical Calculations receive from the Modern Analysis, in a Subject that has been of so Engaging a Nature, as to have employ'd the Minds of the most Eminent Mathematicians, in all Ages, to the Consideration of it. For as the exact Proportion between the Diameter and the Circumference can never be express'd in Numbers; so the improvements of those Enquirers the more plainly appear'd, by how

---

[71] The overbars are Jones's method for enclosing expressions within parentheses.
[72] [Jones 1706, 263].

much the more Easie and Ready, they render'd the Way to find a Proportion to the nearest possible. But the Method of Series (as Improv'd by Mr. Newton, and Mr. Halley) performs this with great Facility, when compared with the Intricate and Prolix Ways of Archimedes, Vieta, van Ceulen, Metius, Snellius, Lansbergius, &c., Tho' some of them were said to have (in this Case) set Bounds to Human Improvements, and to have left nothing for Posterity to boast of; But we see no reason why the indefatigable Labor of our Ancestors should restrain us to those Limits, which by means of the Modern Geometry, are made so easie to Surpass.[73]

One can hardly imagine a more evocative statement of the awed reaction of the mathematical community to the powerful new methods that were opening up to them.

Machin, eventually to become professor of astronomy at Gresham College, never published a derivation of his series, which led to some confusion about his methods.[74] It was finally reconstructed in 1758 as an appendix to an unrelated work by Francis Maseres (1731–1824). The method relies on the tangent sum and difference formula[75]

$$\tan(\alpha \pm \beta) = \frac{\tan \alpha \pm \tan \beta}{1 \mp \tan \alpha \tan \beta}. \tag{3.23}$$

The idea is to use the arc tangent series (3.10) with two different arguments. Machin chooses the first to be $x = AB = \frac{1}{5}$, which gives arc $\theta = \widehat{AE} = \arctan \frac{1}{5} \approx 11.3099°$ in figure 3.15.[76] By coincidence 4θ is very slightly larger than $45°$, which corresponds to an arc length of $\widehat{AK} = \frac{\pi}{4}$. Let $\widehat{AF} = 2\theta$ and $\widehat{AG} = 4\theta$; then from the tangent double-angle formula[77] $AC = \tan 2\theta = \frac{5}{12}$ and $AD = \tan 4\theta = \frac{120}{119}$, very slightly larger than $AL = 1$. Draw KO tangent to the circle from K to MD (magnified in figure 3.16); then using the tangent difference formula,

$$KO = \tan\left(4\theta - \frac{\pi}{4}\right) = \frac{\tan 4\theta - 1}{1 + \tan 4\theta} = \frac{1}{239}. \tag{3.24}$$

---

[73] [Jones 1706, 243–244].

[74] On Machin and his series see [Tweddle 1991] and [Beckmann 1971, 144–145]. Maseres's reconstruction of Machin's argument is in [Tweddle 1991, 13–14].

[75] The tangent sum formula appeared in the *Acta Eruditorum* in 1706 in a paper by Jakob Hermann; see [Tweddle 1991, 8].

[76] It is a sign of the continued importance of geometry at the time that the diagram was drawn at all; from a modern point of view it is unnecessary.

[77] Let $\alpha = \beta$ in (3.23).

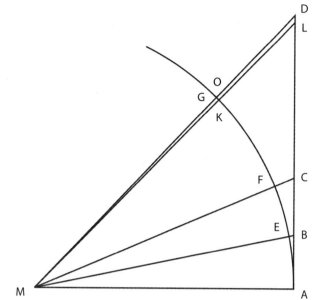

Figure 3.15
Machin's
computation of π.

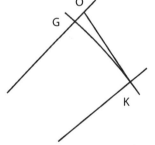

Figure 3.16
Magnification of part of the previous figure.

Therefore

$$\text{arc tan}\frac{1}{239} = 4\theta - \frac{\pi}{4},$$

and so

$$\frac{\pi}{4} = 4\,\text{arc tan}\frac{1}{5} - \text{arc tan}\frac{1}{239}. \qquad (3.25)$$

leading to (3.22). This result makes for effective π calculations for two reasons: using the arc tangent series with $x = \frac{1}{5}$ is computationally easy in base 10, and applying $x = \frac{1}{239}$ produces a quickly converging series.

Successors to Machin broke his records quickly due more to industry than to creativity. Thomas Fantet de Lagny calculated 120 places in 1719 using Sharp's method, and an unknown author in Philadelphia calculated 154 places in 1721 with a less clever application of the arc tangent series.[78] The record held until 1844, and arc tangent series methods remained dominant in the digit chase even up to the use of mechanical computers in the early twenty-first century.

The new approaches to computing tables (and related quantities like π) with series reflects significant changes in how trigonometric functions were conceived. Firstly, the distinction between methods based entirely on finite geometric constructions and those requiring some form of approximation had vanished. The uses of truncated infinite series and of interpolation passed without further justification as arguments in the style of the new calculus became more familiar and trusted. Secondly, series and interpolation methods were applied equally to a variety of types of function, of which trigonometric was only one subcategory. Although (as we shall see) there was some resistance to the notion of trigonometric functions as analytic rather than geometric entities, within the realm of table computation they were no longer a special category with their own methods.

Of course, once one has a polynomial approximation to a curve over a given interval, such as the ones that were being generated for interpolation, one may use it to find the definite integral. In the *Methodus differentialis* Newton takes advantage of this. He uses a cubic polynomial to approximate the area under a curve, arriving at Simpson's Three-Eighths Rule:

$$\int_a^{a+3h} f(x)\ dx \approx \frac{3h}{8}(f(a)+3f(a+h)+3f(a+2h)+f(a+3h)). \qquad (3.26)$$

Eventually Roger Cotes would extend Newton's idea further, integrating numerically with approximating polynomials of degree up to 11; today they are known as the Newton-Cotes formulas.[79]

Newton was not the first to think in this way; already in 1639 Cavalieri had approximated the definite integral of a given function using a quadratic polynomial. Gregory also had done so in his 1668 *Exercitationes geometricae* and had used the results to tabulate

$$\int_0^x \tan\theta\ d\theta = \ln\sec x. \qquad (3.27)$$

---

[78] See [Wardhaugh 2015] and [Wardhaugh 2016].

[79] This appears as the short treatise *De methodo differentiali Newtoniana*, in [Cotes 1722, 23–33].

Incidentally, in the same book Gregory had been the first to demonstrate that

$$\int_0^x \sec\theta \; d\theta = \ln\tan\left(\frac{x}{2}+\frac{\pi}{4}\right), \qquad (3.28)$$

the integral that had been so crucial to the construction of the Mercator map projection. His solution involved six geometric transformations and was so complicated that Edmund Halley, in a 1695 article describing a simpler solution, described Gregory's solution as "not without a long train of consequences and complication of proportions, whereby the evidence of the demonstration is in a great measure lost, and the reader wearied before he attained it." Nevertheless, given its nautical importance, a solution to this integral was still a major accomplishment.[80]

Geometric Derivatives and Integrals of Trigonometric Functions

The conquest of the integral of the secant, involving some of the biggest names in seventeenth-century English mathematics (including John Collins, Barrow, Newton, and Wallis), was a spectacular accomplishment. We have seen that derivatives and integrals of other trigonometric functions had been determined before Newton and Leibniz. However, here the word "function" must be treated with extra caution. To modern ears the word conjures the image of an independent variable, a rule of calculation, and a dependent variable. While one might make a case that such concepts made sense in the late seventeenth century for polynomials or even exponential and logarithmic functions, they fit less well for trigonometric quantities. Magnitudes such as arcs, angles, and lengths in a circle would relate to each other according to the structures of the geometric diagrams in which they resided, but there was usually no clear distinction between what we would call input and output variables. The derivations we have seen thus far, then, are *geometric* derivatives and integrals rather than modern *analytic* processes.

The history of mathematics frequently deals with subtle shades of meaning like this. For instance, there is a sense in which the statement $\frac{d}{dx}\sin x = \cos x$ may be said to have been known already in medieval India: Āryabhaṭa's difference method of computing sines reveals an intuitive

---

[80] [Halley 1695, 203]. Gregory's solution is in [Gregory 1668a, 14–24]; see [Baron 1969, 235–238] for a modern account of Gregory's argument. On Halley's solution and Cotes's improvement of it, see [Gowing 1995].

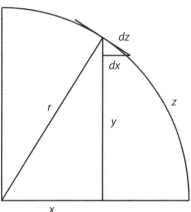

Figure 3.17
Leibniz's derivation of the derivative of the
arc cosine.

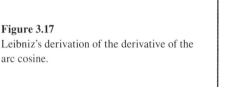

awareness of what we call the first and second derivatives of the sine,[81] and the Keralite mathematicians of the fourteenth and fifteenth centuries even generated Taylor series for the sine and cosine using a method that we might say uses infinitesimal quantities.[82] However, the term *calculus* usually refers to a collection of algorithms that evaluates derivatives or integrals in a uniform way that applies to a group of functions rather than to specific instances of them. Medieval Indian methods anticipate some of the *results* of calculus, but that does not imply that they are a realization of the calculus itself.[83] The same may be said of some of the seventeenth-century precursors of calculus, especially the quadratures and methods of finding tangent lines. We must conclude, then, that a calculus of the trigonometric functions did not exist prior to Newton and Leibniz, even though a number of derivatives and integrals were known. Depending on one's interpretation of the texts, one might even say that there was no calculus of the trigonometric functions *after* Newton and Leibniz. One might test this claim by considering to what extent the works of these two men can be called a "calculus." We turn to Leibniz now.

Recall that Leibniz wrote of his inspiration for the differential calculus, and the differential triangle in particular, in the work of Blaise Pascal. The differential triangle was hardly unique to Newton and Leibniz; we find it also in the writings of Isaac Barrow. It has even been suggested that Barrow may have been the key influence on Leibniz.[84] In any case, once the idea of the differential triangle had arisen, Leibniz found it straightforward to differentiate trigonometric functions. For instance, in a letter written in June 1676

---

[81] [Van Brummelen 2009, 99–102].
[82] [Van Brummelen 2009, 113–121].
[83] Arguments like this have made, for instance, in [Plofker 2001], especially pp. 293–294.
[84] [Leibniz (Child) 1920, 13ff].

while dealing with another mathematical function, Leibniz finds himself in need of an expression for the derivative of the arc cosine. In figure 3.17, where x is the Cosine of arc z (in a circle of radius r), he increases x by dx and decreases z by dz. Then, by similar triangles,

$$\frac{dz}{dx} = \frac{r}{\sqrt{r^2 - x^2}}, \text{ so } dz = \frac{dx \cdot r}{\sqrt{r^2 - x^2}}, \tag{3.29}$$

which gives us the derivative of the arc Cosine almost immediately.[85] Since derivations like this were so simple, Leibniz simply derived the relation he needed when it arose. Does this qualify as a calculus? It generates derivatives fairly easily, although it requires a small new geometric argument for each new situation. The question is left to the reader to decide.

Trigonometric functions also arise as solutions to differential equations, which in this geometric context are not much different from integrals. In 1693 Leibniz, dealing with essentially the same figure 3.17, derives

$$\frac{dz}{dy} = \frac{r}{\sqrt{r^2 - y^2}}$$

in a manner similar to the above. Cross-multiplying and eliminating the square root, he has

$$r^2 dz^2 = r^2 dy^2 + y^2 dz^2.$$

Assuming that the arc z is increasing at a constant rate (so that $d(dz) = 0$), Leibniz takes the differential of both sides, arriving at

$$0 = r^2 dy\, d^2 y + y\, dy\, dz^2.$$

He works out the coefficients of the power series and finds

$$y = z - \frac{z^3}{3!r^2} + \frac{z^5}{5!r^4} - \frac{z^7}{7!r^6} + \cdots, \tag{3.30}$$

correctly identifying the series for the sine function. Again, the series provided Leibniz the access required for analysis of the trigonometric functions.[86]

A Transition to Analytical Conceptions

One of the first steps that transitioned trigonometric functions away from geometry was the recognition of their periodicity. If the sine is considered strictly as a geometric quantity (the length of a line segment), there is no meaning to or

---

[85] [Leibniz (Child) 1920, 116–117]. The modern derivative of the arc cosine has a negative sign in the numerator. In this geometric context, one notes simply that dz is in a downward direction.
[86] [Leibniz 1693, 178–180]; see also [Katz 1987, 313].

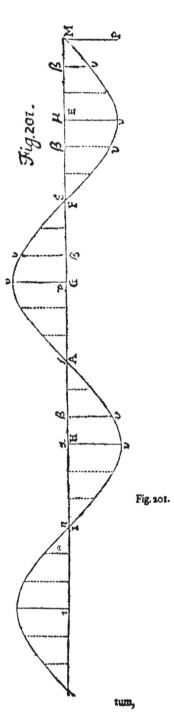

Figure 3.18
Two periods of the sine curve from Wallis's
Mechanica.

necessity for arguments greater than 360°. But if the sine is to be used to model periodic phenomena, then one must permit its arguments to be as large as one needs. A sine curve with two complete periods may be found as early as John Wallis's *Mechanica* (figure 3.18).[87] However, this seems to be an isolated appearance; Wallis's successors did not pursue this idea for several decades.

We find another reference to periodicity, this time with respect to the tangent function, in Thomas Fantet de Lagny's work on formulas for $\tan n\theta$ and $\sec n\theta$ in terms of $\tan\theta$ and $\sec\theta$. De Lagny considers explicitly what happens to the formulas when $n\theta > 90°$. He correctly concludes that when the angle is in the first and third quadrant then the tangent is positive, and if it is in the second and fourth quadrant then the tangent is negative. Above these four quadrants, the sequence of positive and negative continues in the same way ad infinitum.[88]

Awareness of the periodicity of the trigonometric functions seems gradually to have become more explicit in the early decades of the eighteenth century. For instance, we find a clear image of Cartesian graphs of the secant and tangent functions, both drawn with multiple periods (figure 3.19), in a text by English scientist and mathematician Roger Cotes (1682–1716). In his short life Cotes's impact was significant. He is perhaps best known for working with Newton to produce a polished second edition of the *Principia Mathematica*; Newton was so impressed with Cotes that he once remarked "had Cotes lived we might have known something." Cotes's own work was limited in magnitude, although not in depth. He published only one paper during his life and is known today mostly through a collection of his works, *Harmonia mensuram*, gathered after his death.[89] More than half of this book is devoted to a systematic approach to the integrals of no less than 94 types of logarithmic and trigonometric functions. Not long before this, Cotes integrates the tangent and secant functions. In his diagrams one sees coordinate graphs for both functions for the first time, drawn for multiple periods of the argument and clearly intended to be reproduced indefinitely to the left and right.[90]

The *Harmonia mensuram* contains several other interesting passages. It opens with Cotes's only published essay, the "Logometria," which opens by "measuring" (i.e., finding the logarithm of) a given ratio. The measure of

---

[87] [Wallis 1670].

[88] [von Braunmühl 1900/1903, vol. 2, 71–72].

[89] [Cotes 1722]. For a summary of this work see [Anonymous 1722–23]. The first of the four essays ("Logometria") is translated in [Gowing 1983], the standard biography of Cotes. See also [Gowing 1992] on Cotes's work on the Archimedean spiral.

[90] [Cotes 1722, 78–80]. After this passage Cotes notes the importance of $\int \sec x\, dx$ for navigation with the Mercator chart and presents a solution to the Mercator problem. See also [Cotes 1722, *Opera miscellanea*, 110–111] for another appearance of these graphs.

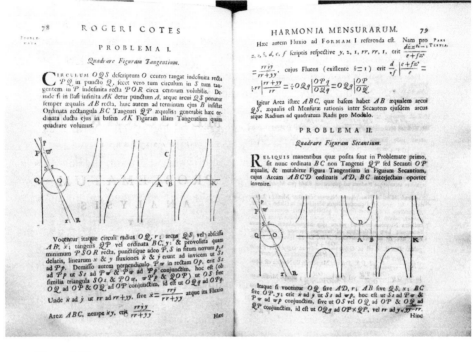

Figure 3.19
Graphs of the tangent and secant functions in Cotes's 1722 *Harmonia mensuram.*

$(1 + x) : 1$ for small values of x might be any number, if (in modern terms) one does not specify a base for the logarithm. Cotes sets the measure of this ratio equal to x itself, which leads him to a base of e—what we call today the natural logarithm. He goes on to determine a scaling factor between it and the more common base 10 logarithms. Next he calculates e using continued fractions, arriving at 2.718281828459. . . .[91] At the end of the book, Cotes's cousin Robert Smith (inspired, he says, by a now-lost paper of Cotes) considers analogously what the natural measure should be for an angle. He concludes that it should be the radius of the base circle, in other words, 57.2957795130°. Cotes and Smith thus quantified radian measure for angles for the first time (without using the word), parallel to Cotes's determination of the natural base of a logarithm.[92] In a sense, much of the calculus of trigonometric functions

---

[91] [Cotes 1722, 4–10]; see also [Gowing 1983, 23–25]. Here Cotes was pursuing an idea originally discussed by Edmund Halley.

[92] [Cotes 1722, 44–45; *Opera miscellania*, 94–96]; see also [Jones 1953, 423–424] and [Gowing 1983, 38–39]. The word "radian" for this measure took another century and a half to appear.

earlier in this chapter had already effectively employed radians by working with the lengths of arcs of circles; Cotes's advance here is to justify the use of radians to measure *angles*.

Finally, in the essay "Aestimatio errorum in mixta mathesi," Cotes considers the question of estimating errors in calculated values of sides and angles of triangles when the given quantities are known only to some level of precision. He begins with three lemmas that state the derivatives of the three standard trigonometric functions: $\frac{d}{dx}\sin x = \cos x$, $\frac{d}{dx}\tan x = \sec^2 x$, and $\frac{d}{dx}\sec x = \sec x \, \tan x$, assembling in one place for the first time a foundation for the systematic differentiation of trigonometric functions. The demonstrations are geometric and are expressed as ratios of infinitesimals; for example, the derivative of the sine is expressed in words corresponding to the equation

$$\frac{dx}{d(\mathrm{Sin}\,x)} = \frac{R}{\mathrm{Cos}\,x}. \tag{3.31}$$

The remaining 28 theorems build on these lemmas, proving various results concerning variations of sides and angles of plane and spherical triangles, to be used by astronomers for error analysis. We provide the first example here.

Text 3.4
Cotes, Estimating Errors in Triangles
(from *Aestimatio errorum in mixta mathesi*)

Theorem 1. In a plane triangle, with any one angle and one side adjacent to it fixed, the ratio of the small[93] variation of the other adjacent side to the small variation of the opposing side is as the ratio of the radius to the complement of the sine of the angle opposite the fixed side.

In triangle *ABC*, take any angle *B* and one adjacent side *AB* to be fixed; change to triangle *ABD*. In *AD* take *AE* equal to *AC*; join *CE*. I say that the ratio of the variation *CD* of the other adjacent side is to the variation *DE* of the opposing side *AC* is as the ratio of the radius to the complement of the sine of angle *C*, opposite the fixed side *AB*, when the variations are small. For in this case we have right angles *CED* and *ACE*. Therefore, the ratio *CD* to *DE* is as the radius to the sine of angle *DCE*, whose complement is angle *C*. Q.E.D.[94]

The complicated story of its origin, involving Thomas Muir and James Thomson in the late 1860s and early 1870s, is told in [Cooper 1992] and in chapter 5 of this book.

[93] The Latin word is "minima."

[94] [Cotes 1722, *Opera Miscellanea*, 4–5].

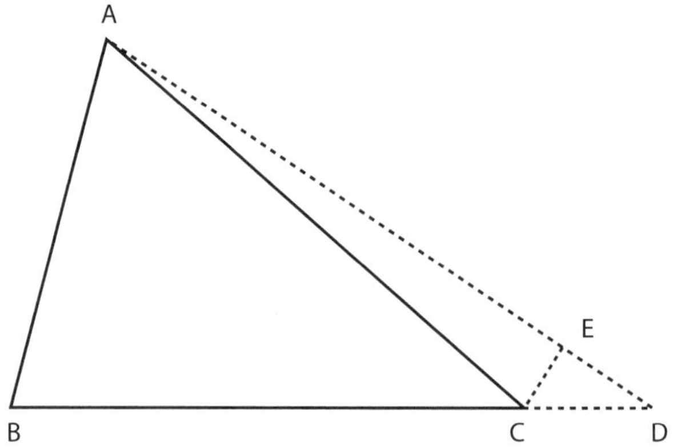

Figure 3.20
Cotes's triangle
error analysis.

Explanation: In figure 3.20, side *BC* in triangle *ABC* is extended by an infinitesimal length *CD*, forming triangle *ABD*. The goal is to determine the change in *AC* in terms of *CD*. Since the changes are infinitesimal, the angles at the base of isosceles triangle *ACE* are both right. Therefore

$$\frac{d(BC)}{d(AC)} = \frac{CD}{DE} = \frac{R}{\operatorname{Sin} \angle DCE} = \frac{R}{\operatorname{Cos} C}$$

since $\angle C = 90° - \angle DCE$.

Cotes's work was greatly expanded and given explicit applications to spherical astronomy by Nicolas-Louis de La Caille (1713–1762) in 1741, and the expanded theory would be adopted by others, including Euler.[95]

Cotes, working so closely with the calculus of logarithmic and trigonometric functions, inevitably noticed the symmetries between the two groups that had been hinted at several times previously. He was able to make the relationship explicit in his paper "Logometria," the only work he published during his life.[96] The essay consists of a number of quadratures, areas under parts of conic sections and various other curves, surface areas of these curves revolved around axes, and related problems in physics. In his massive table of integrals he finds such a striking pattern between the integrals of trigonometric and of logarithmic functions that his book *Harmonium mensuram* is named after it: "That harmony of measures, which is so strong that I propose a single notation serve to designate measures, whether of ratios or of angles."[97]

---

[95] [Delambre 1827, 457].
[96] [Cotes 1714] and [Cotes 1722, 1–41].
[97] [Cotes 1722, 44]; translation in [Gowing 1983, 35].

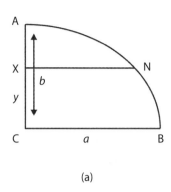
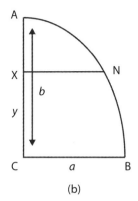

(a)   (b)

Figure 3.21
Cotes on the surface area of an ellipsoid.

Cotes's combined notation involves the specification of a certain constant as the square root of either a positive or a negative quantity. Whether or not that quantity is positive implies whether the meaning of another of Cotes's symbols is trigonometric or logarithmic. This allows Cotes to write the formula for a trigonometric integral and its logarithmic counterpart simultaneously in a single expression. Cotes's harmony turns out to be equivalent to the equation

$$i\phi = \ln(\cos\phi + i\sin\phi). \tag{3.32}$$

This relation is stated explicitly only once in the "Logometria," arising in the context of the determination of the surface area of part of an ellipsoid of revolution. In figure 3.21(a), let $BC=a$ be the semimajor axis of the ellipse, and let $AC=b$ be the semiminor axis as well as the axis of rotation. Let $XC=y$ be any distance along the semiminor axis, and let XN be parallel to CB. Arc BN is rotated around the axis; the surface area is required. Cotes applies one of his quadrature formulas and arrives at

$$\pi a\left\{y\sqrt{1+\frac{a^2-b^2}{b^4}}+\frac{b^2}{\sqrt{a^2-b^2}}\ln\left(y\sqrt{\frac{a^2-b^2}{b^4}}+\sqrt{1+y^2\frac{a^2-b^2}{b^4}}\right)\right\}. \tag{3.33}$$

However, if one considers $b>a$ as in Figure 3.21(b), then the integral changes and the surface area becomes

$$\pi a\left\{y\sqrt{1-\frac{b^2-a^2}{b^4}y^2}+\frac{b^2}{\sqrt{b^2-a^2}}\phi\right\}, \text{ where } \sin\phi = y\sqrt{\frac{b^2-a^2}{b^4}}. \tag{3.34}$$

If we combine these two results and allow $i = \sqrt{-1}$ to be a constant, we eventually get (3.32). (The integrals involved are related to

$$\int \frac{1}{\sqrt{x^2 - a^2}}\, dx = \ln\left| x + \sqrt{x^2 - a^2} \right|$$

and

$$\int \frac{1}{\sqrt{a^2 - x^2}}\, dx = \arcsin\frac{x}{a},$$

distinguished in elementary calculus by whether $x > a$ or $x < a$.) Cotes's algebraic combination, much more than a unification of trigonometry and logarithms, was a strong confirmation of the validity of the use of imaginary quantities and an early step to the legitimation of what was to become complex analysis.[98]

Meanwhile, Abraham de Moivre (1667–1754) was arriving at similar conclusions but from a different direction. A French Protestant, de Moivre may have been threatened with imprisonment for his beliefs due to the 1685 revocation of the Edict of Nantes. The loss of this protective measure caused several hundred thousand Protestants to abandon France for other countries; de Moivre ended up in England. He quickly became connected with the greatest English scientists and mathematicians, among them Halley and Newton. His focus was probability theory, and he is most known for *The Doctrine of Chances*. But he also played a role in the development of trigonometry, logarithms, and algebra with complex numbers, some of which was motivated by his interest in certain problems in probability.[99]

De Moivre's interest in these latter topics seems to have begun as early as 1698, when he derived Newton's formula (3.9) for $\sin n\theta$ in terms of $\sin \theta$,

$$\sin n\theta = n\sin\theta - \frac{n(n^2 - 1^2)}{3!}\sin^3\theta + \frac{n(n^2 - 1)(n^2 - 3^2)}{5!}\sin^5\theta - \cdots.$$

In a paper in the *Philosophical Transactions*,[100] he pursues equations like this one but expressed as polynomials, in this case setting $y = \sin\theta$:

$$ny - \frac{n(n^2 - 1)}{3!}y^3 + \frac{n(n^2 - 1)(n^2 - 9)}{5!}y^5$$
$$- \frac{n(n^2 - 1)(n^2 - 9)(n^2 - 25)}{7!}y^7 - \cdots = a. \tag{3.35}$$

[98] [Gowing 1983, 35–38]. The explicit statement is in [Cotes 1722, 28] and contains a minor error, noted in [Schneider 1968, 236]. The argument, converted to modern notation as found here, is in [von Braunmühl 1904, 361].

[99] On de Moivre's life and work see especially [Walker 1934], [Schneider 1968], [Bellhouse/Genest 2007] (which reproduces a biography of de Moivre written by Matthew Maty the year after de Moivre's death), and [Bellhouse 2011].

[100] [de Moivre 1707].

When n is odd, the series terminates. In this case de Moivre identifies four related solutions, one of which is

$$y = \frac{1}{2}\sqrt[n]{a + \sqrt{a^2 - 1}} + \frac{1}{2}\sqrt[n]{a - \sqrt{a^2 - 1}}. \tag{3.36}$$

He gives an example for a value of $a < 1$, namely $5y - 20y^3 + 16y^5 = \frac{61}{64}$, which produces a solution containing imaginary numbers. This example may be resolved algebraically to a real-valued solution, but de Moivre concludes the essay by giving an alternate solution of the same equation that refers back to its trigonometric origin. We set $\sin n\theta = \frac{61}{64} = a$; then, according to (3.9), the solution is $y = \sin \theta$. Substituting into (3.36), we get

$$\sin\theta = \frac{1}{2}\sqrt[n]{\sin n\theta + i\cos n\theta} + \frac{1}{2}\sqrt[n]{\sin n\theta - i\cos n\theta} \tag{3.37}$$

for odd n.

De Moivre pursues these ideas further in several papers, explaining how he came to consider these questions and carrying his thoughts further. In *De sectione anguli*,[101] he does not explicitly state the formula that bears his name ((3.39) below), but it lurks just below the surface. His goal is to determine a multiple-angle formula for the versed sine. Let $x = $ vers θ and $t = $ vers $n\theta$. De Moivre generates the pair of equations

$$1 - 2z^n + z^{2n} = -2z^n t \text{ and } 1 - 2z + z^2 = -2zx. \tag{3.38}$$

The goal is to eliminate the z variable to obtain the desired multiple-angle formula for the versed sine. The solution to the former turns out to be $z = \sqrt[n]{\cos n\theta \pm i \sin n\theta}$ while the solution to the latter is $z = \cos \theta \pm i \sin \theta$. If we combine the two, we get de Moivre's formula

$$(\cos \theta + i \sin \theta)^n = \cos n\theta + i \sin n\theta, \tag{3.39}$$

again limited to odd values of n. Finally, in another paper entitled *Miscellanea analytica*,[102] de Moivre calls once more upon an equivalent of (3.39), this time within a factoring problem that led to the solution of a particular indefinite integral.[103]

Although Cotes, de Moivre and others were glimpsing the new world of analytic trigonometry, as yet there was no pressing need to leave the old world

---

[101] [de Moivre 1722].

[102] [de Moivre 1730].

[103] English translations of extracts from these papers may be found in [Smith 1929, 440–450]. See also [von Braunmühl 1901a], [Schneider 1968], [Glushkov 1982], and [Bellhouse 2011, 65, 141–143].

behind. Derivatives and integrals of sines, tangents, and what we call their inverse functions were simple enough to find with geometric arguments. This left trigonometric quantities initially outside of the *calculus* of calculus: that is, they were not part of the standard collection of functions for which a standard set of techniques was developed. Cotes's tables of integrals would have been a first step, but his early demise may have slowed progress. Certainly, momentum was absent: almost none of the new calculus texts from l'Hôpital's 1696 *Analyse des Infiniment Petits*[104] onward until Thomas Simpson's 1737 *New Treatise of Fluxions*[105] went beyond Newton. Even Simpson simply repeated the method of the differential triangle applied to a circle that had gone back at least to Leibniz (see figure 3.17). This is in stark contrast to how other functions, particularly exponential and logarithmic, were incorporated quickly into the new analysis. Why was trigonometry the last holdout? Two related answers have been suggested: the sine and other trigonometric quantities were still linked with geometric rather than arithmetic conceptions, and no compelling applications had provided big enough temptations to trigger a crossing of the geometry-analysis divide.[106] A radical transformation would be needed to move forward.

Simpson's trigonometry textbook, *Trigonometry, Plane and Spherical* (1748),[107] hadn't made it across the chasm either. The book opens with the standard geometric account of how to compute tables of sines, tangents, and secants and goes on to treat trigonometry using a Euclidean approach. Most other early eighteenth-century textbooks containing trigonometry were similar. Some presented more elegant constructions and perhaps proved a new theorem or two, but the genre as a whole remained impervious to real change. There were exceptions. John Keill's popular *Elements of Plain and Spherical Trigonometry* was on the whole geometrically conservative—no surprise, since he had been a vocal defender of geometric purity in his edition of Euclid's *Elements*.[108] But after a traditional geometric treatment of the computation of sines, he included a section on the use of series containing some new series he had constructed himself to improve computations.[109]

[104] [l'Hôpital 1696].

[105] [Simpson 1737], a popular textbook that would continue to be published and revised for nearly a century.

[106] [Katz 1987, 314–317].

[107] [Simpson 1748].

[108] [Keill 1726]. See [Malet 2006], [Goldstein 2000], and [Cajori 1928] (among others) for discussions of controversies on the issue of arithmetization in English editions of the *Elements* from the sixteenth to the eighteenth centuries.

[109] [Keill 1726, 15–18].

A more substantial break with the past is found in *Analyis speciosa trig-onometriae* (1720) by Jakob Kresa (1648–1715), a Jesuit and professor at the University of Olomouc, Charles University in Prague, and the Colegio Impe-rial de Madrid.[110] Kresa's novel approach is clear from the very outset of his chapter on plane trigonometry.

Text 3.5
Jakob Kresa, Relations Between the Sine and the Other Trigonometric Quantities
(from *Analysis speciosa trigonometriae*)

Proposition 1. Suppose that the radius of the circle $= r$, and the sine of what-ever arc $= x$. To determine the remaining trigonometric lines, in terms of the same letters r & x.

The sine of arc *FD* is $FE = x$. Therefore *AE*, the cosine[111] of the same arc, $= \sqrt{rr - xx}$. 47.P.I.

And with the proportion $AE .. EF :: AD .. DG$,

this is by analysis $\sqrt{rr - xx} .. x :: r .. \dfrac{rx}{\sqrt{rr - xx}}$.

So, the tangent of arc $FD = \dfrac{rx}{\sqrt{rr - xx}}$.[112]

Figure 3.22
Kresa's definition of the six trigonometric quantities.

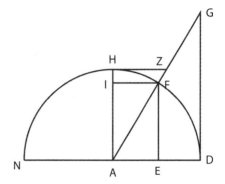

Explanation: Figure 3.22 is the standard diagram defining the six standard trigonometric quantities. Let $\theta = \widehat{FD}$. Kresa's first proposition determines values for the other five in terms of $x = EF = \operatorname{Sin} \theta$. Here $\operatorname{Cos} \theta$ is $\sqrt{r^2 - x^2}$ (where r is the radius of the base circle) according to the Pythagorean Theorem

---

[110] [Kresa 1720]. On Kresa see [Komenský/Marci 2008], [Komenský/Marci 2009], [Vetter/Archibald 1923, 52], and [von Braunmühl 1900/1903, vol. 2, 94–95].
[111] "Sinus secundus."
[112] [Kresa 1720, 120].

(*Elements* I.47), and $\mathrm{Tan}\,\theta = DG = \dfrac{rx}{\sqrt{r^2 - x^2}}$ by similar triangles. Kresa

goes on to find the cotangent $\left(\mathrm{Cot}\,\theta = HZ = \dfrac{r\sqrt{r^2 - x^2}}{x}\right)$, the secant

$\left(\mathrm{Sec}\,\theta = AG = \dfrac{r^2}{\sqrt{r^2 - x^2}}\right)$, and the cosecant $\left(\mathrm{Csc}\,\theta = AZ = \dfrac{r^2}{x}\right)$. This mathe-

matics itself is elementary. However, a proposition that would have been worth hardly a mention in earlier texts is elevated here to the beginning, its conclusions considered fundamental to what follows.

Kresa's introduction of analytical language was followed by a couple of authors before Euler: for instance, Friedrich Christian Maier's *Trigonometria* (1727) and especially Friedrich Wilhelm von Oppel's *Analysis triangulorum* (1746). Although Oppel's original definitions and earliest theorems are all geometric, he quickly transitions to solving problems algebraically. Consider for instance his first problem: find the length of the third side of a triangle when two sides and the diameter of the circumscribed circle are known. In figure 3.23 let a be the diameter (not drawn here), and let b and c be the known sides. From a previous theorem Oppel knows that $AF = bc/a$, where AF is the perpendicular from A to BD. Then by Pythagoras

$$FB = \sqrt{c^2 - \frac{b^2 c^2}{a^2}} = \frac{c\sqrt{a^2 - b^2}}{a} \quad \text{and} \quad FD = \sqrt{b^2 - \frac{b^2 c^2}{a^2}} = \frac{b\sqrt{a^2 - c^2}}{a}, \text{ so}$$

$$BD = FD + FB = \frac{b\sqrt{a^2 - c^2} + c\sqrt{a^2 - b^2}}{a}.\ {}^{113}$$

Neither the nature of the solution nor even the problem itself would have been natural in previous texts. Even stating that $AF = bc/a$ requires a new conception of what "AF" means. However, the fundamental transformation—working with sines, cosines, and so on analytically as functions themselves—was yet to be made.

But even as geometry was about to be abandoned in one direction, it was providing inspiration in a couple of others. One of these, an entirely new approach to deriving the standard theorems of spherical trigonometry, is also found in Oppel's *Analysis triangulorum*.[114] Rather than working directly with the sides of the spherical triangle (*ABC* in figure 3.24), Oppel turns his atten-

---

[113] [von Oppel 1746, 3].

[114] Oppel's approach was anticipated in 1737 by Roger Boscovich (1711–1787) in a method related to the ancient analemma and especially by John Caswell in 1690; see [von Braunmühl 1901b]. In the following exposition, based on [von Oppel 1746, 51], we have relabeled some of the points for clarity. We have also set the sphere's radius $R = 1$.

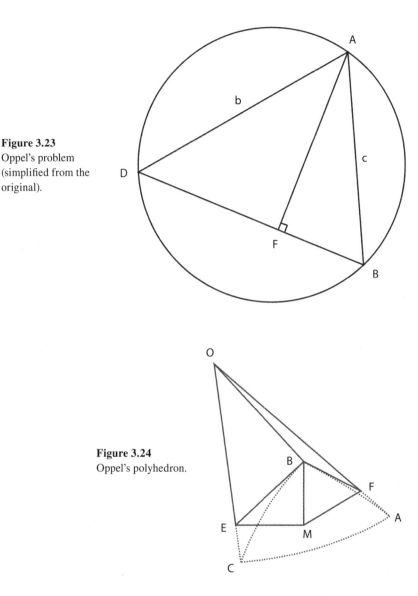

Figure 3.23
Oppel's problem
(simplified from the
original).

Figure 3.24
Oppel's polyhedron.

tion to the trihedral angle at the center O of the sphere. The three edge angles
at O correspond to the triangle's sides a, b, and c. From vertex B, drop perpendiculars BE and BF onto OA and OC, and then draw perpendiculars EM
and FM from E and F respectively, intersecting at M. Finally, join M to B,
forming the polyhedron $OEMFB$ in figure 3.24. We see in the polyhedron that
$\angle MFB = \angle A$ and $\angle MEB = \angle C$. Oppel does not work with this polyhedron directly; rather, he unfolds it into its ***development*** on the plane in figure 3.25,

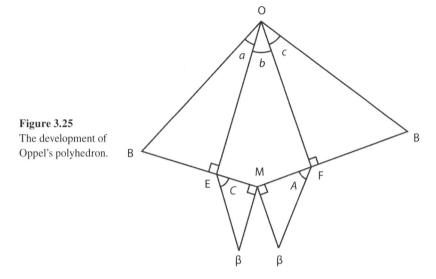

Figure 3.25
The development of
Oppel's polyhedron.

as if it were a paper cutout model. He achieves this by cutting the polyhedron along *OB* and laying out the triangles onto the plane of *OEMF*. This causes *B* to separate into four different points in the development, two of which Oppel relabels as *β*. Then, from the development, we see that

$$\frac{\sin a}{\sin c} = \frac{BE}{BF} = \frac{\beta E}{\beta F} = \frac{\csc \beta EM}{\csc \beta FM} = \frac{\csc C}{\csc A} = \frac{\sin A}{\sin C}, \tag{3.39}$$

and without further ado, we have the spherical Law of Sines. Oppel continues in this way, systematically unfolding polyhedra and representing sides of spherical triangles as angles in the development, and the standard theory of spherical trigonometry unfolds rather easily. This elegant geometric approach, eminently suitable for pedagogical purposes, might have generated a following if not for its misfortune of being invented only two years before Euler's *Introductio in analysin infinitorum* was published. As one historian commented wryly, "the better is the enemy of the good."[115]

The second new geometric inspiration originated in textbooks. Projections of the sphere, especially stereographic and orthographic, had never been far removed from spherical trigonometry; we have seen them already in this volume in the writings of Clavius, Napier, and Harriot. Several English authors felt that projections were important enough that they should

[115] [von Braunmühl 1900/1903, vol. 2, 100].

be treated before spherical trigonometry itself; presumably they were inspired by the usefulness of projections in astronomy and navigation. From the preface to John Harris's *Elements of Plain and Spherical Trigonometry* (1706):

> The reason of its publication was, that it might be in the hands of my auditory at the publick mathematick lecture in Birchin Lane, when I am explaining the subject to them; and where as soon as I have finished plain trigonometry, I shall proceed to read spherick geometry, and projection of the sphere in plano,[116] and then spherick trigonometry; which last science cannot be well understood, without the learner go in the order and method I have now mentioned. And this hint may serve to show the great defect of such treatises of the whole art of trigonometry, as have nothing of spherick projection introductory to the doctrine of spherick triangles.[117]

Harris is true to his word: he considers projections before tackling spherical trigonometry, although his integration of the two subjects does not quite live up to his advertisement. However, William Hawney's *Doctrine of Plain and Spherical Trigonometry* (1725) actually uses not only the mathematical idea but also the physical process of projection to solve triangles.[118] Space precludes a full explanation, but we provide here a precis of one of Hawney's examples.[119] In figure 3.26 Hawney intends to solve right-angled triangle ABC with the right angle at B, where $\angle BAC = 40°$ and $\overset{\frown}{AC} = 46°31'$. What we see in the figure is the plane of a stereographic projection of the sphere containing the triangle with the sphere oriented so that A is at the north pole and AB moves horizontally to the right. The projected triangle must be drawn first. We are instructed to begin by drawing AC inclined at a $40°$ angle to AB and then to measure $AC = \tan\frac{1}{2}(46°31')$, which turns out to be the correct length of the projection of $\overset{\frown}{AC}$. Finally, the projection of the third side $\overset{\frown}{CB}$ of the triangle may be drawn by joining C, and the top and bottom points of the circle, with a circular arc. Hawney constructs the projection a of the pole of this arc and then extends aC to b. It turns out that $\overset{\frown}{bD}$ is

---

[116] That is, "onto a plane."

[117] [Harris 1706, fifth and sixth unnumbered pages of "To the Reader"]. Henry Wilson's 1720 *Trigonometry Improv'd* [Wilson 1720] expresses similar sentiments. See also [Jones 1706, 279–282].

[118] [Hawney 1725, 168–197]. This is reminiscent of Clavius's *Astrolabium*, discussed earlier in this chapter.

[119] [Hawney 1725, 181–182]. An equivalent example, taken from Benjamin Martin's *The Young Trigonometer's Compleat Guide* [Martin 1736, vol. 2, 150–152], is explained thoroughly in [Van Brummelen 2013, 133–139].

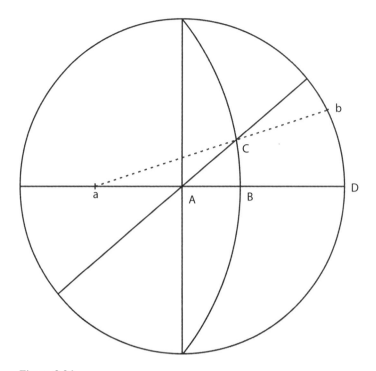

Figure 3.26
Hawney's use of stereographic projection to solve a right-angled spherical triangle.

equal to the original side $\widehat{CB}$ (not the projection of $\widehat{CB}$ in the figure); we measure this arc with an instrument to find $\widehat{CB} = 27°48'$. The other sides and the third angle are found in similar ways.

This curious method, requiring that the student perform geometric constructions on paper and measure the results with rulers and protractors, endured for a surprisingly long time. It is found soon after Hawney in Benjamin Martin's *Young Trigonometer's Compleat Guide* (1736).[120] In fact it lived on well into the nineteenth century and is present for instance in Thomas Keith's *Introduction to the Theory and Practice of Plane and Spherical Trigonometry*.[121]

[120] [Martin 1736].
[121] The book went through a number of editions through 1854. Here we rely on the second edition [Keith 1810]. Keith was familiar with Hawney's work, having published a revised edition of Hawney's *Complete Measurer*. For other constructive graphical methods of solving spherical triangles, see [Valette 1757] and [Mauduit 1765, 87–109].

▨   Euler on the Analysis of Trigonometric Functions

The final step in the conversion of trigonometry to the new analysis and the full admittance of trigonometric quantities to the family of functions was accomplished by perhaps the greatest mathematician of the past three centuries, Leonhard Euler (1707–1783). It is simply impossible to encapsulate Euler's contributions briefly, except to say that the history of most eighteenth-century mathematical disciplines may be divided into two periods: before Euler and after him. Among the many transformations for which he is responsible, one might make an argument that his approach to calculus had the most impact. Starting with his characterization of a function and working through to differential equations and the calculus of variations, there is hardly a topic in the university curriculum in these areas that does not rely in some fundamental way on Euler's analysis. His mathematics interacted strongly with the physical sciences and with what we would now call engineering, including topics in mechanics and astronomy. However, he also transformed purely mathematical subjects such as algebra and number theory and even anticipated subjects that would not come into existence much later, such as graph theory and topology. His prodigious output, well over 800 publications, is still being compiled in an *Opera omnia* consisting of over 80 volumes thus far in a project that has already lasted over a century.[122]

From a young age Euler had the benefit of contact with outstanding mentors. Even at 14 he was able to meet weekly with Johann Bernoulli at the University of Basel. At 20 he received his first academic posting, curiously, in physiology at the new St. Petersburg Academy of Sciences. Among his illustrious senior colleagues there were Daniel Bernoulli, Christian Goldbach, Jakob Hermann (who had worked on multiple-angle formulas), and Friedrich Christian Maier, whose *Trigonometria* (published in 1727, the year that Euler arrived) was beginning to convert trigonometry to analytic representations. Perhaps within months of his arrival in St. Petersburg he authored a manuscript entitled *Calculus differentialis*, which was never published and presumably

---

[122] Publication of *Leonhardi Euleri Opera omnia* began in 1911 and is currently being published by Birkhäuser, supervised by the Euler Committee of the Swiss Academy of Sciences. Many of Euler's publications are also available (some with English translations) at the Euler Archive (eulerarchive.maa.org). We cannot hope to refer to the entire literature on Euler here. One may begin with the biographies [Calinger 2016] (partly based on the papers [Calinger 1983] and [Calinger 1996]), [Fellmann 2007], and [Thiele 1982] (the latter in German). On Euler's mathematics a good starting point is a recent five-volume set published by the Mathematical Association of America ([Sandifer 2007a], [Dunham 2007], [Sandifer 2007c], [Bogolyubov/Mikhaïlov/ Youschkevitch 2007], and [Bradley/D'Antonio/Sandifer 2007]); see also [Sandifer 2015], [Debnath 2010], [Baker 2007], [Bradley/Sandifer 2007], [Dunham 1999], [Thiele 2005], [Fellmann et al. 1983], and [Engel 1983].

was intended for teaching purposes. He begins the manuscript with the definition of a function taken from Johann Bernoulli ("a quantity composed somehow from one or a greater number of quantities is called its, or their, function"), but the trigonometric quantities are not to be found anywhere. Indeed, they couldn't be. By "composed," Euler meant via the arithmetic operations, including extraction of roots and the taking of logarithms.[123] No finite expression of this sort is equal to the sine. In the middle of the next decade Euler unwittingly came closer to an analytic understanding of the sine in a paper on oscillations of rigid and flexible bodies that had emerged from his correspondence with Daniel Bernoulli. In it he solves the differential equation $k^4 y^{(iv)} = y$. His method invokes power series, but since he had used information about the initial values in his solution, he did not notice that trigonometry lurks inside the general solution to this equation.[124]

Euler's continued work on differential equations brought him closer in 1739, when he studied a harmonic oscillator driven sinusoidally. Along the way he considers the differential equation

$$dt = \frac{a\,dy}{\sqrt{a^2 - y^2}},$$

where y is the sine in a circle of radius a, and t is time measured as one travels at a constant rate along that circle. We have seen this equation before (for instance in the work of Leibniz), and Euler quotes the well-known solution $t = a \arcsin \frac{y}{a}$. But since for him t is the independent variable, Euler is able to take the extra step of expressing y in terms of t $\left(y = a\sin\frac{t}{a}\right)$. This allows him to proceed to the differential equation for the oscillator,

$$2ad^2 s + \frac{sdt^2}{b} + \frac{adt^2}{g}\sin\frac{t}{a} = 0, \tag{3.40}$$

where s is the oscillator's position. The sine (notably, stated explicitly for a unit circle) appears directly within the equation. However, Euler's method of attack on the problem uses his knowledge of the differentials of sines and cosines, and trigonometric functions also appear in the solution. And mere weeks later, in a letter to Johann Bernoulli, he reported that he had solved the equation $a^3 y''' = y$ as

---

[123] [Youschkevitch 1983]; see p. 161 for the quotation. See also [Calinger 2016, 73–75] and [Thiele 2007a, 377–378].

[124] [Euler 1740a (E40)], written in 1735; see also [Katz 1987, 318]. This and the next two paragraphs are based on the latter article. We shall refer to Euler's publications both by date and by the commonly used "Eneström numbers"; this paper's number is E40.

$$y = be^{x/a} + ce^{-x/2a} \sin\frac{(f+x)\sqrt{3}}{2a} \qquad (3.41)$$

(with the three integration constants b, c, and f). Both exponential and trigonometric terms were on the same playing field and arose in similar ways from the same equation. The separation between trigonometry and analysis in Euler's mind was fading rapidly.[125]

Euler went further only a few months later. In another letter to Johann Bernoulli, he was working with a natural generalization to the equations above: namely, nth order linear differential equations with constant coefficients:

$$y + ay' + by'' + cy''' + \ldots = 0. \qquad (3.42)$$

He discovered "unexpectedly" that the key to solving these equations was to consider what is now called the "characteristic equation":

$$1 + ap + bp^2 + cp^3 + \ldots = 0. \qquad (3.43)$$

He found that linear factors $1 - \alpha p$ of this polynomial correspond to exponential solutions $y = Ce^{x/\alpha}$ of the differential equation, but irreducible quadratic factors $1 + \alpha p + \beta p^2$ correspond to solutions of the form

$$e^{-\alpha x/2\beta}\left(C\sin\frac{x\sqrt{4\beta - \alpha^2}}{2\beta} + D\cos\frac{x\sqrt{4\beta - \alpha^2}}{2\beta} \right). \quad \text{[126]} \qquad (3.44)$$

The sine and cosine were becoming essential to Euler's analytic toolbox. In addition, the connection between exponential and trigonometric functions through complex numbers was becoming more and more obvious to Euler.

Later in 1740 Euler, continuing the discussion with Bernoulli, showed him that both $y = e^{ix} + e^{-ix}$ and $y = 2\cos x$ are solutions to the differential equation $y'' + y = 0$ with initial values $y(0) = 2$ and $y'(0) = 0$.[127] Setting these two expressions equal to each other, we have the first explicit appearance of an equality between trigonometric and imaginary exponential quantities:

$$\cos x = \frac{e^{ix} + e^{-ix}}{2}. \qquad (3.45)$$

From this point onward, Euler used trigonometric functions routinely, both in his mathematical analysis and in physical applications.

---

[125] Euler's published paper is [Euler 1750b (E126), esp. 133–134], written in 1739; the letter appears in [Eneström 1905, 24–33]. See also [Katz 1987, 318–320].

[126] The letter appears in [Eneström 1905, 33–38, esp. 37–38]; see also [Katz 1987, 320–321], [Eneström 1897], and [Roy 2011, 263]. Euler published his solution in [Euler 1743a].

[127] [Eneström 1905, 76–77]; see also [Nahin 1998, 143].

However, matters were hardly routine for Euler, either in life or in mathematics. The political situation in St. Petersburg had grown unstable, and he answered a call from Frederick the Great to become a member of the Academy of Sciences in Berlin in 1741. In addition, the interactions among exponential functions, trigonometric quantities, and imaginary numbers were hardly settled among his colleagues. For decades disagreements had arisen among such luminaries as Leibniz, Johann Bernoulli, and Euler himself. One of the points of contention arose over the question of whether there is such a thing as the logarithm of a negative number, and if so, what is its value. Arguments based on geometrical principles and analytic solutions to differential equations did not produce consistent results. A well-known debate between Euler and Jean le Rond d'Alembert from 1746 to 1748 has d'Alembert claiming that $\log(-x) = \log x$ for $x > 0$ while Euler counters with the modern definition of the logarithm of a complex number, $\log z = \ln z + i(\arg(z) + 2\pi n)$ for $n \in \mathbb{Z}$.[128]

At the time of his argument with d'Alembert, Euler had already submitted to the publisher the book that would become the foundation of modern analysis, *Introductio in analysin infinitorum* (1748). This two-volume work, one of the most influential books in the history of mathematics, impacted two overlapping audiences. Firstly, the mathematical research community quickly adopted the *Introductio* as the primer on how analysis was to be done. Many of its innovations, notations, and concepts became standards, so much so that eighteenth-century analysis is often broken into the period before the *Introductio* and the period after it. Secondly, the *Introductio* became a leading textbook, establishing analysis with infinitesimal quantities alongside geometry and algebra as one of the standard mathematical disciplines that serious students needed to learn. Even more, it attempted to unify all of mathematics within an analytical framework. The next generation of mathematical leaders grew into their profession reading the *Introductio*, and many of its innovations are still taught in our high school and university classrooms today.[129]

The book opens with a general definition of a mathematical function: "a function of a variable quantity is an analytic expression composed in any way

---

[128] On this debate see [Bradley 2007], [Calinger 2016, 269–270], and [Cajori 1913b]. For Euler's views on the subject see [Euler 1932 (E807)] (with an English translation by Todd Doucet at eulerarchive.maa.org) and [Euler 1751 (E168)]; for d'Alembert's views see [d'Alembert 1761]. In [Euler 1932 (E807), paragraph 26], he states explicitly that $\log(\cos \theta + i \sin \theta) = i(\theta \pm 2\pi n)$.

[129] The *Introductio* ([Euler 1748 (E101, E102)]) is available in a number of editions and translations. In English see [Euler 1988 (E101)] and [Euler 1990 (E102)], and [Euler 2013a (E101, E102)]. Translations have also appeared in French, German and Spanish. Passages of the *Introductio* related to trigonometry appear in English translation in [Struik 1969, 345–351]. On the *Introductio* see [Katz 2007, 214–222], [Ferraro 2007b, 41–56], [Panza 2007], [Reich 2005], and [Boyer 1951].

whatsoever of the variable quantity and numbers or constant quantities."[130] This seems little changed from the definition in his 1727 manuscript *Calculus differentialis*. However, the impact of this seemingly innocuous statement was substantial. Prior to Euler, analysis had been about *curves* that could be represented by equations containing *variables*. Euler's primitive notion is not the geometric curve but the analytic *function*. What precisely the various terms in this definition meant has been debated, but Euler himself interpreted his own expression "composed in any way whatsoever" very broadly, permitting the use of infinite processes such as series and continued fractions within its ambit.[131]

Of special interest to us is that Euler divides functions into algebraic and transcendental, and he explicitly extends his description of transcendental functions from his *Calculus differentialis* (which had included only exponentials and logarithms) to also include "quantities which arise from the circle." This phrase is the title of the *Introductio*'s eighth chapter, which contains a number of innovations already in its first page.[132]

Text 3.6
Leonhard Euler, On Transcendental Quantities Which Arise from the Circle
(from *Introductio in analysin infinitorum*)

After having considered logarithms and exponentials, we must now turn to circular arcs with their sines and cosines. This is not only because these are further genera of transcendental quantities, but also since they arise from logarithms and exponentials when complex values are used. This will become clearer in the development to follow.

We let the radius, or total sine,[133] of a circle to be equal to 1, then it is clear enough that the circumference of the circle cannot be expressed exactly as a rational number. An approximation of half of the circumference of this circle is 3.14159265358979323846264338327950288419716939937510582097494459 23078164062862089986280348253421170679821480865132723066470938446+.

---

[130] [Euler (Blanton) 1988 (E101), 3].

[131] The development of the concept of function from the seventeenth through the nineteenth century is surveyed in [Youschkevitch 1976–77]; on Euler see pp. 61–63. See also [Thiele 2007a], [Thiele 2007b], and [Kleiner 1989]. On specifically eighteenth-century notions of function see [Panza 1996] and [Ferraro 2000] and [Ferraro 2001].

[132] We shall follow chapter 8 for the next several pages. See [Euler 1748, vol. 1 (E101), 93–107], and an English translation in [Euler (Blanton) 1988 (E101), 101–115].

[133] This is the last time the term "total sine," or "sinus totus," appears in Euler's work as a reference to the radius of the base circle.

For the sake of brevity we will use the symbol π for this number. We say, then, that half of the circumference of a unit circle is π, or that the length of an arc of 180 degrees is π.

We always assume that the radius of the circle is 1 and let z be an arc of this circle. We are especially interested in the sine and cosine of this arc z. Henceforth we will signify the sine of the arc z by sin z. Likewise, for the cosine of the arc z we will write cos z.[134]

Explanation: The fact that this passage may be so clear to us as to require no explanation at all is itself instructive. The reader may wish to look back at some of the texts earlier in this chapter to get a sense of the contrast in its style with some of its predecessors. One feature that immediately attracts the eye is the lack of even a single geometric diagram not only within this text but also within the entire first volume.

For eighteenth-century readers several other features of this passage would have been noteworthy or even shocking:

- The reference to the emergence of sines and cosines from exponential quantities would have been new to many readers, and the use of complex numbers as legitimate values for variables (especially in the context of trigonometry) was startling.
- We have seen that Euler had been using the unit circle in previous work, and it had come up occasionally in other authors' writings. However, it is here in the *Introductio* that it became established as a standard within the mathematical community. With the adoption of the unit circle, for the first time the sine and cosine are considered to be *ratios* of line segments rather than their *lengths*. Indeed, at the end of the following paragraph, tan z and cot z are introduced directly as sin z/ cos z and cos z/ sin z respectively.[135]
- Similarly, the notation "sin" and "cos," while appearing in various forms in the works of earlier authors, became definitive with its appearance here.

Euler does not *define* the sine and cosine at all in this text; the primitive notions remain geometrical. The arguments of the new functions (and now, for the first time, we may use the word "function" without much fear of anachronism) are chosen explicitly to be arcs of the unit circle, not an angle. Thus, while Euler's sine function is mathematically equivalent to the modern function with the argument measured in radians, the final step of measuring the angles themselves in radians is not yet taken. Whereas in earlier papers he had referred to the sine of arc z as "sin. A.z," here he immediately suppresses the "A" and writes only "sin. z."

---

[134] [Euler (Blanton) 1988 (E101), 101].

[135] Euler's abbreviation for the tangent is "tang"; abbreviations of the other trigonometric quantities correspond to modern usage.

Now that Euler has established the basic terms, he begins his analysis of the sine and cosine by stating (not proving) the basic identities $\cos z = \sin\left(\frac{\pi}{2} - z\right)$, $\sin z = \cos\left(\frac{\pi}{2} - z\right)$, $(\sin z)^2 + (\cos z)^2 = 1$ (in that notation), and the sine and cosine sum and difference laws. The absence of proofs is due presumably to the fact that Euler sees himself to be doing analysis here, not geometry; he takes from geometry here what he must to begin his analytical work.[136] As we shall see, he is able to grow spectacular flowers from such meager seeds.

Euler's first collection of theorems might appear today as ordinary statements that would not be out of place in a high school classroom, but this is only because our students learn Euler's system from the beginning. Substituting into the sum and difference formulas, he begins with the routine formula $\sin\left(\frac{\pi}{2} + z\right) = +\cos z$, building further and further until he ends up with formulas such as

$$\sin\left(\frac{4n+4}{2}\pi - z\right) = -\sin z \text{ and } \cos\left(\frac{4n+4}{2}\pi - z\right) = -\cos z \quad (3.46)$$

for any integer n. What Euler's contemporaries would have noticed here is that Euler implicitly permits the arguments of trigonometric functions to take on any real value, positive or negative.

The heart of chapter 8, on computing trigonometric quantities, is a potent mixture of traditional and revolutionary. The strategy is traditional—it goes back all the way back to Ptolemy and was still present in Euler's day: develop the basic trigonometric identities and use them to generate new sine values from given ones, gradually building a sine table. However, the treatment is almost entirely new, including theorems not seen before in this context and eventually branching into complex numbers. Euler begins by demonstrating the following pair of identities derived from the sine/cosine sum formulas and the Pythagorean Theorem:

$$\sin((n+1)y+z) = 2 \cos y \sin(ny+z) - \sin((n-1)y+z), \text{ and} \quad (3.47)$$

$$\cos((n+1)y+z) = 2 \cos y \cos(ny+z) - \cos((n-1)y+z).[137] \quad (3.48)$$

These formulas offer an immediate benefit to any table compiler: as soon as one has values for the sine or cosine of any two arcs, say, $3°$ and $6°$, one can

---

[136] Indeed, Euler has no notion of analytical geometry, distinguished from synthetic geometry. He has simply geometry, followed by analysis. See [Ferraro 2007b, 51].

[137] Euler does not express these formulas for arbitrary n; rather, he derives them up to $n = 3$ and ends with "etc."

quickly compute the sine or cosine of the arcs in the arithmetic sequence that follows from them: 9°, 12°, 15°, . . .

Next Euler develops the sine and cosine half-angle formulas

$$\sin\frac{v}{2} = \sqrt{(1-\cos v)/2} \text{ and } \cos\frac{v}{2} = \sqrt{(1+\cos v)/2}, \qquad (3.49)$$

standard fare for table compilers. But he derives them smoothly and entirely algebraically using the product-to-sum formulas

$$\begin{matrix} \sin \\ \cos \end{matrix} y \begin{matrix} \cos \\ \sin \end{matrix} z = \frac{\sin(y+z) \pm \sin(y-z)}{2} \qquad (3.50)$$

and

$$\begin{matrix} \cos \\ \sin \end{matrix} y \begin{matrix} \cos \\ \sin \end{matrix} z = \frac{\cos(y+z) \pm \cos(y-z)}{2}, \qquad (3.51)$$

well known through their applications in prosthaphaeresis (the predecessor to logarithms) a century and a half before. Before proceeding, Euler takes the opportunity to derive some corollaries, including the sum-to-product identities

$$\sin a \pm \sin b = 2 \begin{matrix} \sin \\ \cos \end{matrix} \frac{a+b}{2} \begin{matrix} \cos \\ \sin \end{matrix} \frac{a-b}{2} \qquad (3.52)$$

and

$$\cos a \pm \cos b = 2 \begin{matrix} \cos \\ \sin \end{matrix} \frac{a+b}{2} \begin{matrix} \cos \\ \sin \end{matrix} \frac{a-b}{2}. \qquad (3.53)$$

Dividing these into each other, he comes up with several related theorems, such as

$$\cot\frac{a+b}{2} = \frac{\sin a - \sin b}{\cos b - \cos a}. \qquad (3.54)$$

None of these results were new (for instance, (3.54) goes back to Viète[138]), but Euler's algebraic process of discovery would have presented itself as breathtakingly simple to his colleagues.

Trigonometers would expect Euler to demonstrate next how to calculate one or more sines inaccessible by the usual identities, and he does not disappoint—but his approach is completely novel. He begins by introducing complex numbers into his analysis:[139]

$$(\cos y + i \sin y)(\cos z + i \sin z) = \cos y \cos z - \sin y \sin z$$
$$+ i (\cos y \sin z + \sin y \cos z) = \cos(y+z) + i \sin(y+z).$$

---

[138] [Viète 1579]; see also [Ritter 1895, 51].
[139] Although Euler would later invent the i notation, in the *Introductio* Euler uses $\sqrt{-1}$.

Setting $y = z$, we have

$$(\cos z + i \sin z)^2 = \cos 2z + i \sin 2z. \tag{3.55}$$

After a similar derivation for $(\cos z + i \sin z)^3$, Euler concludes that

$$(\cos z + i \sin z)^n = \cos nz + i \sin nz. \tag{3.56}$$

This is de Moivre's formula (3.39), this time demonstrated for any whole number n (not just odd n). Similarly,

$$(\cos z - i \sin z)^n = \cos nz - i \sin nz. \tag{3.57}$$

From (3.56) and (3.57),

$$\cos nz = \frac{(\cos z + i \sin z)^n + (\cos z - i \sin z)^n}{2}. \tag{3.58}$$

Using the binomial theorem to expand the terms in the numerator, Euler gets

$$
\begin{aligned}
\cos nz = {} & (\cos z)^n - \frac{n(n-1)}{1 \cdot 2}(\cos z)^{n-2}(\sin z)^2 \\
& + \frac{n(n-1)(n-2)(n-3)}{1 \cdot 2 \cdot 3 \cdot 4}(\cos z)^{n-4}(\sin z)^4 \\
& - \frac{n(n-1)(n-2)(n-3)(n-4)(n-5)}{1 \cdot 2 \cdot 3 \cdot 4 \cdot 5 \cdot 6}(\cos z)^{n-6}(\sin z)^6 \\
& + \cdots.
\end{aligned}
\tag{3.59}
$$

How does this help? Euler uses the formula in a shocking way:

> "Let the arc z be infinitely small; then $\sin z = z$ and $\cos z = 1$. If n is an infinitely large number, so that nz is a finite number, say $nz = v$, then since $\sin z = z = \dfrac{v}{n}$, we have

$$\cos v = 1 - \frac{v^2}{1 \cdot 2} + \frac{v^4}{1 \cdot 2 \cdot 3 \cdot 4} - \frac{v^6}{1 \cdot 2 \cdot 3 \cdot 4 \cdot 5 \cdot 6} + \cdots,\text{"[140]} \tag{3.60}$$

the Taylor series for the cosine. A similar derivation produces the Taylor series for the sine. With these series, Euler can now find the sine and cosine of any arc. Of course, these results were not new; Taylor series had been used to compute sines in Europe at least since Abraham Sharp several decades earlier. However, the steps Euler took to get there were utterly novel—although, a little later when Euler describes the practical business of calculating sines and cosines reliably, we essentially find Sharp's approach in (3.19), a formula that

---

[140] Euler does not express these formulas for arbitrary n; rather, he derives them up to $n = 3$ and ends with "etc."

goes back as far as Viète.[141] One novelty here is that Euler considers separately how to compute tangents and cotangents using series. Generally, the tangent had been computed simply by dividing the sine by the cosine.[142]

Chapter 8 concludes with two tangential topics. Using methods similar to the above, Euler quickly derives his famous theorem

$$e^{\pm iv} = \cos v \pm i \sin v.\tag{3.61}$$

Then he derives the series

$$\arctan t = \frac{t}{1} - \frac{t^3}{3} + \frac{t^5}{5} - \frac{t^7}{7} + \frac{t^9}{9} - \cdots.\tag{3.62}$$

With the substitution $t = 1$, we get Leibniz's series for π, which Euler shows how to manipulate to calculate approximations to π more quickly. With this, one of the monumental chapters in the history of trigonometry ends.

But the *Introductio* has much more in store. The next chapter begins by appearing to branch into a different topic, techniques for factoring polynomials, but this is only a preparation for something bigger. Earlier in the book Euler had established the now-standard definition of the exponential function e^x, using the expression $\left(1 + \dfrac{x}{j}\right)^j$. Today we would take its limit as $j \to \infty$; Euler simply says that e^x is its value when $j = \infty$. From series he already knows that

$$e^x = 1 + x + \frac{x^2}{2!} + \frac{x^3}{3!} + \cdots;\tag{3.63}$$

therefore,

$$\frac{e^x - e^{-x}}{2} = x + \frac{x^3}{3!} + \frac{x^5}{5!} \cdots.\tag{3.64}$$

But $(e^x - e^{-x})$ is also equal to $\left(1 + \dfrac{x}{j}\right)^j - \left(1 - \dfrac{x}{j}\right)^j$ with $j = \infty$. Euler applies his newly developed factoring technique to the latter expression and quickly arrives at

$$\frac{e^x - e^{-x}}{2} = x\left(1 + \frac{x^2}{\pi^2}\right)\left(1 + \frac{x^2}{4\pi^2}\right)\left(1 + \frac{x^2}{9\pi^2}\right)\left(1 + \frac{x^2}{16\pi^2}\right)\left(1 + \frac{x^2}{25\pi^2}\right)\cdots;\tag{3.65}$$

---

[141] Whether by intent or by a slip of the tongue, here for the first time Euler talks of the sines and cosines of *angles* rather than of *arcs*. See [Euler (Blanton) 1988 (E101), 110].

[142] Euler explores these series in more detail in [Euler 1750a (E128), 204–217], written in 1739, including sample calculations of sin 9° and cos 9° to 28 places. (The sine value is correct; the cosine value is in error only by one unit in the last place.) On this paper see also [Sandifer 2007a, 323–327].

similarly, he finds

$$\frac{e^x + e^{-x}}{2} = \left(1 + \frac{4x^2}{\pi^2}\right)\left(1 + \frac{4x^2}{9\pi^2}\right)\left(1 + \frac{4x^2}{25\pi^2}\right)\left(1 + \frac{4x^2}{49\pi^2}\right)\cdots. \quad (3.66)$$

Eventually these quantities would be named the hyperbolic sine and cosine, but not here and never by Euler. Next, setting $x = zi$ in (3.65) he has

$$\sin z = \frac{e^{zi} - e^{-zi}}{2} = z\left(1 - \frac{z^2}{\pi^2}\right)\left(1 - \frac{z^2}{4\pi^2}\right)\left(1 - \frac{z^2}{9\pi^2}\right)$$

$$\left(1 - \frac{z^2}{16\pi^2}\right)\left(1 - \frac{z^2}{25\pi^2}\right)\cdots, \quad (3.67)$$

and similarly

$$\cos z = \left(1 - \frac{4z^2}{\pi^2}\right)\left(1 - \frac{4z^2}{9\pi^2}\right)\left(1 - \frac{4z^2}{25\pi^2}\right)\left(1 - \frac{4z^2}{49\pi^2}\right)\cdots. \text{[143]} \quad (3.68)$$

Of course, products are much harder to calculate with than sums are, so one might wonder why Euler bothered to develop these formulas. In fact he had worked with them already more than a decade earlier; (3.67) had been key to his famous solution of the Basel problem, namely, to show that

$$1 + \frac{1}{4} + \frac{1}{9} + \frac{1}{16} + \frac{1}{25} + \cdots = \frac{\pi^2}{6}. \text{[144]} \quad (3.69)$$

The benefits of the product formulas in the *Introductio* reveal themselves with some algebraic manipulation. Substituting $z = \frac{m\pi}{2n}$ into (3.67) and $z = \frac{\pi}{2} - \frac{m\pi}{2n}$ into (3.68), he generates two formulas for $\sin\frac{m\pi}{2n}$. Placing them in ratio to each other and simplifying, he gets

$$1 = \frac{\pi}{2} \cdot \frac{1}{2} \cdot \frac{3}{2} \cdot \frac{3}{4} \cdot \frac{5}{4} \cdots,$$

or the striking expression

$$\frac{\pi}{2} = \frac{2 \cdot 2 \cdot 4 \cdot 4 \cdot 6 \cdot 6}{1 \cdot 3 \cdot 3 \cdot 5 \cdot 5 \cdot 7} \cdots, \quad (3.70)$$

[143] [Euler 1748, vol. 1 (E101), 117–121], [Euler (Blanton) 1988 (E101), 124–128]. Euler goes on to consider infinite products for the other four standard trigonometric functions, which may be obtained easily from these. In correspondence with Euler, Nicholas I Bernoulli questioned the convergence of these series, but Euler continued to use them in subsequent work. See [von Braunmühl 1900/1903, vol. 2, 111], [Euler 1750c (E130)], and [Euler 1743b (E59)].

[144] [Euler 1740b (E41)].

Wallis's infinite product for π. Variants of the same process lead to other infinite products, for instance,

$$\sqrt{2} = \frac{2 \cdot 2 \cdot 6 \cdot 6 \cdot 10 \cdot 10 \cdot 14 \cdot 14 \cdot 18 \cdot 18}{1 \cdot 3 \cdot 5 \cdot 7 \cdot 9 \cdot 11 \cdot 13 \cdot 15 \cdot 17 \cdot 19} \cdots.^{145} \tag{3.71}$$

But these products are bagatelles compared with Euler's real quarry. Returning to (3.67) and again substituting $z = \dfrac{m\pi}{2n}$, he takes logarithms to obtain

$$\ln \sin \frac{m\pi}{2n} = \ln \pi + \ln \frac{m}{2n} + \ln \left(1 - \frac{m^2}{4n^2}\right) + \ln \left(1 - \frac{m^2}{16n^2}\right) + \cdots. \tag{3.72}$$

But he already knows how to expand logarithms as series themselves. After expanding and massaging the result a bit, he arrives at

$$\begin{aligned}
\ln \sin \frac{m\pi}{2n} = {} & \ln m + \ln(2n - m) + \ln(2n + m) + \ln \pi - \ln 8 \\
& - \frac{m^2}{n^2}\left(\frac{1}{4^2} + \frac{1}{6^2} + \frac{1}{8^2} + \cdots\right) \\
& - \frac{m^4}{2n^4}\left(\frac{1}{4^4} + \frac{1}{6^4} + \frac{1}{8^4} + \cdots\right) \\
& - \frac{m^6}{3n^6}\left(\frac{1}{4^6} + \frac{1}{6^6} + \frac{1}{8^6} + \cdots\right) \\
& - \frac{m^8}{4n^8}\left(\frac{1}{4^8} + \frac{1}{6^8} + \frac{1}{8^8} + \cdots\right) \\
& - \cdots.
\end{aligned} \tag{3.73}$$

Although this looks complicated, Euler has already computed the constants and the series within this expression. Substituting these values, he ends up with a simple and quickly converging series for $\ln \sin \dfrac{m\pi}{2n}$. The benefit of this to the table makers was immediate. As we have seen, it was common to compose tables of logarithms of trigonometric functions. Here Euler had found an efficient method to compute logarithms of sines—without having to compute the sines at all.[146] His methods were picked up by the table makers; they were a standard part of their repertoire well into the twentieth century.[147]

---

[145] [Euler 1748, vol. 1 (E101), 145–147] and [Euler (Blanton) 1988 (E101), 160–165]; see also [Euler 1750a (E128), 202].

[146] [Euler 1748, vol. 1 (E101), 151–158] and [Euler (Blanton) 1998 (E101), 160–165]; see also [Euler 1750a (E128), 217–223]. Euler has a parallel calculation for the cosine. Given the properties of logarithms and the definitions of the other trigonometric functions, it is easy for him to extend his methods to co/tangents and co/secants.

[147] Perhaps the most heavily used set of trigonometric tables was that first published by François Callet in 1795 and issued in various editions through the nineteenth century [Callet 1795]. Its introduction of more than 100 pages describes the functions tabulated, the tables' methods of

Euler returns to trigonometry in chapter 14, "On the Multiplication and Division of Angles." Once again the topic is inspired by the concerns of table makers, and once again he goes well beyond them. The chapter begins by considering formulas for sin nz in terms of sin z. The formula for $n = 3$ had been used since the fifteenth century in Persia (al-Kāshī) and the late sixteenth century in Europe (Viète) to find $\sin\dfrac{\theta}{3}$ when sin θ is known. Viète had extended this to $n = 5$ to help build tables, and he had found recursive formulas to find sin nz for larger values of n. Euler's general formulas

$$\sin nz = nx - \frac{n(n^2 - 1)}{3!}x^3 + \frac{n(n^2 - 1)(n^2 - 9)}{5!}x^5 - \cdots \tag{3.74}$$

(where $x = \sin z$) for odd n and

$$\sin nz = \left(nx - \frac{n(n^2 - 4)}{3!}x^3 + \frac{n(n^2 - 4)(n^2 - 16)}{5!}x^5 - \cdots \pm 2^{n-1}x^{n-1} \right)\sqrt{1 - x^2} \tag{3.75}$$

for even n[148] were not new, going back to Newton and Jakob Bernoulli. But Euler goes further, generating multiple-angle formulas not just for the sine and cosine but for all six trigonometric functions. Before wrapping up, Euler deals with two further challenges. The first is the sum of a series of sines (or cosines) whose arcs are in an arithmetic progression: first an infinite series, and then (by subtracting two infinite series) a finite one. As a final flourish, Euler derives formulas for powers of sin z and cos z in terms of sines and cosines of multiples of z; for instance,

$$32(\cos z)^6 = 10 + 15\cos 2z + 6\cos 4z + \cos 6z. \tag{3.76}$$

These are obtained by iterated use of the sine and cosine product-to-sum formulas.[149]

Although these latter expansions appear to be inspired simply by mathematical curiosity, this is hardly the case. In a paper written in the year the *Introductio* was published, Euler had been using formulas like this within his

---

construction, and various examples of their use. The presentation on methods for trigonometric tables refers to Euler's *Introductio* and his series on pp. 26–29 and to logarithms of trigonometric functions on pp. 48–57. The tables that became standard in the early twentieth century, Henri Andoyer's *Nouvelles Tables Trigonométriques Fondamentales*, also rely on formulas quoted from the *Introductio* [Andoyer 1915–1918, vol. 1, ix].

[148] [Euler 1748, vol. 1 (E101), 198–204] and [Euler (Blanton) 1988 (E101), 204–211].

[149] [Euler 1748, vol. 1 (E101), 217–220] and [Euler (Blanton) 1988 (E101), 223–227]. The formulas for sines are given for n up to nine and for cosines for n up to seven. He would return to this topic in [Euler 1760 (E246)].

work in celestial mechanics related to the three-body problem.[150] Specifically, while dealing with inequalities in the motions of Saturn and Jupiter, he found himself needing integrate the expression $(1 - g \cos \omega)^{-3/2}$. For the value of g that he was using (around 0.8), the binomial expansion leads to a series that converges too slowly to be useful. Instead he transforms the expression using his formulas for $(\cos \omega)^n$. Setting the exponent to $-\mu$, he has

$$(1 - g\cos\omega)^{-\mu} = 1 + \frac{\mu}{1!}g\cos\omega + \frac{\mu(\mu+1)}{2!}g^2(\cos\omega)^2$$
$$+ \frac{\mu(\mu+1)\mu(\mu+2)}{3!}g^3(\cos\omega)^3 + \cdots. \tag{3.77}$$

The powers of cos ω may be reduced to terms involving cos $n\omega$ using identities like (3.76). Euler ends up with an expansion of the required function in the form

$$(1 - g \cos \omega)^{-\mu} = A + B \cos \omega + C \cos 2\omega + D \cos 3\omega + E \cos 4\omega + \ldots. \tag{3.78}$$

The constants $A, B, C, \ldots$ are by no means easy to find, but once he has them he may integrate the result easily enough. The series he ends up with turns out to converge more quickly than the one he derived using the binomial expansion.

Euler's fast and loose handling of infinite series raises the question, familiar to those who read his work, of the extent to which he understood questions of their convergence. We have seen here a few examples of liberties apparently being taken; and yet, Euler seldom reaches incorrect conclusions. It appears that Euler (and some of his contemporaries) had a practical sense—if not a full theoretical awareness—of when series converge and when they don't. Behind the published scenes, calculations often were being performed to assure practitioners that the series they were manipulating were in fact well behaved.

The *Introductio* was the first of Euler's three textbooks. The second, *Institutiones calculi differentialis* (1755), corresponds to what we would call the differential calculus.[151] He chose to approach the topic strictly analytically; no diagrams are to be found. Nevertheless, the modern reader experiences a sense of familiarity within its pages. Euler's development of the ideas is similar to what we find in today's textbooks—a sign of his immense impact even today, two-and-a-half centuries later. One dramatic difference be-

---

[150] [Euler 1749b (E120)]. On this paper and Euler's subsequent work on this topic, see [Golland/Golland 1993]; see also [Ferraro 2007a, 76] and [Wilson 1980, 126–130]. Euler returned to this series, generalizing to arbitrary planets, in [Euler 1769b (E414)].

[151] [Euler 1755b (E212)]. [Euler (Blanton) 2000 (E212)] is a translation into English of the first of the book's two parts, and [Euler (Bruce) 2011 (E212)] is a translation of the entire book. On the *Institutiones calculi differentialis* see [Katz 2007, 222–228], [Ferraro 2007b, 57–77], [Sandifer 2007c], [Demidov 2005], [Ferraro 2004], and [Ferzola 1994].

tween Euler and modern university calculus is his calculation not of a derivative but of a "differential"—an evanescent increment dy of one quantity y
that is the function of another quantity x, having undergone its own evanescent increment dx. The ratio of these increments, dy/dx, is what we call a derivative. There are hints of the beginning of the modern notion of a limit in
the preface, but in practice Euler proceeds without much logical concern for
whether dy is equal to zero or infinitely small. His algebra with infinitesimal
quantities is based on analogies to algebra with finite quantities, supplemented
with statements like $x + dx = x$ and $dx + dx^2 = dx$. For example, the following
is Euler's calculation of the derivative of the sine.

Text 3.7

Leonhard Euler, On the Derivative of the Sine
(from *Institutiones calculi differentialis*)

There remain some quantities that arise as inverses of these functions, namely
the sines and tangents of given arcs, and we ought to show how these are dif-
ferentiated. Let x be a circular arc and let sin x denote its sine, whose differential we are to investigate. We let $y = \sin x$ and replace x by $x + dx$ so that y
becomes $y + dy$. Then $y + dy = \sin(x + dx)$ and

$$dy = \sin(x + dx) - \sin x.$$

But

$$\sin(x + dx) = \sin x \cdot \cos dx + \cos x \cdot \sin dx,$$

and since, as we have shown in the *Introductio*,

$$\sin z = \frac{z}{1} - \frac{z^3}{1 \cdot 2 \cdot 3} + \frac{z^5}{1 \cdot 2 \cdot 3 \cdot 4 \cdot 5} - \cdots,$$

$$\cos z = 1 - \frac{z^2}{1 \cdot 2} + \frac{z^4}{1 \cdot 2 \cdot 3 \cdot 4} - \cdots,$$

when we exclude the vanishing terms, we have $\cos dx = 1$ and $\sin dx = dx$ so that

$$\sin(x + dx) = \sin x + dx \cos x.$$

Hence, when we let $y = \sin x$, we have

$$dy = dx \cos x.$$

Therefore, the differential of the sine of any arc is equal to the product of the
differential of the arc and the cosine of the arc.[152]

---

[152] [Euler (Blanton) 2000 (E212), 116].

Explanation: As in the *Introductio*, Euler does not take geometry out of the definitions of the sine and cosine, but he takes the minimum of what he needs from the geometry (the definitions and the sine sum law) and promptly switches into analysis. He also borrows the Taylor series for the sine and cosine within the derivation, but this is the only time he does so in his development of the differential calculus of trigonometric functions. Finally, the opening sentence of this passage hints that he had begun his work with what we call the inverse trigonometric functions and only afterward turned to sines and cosines. This was standard practice, and we turn to it next.

Euler begins the section on trigonometry by considering the function $y = \arc \sin x$ and begins with a derivation that segues from the previous section on exponentials and logarithms.[153] He recalls from the *Introductio* that

$$y = \arc \sin x = \frac{1}{i} \ln\left(\sqrt{1 - x^2} + ix \right). \tag{3.79}$$

He applies the standard rules of differentiation to this expression; simplification leads him to

$$dy = \frac{dx}{\sqrt{1 - x^2}}. \tag{3.80}$$

But this is the only time he relies on the link between trigonometry, logarithms, and complex numbers. He returns immediately to this same function, using infinitesimals but not complex analysis. Since $x = \sin y$, we have

$$x + dx = \sin(y + dy) = \sin y \cos dy + \cos y \sin dy. \tag{3.81}$$

But $\cos dy = 1$, $\sin dy = dy$, and $\cos y = \sqrt{1 - x^2}$; and after substitution and cleanup he concludes with (3.80) as before.

Euler's approach to the calculus of trigonometry had an instant impact, and not just because trigonometric functions were handled on par with other functions. From this point onward, most calculus textbooks abandoned geometry as Euler had done and adopted his analytic approach. Discomfort over Euler's use of infinitesimals would lead gradually to more rigorous treatments and formal definitions of limits, but that was the better part of a century away.

Euler was essentially acting president of the Berlin Academy from 1757 onward under the supervision of Frederick the Great. The two men had vastly different personalities and styles, and the conflicts that arose eventually caused Euler to return to St. Petersburg in 1766. Only about ten days after his arrival, Euler rewarded his new hosts by presenting the first volume of his mon-

---

[153] [Euler (Blanton) 2000 (E212), 110].

umental *Institutionum calculi integralis*.[154] All three volumes of this third
and concluding part of Euler's textbook series were in print by 1770. Continuing the style of the first two parts, this new work contains no geometry
or even applications of the integral calculus to problems in physics. The first
of three sections in volume 1 contains the integration tools needed before he
can continue onward to the solutions of differential equations, and the fifth
and sixth of the nine chapters in this initial section deal with trigonometry.

Chapter 5 is a comprehensive collection of basic antiderivatives involving trigonometric quantities, starting (as usual) with the inverse functions. The
first problem, for instance, deals with

$$\int f(x) \arcsin x \, dx$$

for various functions $f(x)$. For ordinary trigonometric functions, Euler begins
as one might expect by reversing basic results from his *Institutiones calculi
differentialis*, obtaining results such as

$$\int \frac{\cos n\phi}{\sin^2 n\phi} d\phi = -\frac{1}{n \sin n\phi} \tag{3.82}$$

(ignoring, as usual, the additive constant). Next he turns to integrals of powers of $\sin \phi$ and $\cos \phi$ and the products and ratios of these terms. Following
this "we consider fractional formulas, of which the denominator is $a + b \cos \phi$
and powers of this; for such formulas occur most frequently in theoretical
astronomy."[155] Indeed, we just saw these integrals arise in Euler's work on the
perturbations of the orbits of Saturn and Jupiter. He concludes the chapter by
combining trigonometric and exponential functions, considering

$$\int e^{\alpha\phi} \sin^n \phi \, d\phi \text{ and } \int e^{\alpha\phi} \cos^n \phi \, d\phi.$$

Finally, chapter 6 is an exploration of integrals of the form $\int (1 + n \cos \phi)^v \, d\phi$,
this time relying on the method of conversion to trigonometric series that we
saw in his work on planetary astronomy; see (3.77) and (3.78).

▪ Euler on Spherical Trigonometry

The brachistochrone problem, one of the most famous in the history of mathematics, poses the challenge to determine the curve joining two points (one
higher than the other) such that a bead rolls along it from one point to the

---

[154] See [Euler 1768 (E342)], [Euler 1769a (E366)], and [Euler 1770 (E385)]. All three volumes have
been translated into English in [Euler (Bruce) 2010 (E342, E366, E385)]. On the *Institutionum
calculi integralis*, see [Capobianco/Enea/Ferraro 2017], [Katz 2007, 228–231], and [Ferraro
2007b, 77–98].
[155] [Euler (Bruce) 2010, part 1 (E342), chapter 5, 18].

other most quickly under the influence of gravity. It attracted some of the greatest mathematical minds of the last decade of the seventeenth century, including Johann and Jakob Bernoulli and Isaac Newton. It is a classic example of a category of problems that are now grouped together under the title of the **calculus of variations**. This discipline is concerned with finding a function that maximizes or minimizes some functional, usually a definite integral involving the unknown function and its derivatives. In the case of the brachistrochrone, the function is the curve of quickest descent, which turns out to be a cycloid; the functional is the time of travel between the two points. Euler's 1744 book *Methodus inveniendi lineas curvas maximi minimive proprietate gaudentes*,[156] often considered to be the foundation of the calculus of variations, brought together various problems of this sort that had arisen in the previous half century and approached them in a unified way. There is hardly any trigonometry in this work, but the method he invented was soon to become a vital part of our story.

The fundamental tool employed by Euler is a version of what came to be known as the **Euler-Lagrange equation**. In Euler's language, if Z is a function of x, y, and p (where $p = dy/dx$) so that $dZ = M\, dx + N\, dy + P\, dp$, then the value of $\int Z\, dx$[157] reaches a maximum or minimum precisely when

$$N\, dx - dP = 0. \tag{3.83}$$

Euler's general technique, which he called his "method of maxima and minima," would be developed through correspondence with the young Italian-French analyst Joseph-Louis Lagrange. Beginning in 1755, Lagrange had found a way to replace some of the geometric methods Euler had relied on with analysis; eventually Euler would coin the improved theory the "calculus of variations."[158]

However, our attention to the calculus of variations is in the period before Euler's correspondence with Lagrange. One of the most obvious applications of the Euler-Lagrange equation is to find the shortest path—the **geodesic**—between two points on a surface. For instance, on a plane consider the arc length integral $\int \sqrt{1 + (y')^2}\, dx$. To find the curve of minimum distance we set $Z = \sqrt{1 + (y')^2}$; then $M = N = 0$ and $P = y'/\sqrt{1 + (y')^2}$. Substituting into the Euler-Lagrange equation, we have $dP/dx = 0$ so that P is a constant C; setting $P = C$ and solving, we find $y' = C/\sqrt{1 - C^2}$, which is also a constant. Therefore y is a straight line between the endpoints.

---

[156] [Euler 1744 (E65)]. The Euler-Lagrange formula in the next paragraph may be found on pp. 42–43.

[157] Although Euler does not write bounds, integrals are assumed to be definite.

[158] For a detailed history of the calculus of variations, see [Goldstine 1980].

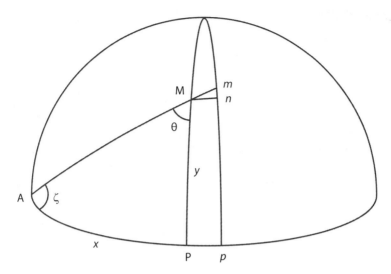

Figure 3.27
Euler's use of the calculus of variations to derive theorems in spherical
trigonometry.

But there is no need to restrict ourselves to operating only in two dimensions, and in 1755 Euler illustrated the power of his new method by converting the playing field from a plane to a sphere in his article "Principes de la
trigonométrie sphérique tirés de la méthode des plus grands et plus petits."[159]
He takes delight in the fact that his method has wider applications than he
had shown previously:

> For, once one has shown that most mechanical and physical prob-
> lems are resolved very quickly by means of this method, it will only
> be very pleasant to see that the same method brings such a great help
> to the resolution of problems in pure Geometry.[160]

We illustrate his method by summarizing part of the first of eleven problems, which along with eight corollaries uses the calculus of variations to
derive a surprising conclusion: the ten fundamental identities of a spherical
right triangle. In figure 3.27, let $\widehat{AP} = x$ and $\widehat{PM} = y$ be at right angles to
each other. The goal is to find the curve on the sphere of least distance s between A and M, which of course in the end will be the hypotenuse of the

---

[159] [Euler 1755c (E214)]; for an English translation see [Euler (Heine) 2014 (E214)], and in German see [Euler (Hammer) 1896 (E214), 1–39]. See also [Papadopoulos 2014, section 3].
[160] [Euler 1755c (E214), 225]; [Euler (Heine) 2014 (E214), 3].

triangle. Extend the curve an infinitesimal amount $ds = \widehat{Mm}$ beyond M, drop a perpendicular to $\widehat{AP}$ through m to p, and drop onto $\widehat{mp}$ another perpendicular $\widehat{Mn}$; then $\widehat{mn} = dy$ and $\widehat{Mn} = \widehat{Pp}\cos y = dx\cos y$. The quantity we wish to minimize is

$$s = \int \sqrt{dy^2 + dx^2 \cos^2 y} = \int \sqrt{p^2 + \cos^2 y} \; dx, \qquad (3.84)$$

where $p = dy/dx$.[161] The quantity inside the integral is Z, and its partial derivatives with respect to x, y, and p are M ($=0$ in this case), N, and P. Now, if we multiply the Euler-Lagrange equation (3.83) through by p, we get $N \, dy = p \, dP$; substituting into $dZ = M \, dx + N \, dy + P \, dp$ we have $dZ = p \, dP + P \, dp$, so $Z = Pp + C$. Substituting and doing some algebra gives us

$$\cos^2 y = C\sqrt{p^2 + \cos^2 y}; \qquad (3.85)$$

and solving for p,

$$\frac{dy}{dx} = p = \frac{\cos y \sqrt{\cos^2 y - C^2}}{C}. \qquad (3.86)$$

Then from (3.84), (3.85), and (3.86) in succession,

$$\widehat{Mm} = ds = dx\sqrt{p^2 + \cos^2 y} = \frac{dx \cos^2 y}{C} = \frac{dy \cos y}{\sqrt{\cos^2 y - C^2}}; \qquad (3.87)$$

and from (3.86),

$$\widehat{Mn} = dx \cos y = \frac{C \, dy}{\sqrt{\cos^2 y - C^2}}; \qquad (3.88)$$

so

$$\sin\theta = \frac{\widehat{Mn}}{\widehat{Mm}} = \frac{C}{\cos y} \qquad (3.89)$$

(where $\theta = \angle AMP$).

To find C Euler uses a clever trick. Imagine that $y = 0$; then we have effectively reduced AMP to an infinitesimal plane triangle, so M coincides with A and $\zeta = \angle MAP = \angle mMn$. Then $dx/ds = \cos \zeta$. But since $\cos y = \cos 0 = 1$, from (3.88) and (3.87) we also have

$$\frac{dx}{ds} = \frac{C \, dy / \sqrt{1 - C^2}}{dy / \sqrt{1 - C^2}} = C. \qquad (3.90)$$

---

[161] Both P and p represent both points in the diagram and quantities.

So $C = \cos \zeta$, and thus

$$\sin \theta = \frac{\cos \zeta}{\cos y}, \tag{3.91}$$

which is none other than Geber's Theorem. Euler does not take a long time to derive eight of the other nine identities of the right-angled spherical triangle in a similar way. He derives the tenth, $\sin x = \sin \theta \sin s$, algebraically from three of the other nine.

The paper goes on to derive further known theorems using the calculus of variations, including formulas for the area of a spherical triangle and the distance between two arbitrary points on the sphere. Euler's work on the latter leads immediately to the Law of Sines and the Law of Cosines. Other theorems include

$$\sin \frac{A}{2} = \sqrt{\frac{\sin \frac{1}{2}(a - b + c) \sin \frac{1}{2}(a + b - c)}{\sin b \sin c}} \tag{3.92}$$

and

$$\cos \frac{A}{2} = \sqrt{\frac{\sin \frac{1}{2}(b + c - a) \sin \frac{1}{2}(b + c + a)}{\sin b \sin c}} \tag{3.93}$$

(already found in Napier's 1614 *Descriptio*).

Of course, all these theorems were already known; Euler's goal is not to generate new mathematics but rather to illustrate the power of his new method. His analysis does introduce a couple of new results, including

$$\tan \frac{A}{2} = \sqrt{\frac{\sin \frac{1}{2}(a - b + c) \sin \frac{1}{2}(a + b - c)}{\sin \frac{1}{2}(b + c - a) \sin \frac{1}{2}(b + c + a)}} \tag{3.94}$$

and

$$\cos a = \frac{1}{4}\cos(A - b + c) + \frac{1}{4}\cos(A + b - c) - \frac{1}{4}\cos(A - b - c)$$
$$- \frac{1}{4}\cos(A + b + c) + \frac{1}{2}\cos(b - c) + \frac{1}{2}\cos(b + c). \tag{3.95}$$

Euler's new approach led the way for others to discover new theorems in spherical trigonometry seemingly unreachable using conventional methods. However, as in so many other areas, Euler's novel notation would make an impression almost as important as the theorems. For the first time, Euler named the angles of a triangle A, B, and C and the sides opposite these angles a, b, and c respectively.[162] This step, obvious once it has been taken, allowed

[162] [von Braunmühl 1900, 73]; [Chemla 1989, 334–335].

Euler to observe parallel structures in spherical trigonometry that had been seen only partially before. His notation supplies the organizing principle for another section of his paper, on solving triangles, as we see in his coverage of the related problems 6 and 7.

Problem 6 considers the case where the three sides of a given triangle are known and the three angles are sought. Euler begins with the Law of Cosines

$$\cos A = \frac{\cos a - \cos b \cos c}{\sin b \, \sin c}. \tag{3.96}$$

He works through a series of corollaries that include (among others) (3.92), (3.93), (3.94), and two of Napier's Analogies:

$$\tan \frac{A+B}{2} = \frac{\cos \frac{1}{2}(a-b)}{\tan \frac{1}{2} C \, \cos \frac{1}{2}(a+b)} \tag{3.97}$$

and

$$\tan \frac{A-B}{2} = \frac{\sin \frac{1}{2}(a-b)}{\tan \frac{1}{2} C \, \sin \frac{1}{2}(a+b)}. \tag{3.98}$$

In problem 7 Euler turns to triangles where the angles are known and the sides are sought. He begins with the Law of Cosines for angles

$$\cos a = \frac{\cos A + \cos B \, \cos C}{\sin B \, \sin C}. \tag{3.99}$$

Another series of corollaries follows, traveling along a path almost perfectly analogous to those that appear after problem 6. Among his results are

$$\sin \frac{a}{2} = \sqrt{-\frac{\cos \frac{1}{2}(A+B+C)\sin \frac{1}{2}(B+C-A)}{\sin B \, \sin C}} \tag{3.100}$$

(compare with (3.92)) and from Napier's Analogies

$$\tan \frac{a+b}{2} = \frac{\cos \frac{1}{2}(A-B)}{\cos \frac{1}{2}(A+B)} \tan \frac{1}{2} c \tag{3.101}$$

(compare with (3.97)) and

$$\tan \frac{a-b}{2} = \frac{\sin \frac{1}{2}(A-B)}{\sin \frac{1}{2}(A+B)} \tan \frac{1}{2} c \tag{3.102}$$

(compare with (3.98)).

The obvious similarities between the consequences of problems 6 and 7, with corresponding sides and angles changing places in the formulas, reflect the duality of sides and angles in spherical triangles. Usually the duality relations of spherical trigonometry are derived through the construction of the polar triangle, but not here; Euler simply follows two parallel logical progressions. His notation reveals the resemblances much more visibly than before. In fact, late in life Euler revisited spherical trigonometry in an attempt to bring out these relationships even more clearly. His 1782 "Trigonometria sphaerica universa"[163] does not rely on calculus, building up the basic theorems instead using geometry and then turning to algebra. His key observation: a person can take any formula, replace references to sides with the corresponding angles and vice versa, and change the signs of the cosine terms, and they will have a new correct formula. One may do this, for instance, to transform (3.96) into (3.99). From this he concludes:

> Theorem: Given any spherical triangle, the angles of which are A, B, C, and the sides of which are a, b, c, another analogous spherical triangle can always be exhibited, the angles of which are the supplements from two right angles of the sides of this one, whereas its sides are the supplements from two right angles of the angles of this one.[164]

So Euler travels backward from the usual development. Rather than working from the polar triangle to the dual formulas, he begins with the dual formulas and asserts the existence of the polar triangle.

Meanwhile, Euler's 1753 paper on his method of maxima and minima had another practical purpose. With only minor alterations, it can be made to work on the surface of a *spheroid*; that is, a sphere that has been stretched away from the plane of its equator (prolate) or compressed toward it (oblate). He says:

> From this, it is understood that this research could well become of great importance; for the surface of the Earth is not spherical, but spheroidal; a triangle formed on the surface of the Earth belongs to the sort of which I just mentioned. To see this, one only needs to imagine three points on the surface of the Earth which are joined by the shortest path which leads from one to the other, or formed by a cord stretched from one to the other; for it is thus that those triangles must be represented, which are used in the operations for the measure of the Earth.[165]

[163] [Euler 1782 (E524)]; for an English translation see [Euler (Bruce) 2013b (E524)] and in German see [Euler (Hammer) 1896 (E524), 40–54]. See [Chemla 1989] and [Chemla/Pahaut 1988] for expositions and comparisons of Euler's two papers, and for the history of the subsequent development of the notion of duality. See also [Chemla 2004] and [Papadopoulos 2014, section 3].

[164] [Euler 1782 (E524), 78]; translation from [Chemla 1989, 341].

[165] [Euler 1755c (E214), 224–225]; [Euler (Heine) 2014 (E214), 2].

It had been a matter of debate up to the 1740s whether the earth is oblate or prolate, but not whether it is a spheroid. For his purposes, Euler does not need to decide the matter in advance. In a follow-up paper immediately after this one, he derives formulas for lengths of arcs on spheroids, again from the calculus of variations.[166] He goes on to analyze data that had been conducted in four surveys of the length of a degree at various latitudes. He concludes that the ratio of the earth's major axis (across the equator) to its minor axis (from pole to pole) is close to 230:229, the value that had been inferred by Isaac Newton in his argument in favor of the oblate spheroid. (The correct value is about 299:298.)[167]

Euler continued to work on spherical trigonometry to the end of his life, sometimes with the aid of calculus, sometimes without.[168] In any case, he certainly can be named as the person who completed the project of bringing trigonometry into analysis. And yet, he continued to work as a geometer. Sometimes, revolutionaries do not have to abandon the old ways entirely.

[166] [Euler 1755a (E215)]; for an English translation see [Euler (Heine) 2015 (E215)]. See also [Heine 2013], [Todhunter 1873, vol. 1, 353], and [Papadopoulos 2014, section 3].

[167] Spheroidal trigonometry would gain a systematic treatment in [Grunert 1833]; for trigonometry on arbitrary surfaces, the fundamental work is [Gauss 1828].

[168] See for instance his papers [Euler 1781 (E514)] and especially [Euler 1797 (E698)] (an English translation by Johan Sten is available at www.17centurymaths.com) on the areas of spherical triangles where he alternates between applying the methods of calculus and "geometry," the latter enhanced considerably by algebraic manipulation. In [Euler 1781 (E514)] he also defines the angles at vertices of polyhedra, which led to what is now called the "polar sine" of a solid angle.

4 ⚞ China

Before the arrival of Jesuit missionaries in the late sixteenth century, Chinese mathematics engaged in only limited interactions with other cultures.[1] Thus early China is a useful case study to ask interesting questions such as: how can mathematics be approached differently from the tradition inspired by Euclid? Are our boundaries among disciplines (both within and outside of mathematics) inevitable, or are there other ways of marking out the intellectual terrain? Is it necessary that a sufficiently developed culture will establish something equivalent to what we call "trigonometry"? These questions will arise as themes in the first half of this chapter.

Textual evidence for mathematics in China begins relatively late, in the second century BC,[2] after Euclid and most of the mathematical contributions of ancient Egypt and Babylon had already passed. There was mathematics—probably a lot of it—before this time, but most direct evidence has been lost. Documents tend to be mixtures of mathematical problem solving with various practical concerns of the time: taxation, surveying, and calendar reckoning, to name a few. Certainly, we see nothing of the barrier between geometry and the physical universe that the first ancient Greek trigonometers had had to pierce. This makes it all the more interesting that, even though astronomy was valued highly by the state, trigonometry as a discipline did not surface in these years. Nevertheless, there is a great deal of originality in ancient Chinese mathematics that might be interpreted as something related to trigonometric activity. We shall report on these episodes and allow the reader to decide how to place them in context.

▨ Indian and Islamic Trigonometry in China

From the earliest times interest in astronomy in China was related to state requirements, especially tracking the motions of the heavens for the purpose of calendar design. Astronomers often were essentially government

---

[1] For references to the extensive literature on Chinese trigonometry in Chinese languages, see [Chen 2010] and [Chen 2015]. Bibliographic articles on Chinese mathematics include [Swetz/Ang 1984], a supplement [Youschkevitch 1986], and [Eberhard-Bréard/Dauben/Xu 2003].

[2] The *Suanshu shu*, dating from the early second century BC, may be the oldest Chinese mathematical manuscript. Discovered in 1984 in the tomb of a Chinese civil servant, the treatise contains a series of problems dealing with practical problems that would have been useful for the owner's career. See [Cullen 2004] and [Dauben 2008] for English translations with commentaries.

officials as their work on creating and maintaining calendars had important political implications. There is early evidence of substantial astronomical work; records of three different cosmographical systems of the heavens (*zhou bi*, *xuan ye*, and *hun tian*), with differing aims and methods, have been around as soon as the second century AD.[3] As in other early cultures there was no clear distinction between the mathematicians and the astronomers; both were covered by the single term *chouren*, meaning "surveyors."[4] The beginning of the Chinese practice of surveying the heavens is mostly outside our scope. However, a relatively early episode of transmission of knowledge brought trigonometry temporarily to China before disappearing again for centuries.

Buddhism began to spread in China starting around the first century AD, and along with it came various aspects of Indian learning. At first the astronomical knowledge that found its way from India to China was somewhat primitive. But eventually, as Indian astronomy grew in sophistication and power, their achievements became known in China as well.[5] Early in the Tang dynasty (the seventh and early eighth centuries), three Indian schools served at the Chinese national astronomical bureau. One of the astronomers working there, Qutan Xita (or Gautama Siddhārtha), was ordered to translate an Indian astronomical treatise. The result, the *Jiuzhi li* ("Nine Planets System"), is preserved today as the 104th volume of the 120-volume work *Da tang kaiyuan suanjing* ("Classic of Astrology from the Kaiyuan Reign of the Tang Period"), which—due to its astrological content—was kept secret for almost a millennium after its composition.[6] The Indian sources of the *Jiuzhi li* are various; it relies heavily on the *Pañcasiddhāntikā* but may also have been influenced by Brahmagupta's *Khaṇḍakhādyaka*. The *Jiuzhi li* is filled with instructions for computing various astronomical quantities related to predictions of the motions of celestial bodies, with an emphasis on predicting eclipses. Within the section on eclipses we find a typically Indian table of sines calculated with a base circle radius of $R = 3,438$ for arguments increasing in steps of $3°45'$ (see figure 4.1). Buried in the middle of calculations dealing with the moon's position, this sine table does not seem to have been used or even noticed by anyone. The *Jiuzhi li*'s subsequent disappearance until the early seventeenth century assured that nothing further would come of this transmission.

---

[3] See [Cullen 1996, 37–66].

[4] [Martzloff 1997], referring to a Japanese publication by Mikami.

[5] See [Gupta 2011, 34–37] for a summary of Indian astronomical works in China.

[6] An English translation of the *Jiuzhi li* is available in [Yabuuti 1979]; the sine table may be found on pp. 35–36. See also [Yabuuti 1963].

Argument	Sine
3°45′	225
7°30′	449
11°15′	671
⋮	⋮
45°	2431
⋮	⋮
90°	3438

Figure 4.1
The sine table in the *Jiuzhi li*.

One dramatic exception to the renewed absence of trigonometry may have occurred several years later. Around AD 724, the astronomical bureau began an intensive observational program. One of its activities was the measurement of lengths of shadows cast by gnomons at the equinoxes and solstices at ten stations whose terrestrial latitudes varied between 29°N and 52°N.[7] At the survey's conclusion, the Buddhist monk and astronomer Yixing analyzed the observations. His *Dayan li*, a new and complete astronomical system, appeared posthumously by the end of the decade.[8] In it Yixing attempted to establish procedures that would produce correct astronomical results whatever the observer's location, not just at the observatory in Yang cheng. Among these procedures was a table for finding the length of the shadow cast by a gnomon of length eight *chi* in terms of the angular displacement of the sun from the zenith.

We have seen similar tables before; they can be found as far back as Ptolemy's *Almagest*.[9] However, the appearance it takes on in Yixing's text is unique and startling. In figure 4.2, the sun's distance from the zenith is φ; the shadow length s is given in units of 1/10,000 *chi*. Clearly the function that Yixing tabulates is mathematically equivalent to $s = 80{,}000 \tan \varphi$.

[7] This survey has been studied in [Beer et al. 1961].

[8] An analysis of some of Yixing's astronomical structures in the *Dayan li* leads [Shah 2012b] to reject a connection with Indian astronomy, and a further study in [Shah 2012a] argues that the geometric models found in Greek, Indian, and Islamic astronomy had almost no impact on the *Dayan li*.

[9] [Van Brummelen 2009, 77–80].

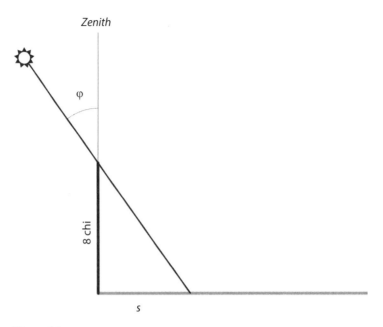

Figure 4.2
The shadow length of a gnomon.

Text 4.1
Yixing, Description of a Table of Gnomon Shadow Lengths
(from the *Dayan li*)

In the south, at the subsolar point, precisely at noon there is no shadow.
    For the first *du* of zenith distance take the initial value 1379.
    Begin the differences here, increasing by one for each *du*.
    End at 25 *du*, reckoning an addition of 26 *fen*.[10]
    Add a further 2 for each *du*.
    End at 40 *du*.
    Add a further 6 for each *du*.
    End at 44 *du*, adding 68.
    (The passage continues in a similar fashion.)[11]

Explanation: The *du*, the unit of angular measurement, is close to but not exactly equal to 1°; it corresponds to the distance that the sun travels along the ecliptic in one day. Often the complete circle was taken to be 365¼ *du*; in this work, the circle is $365 + 779.75/3040$ *du*.

---

[10] One *chi* = 100 *fen*.
[11] The translation is from [Cullen 1982, 4]. See also the reconstruction in [Qu 1997], which questions whether it should be called a table of "tangents," and [Duan/Li 2011].

0	0			
		1379		
1	1379 [+3]		1	
		1380		1
2	2759 [+6]		2	
		1382		1
3	4141 [+9]		3	
		1385		1
4	5526 [+13]		4	
		1389		1
5	6915 [+17]		5	
⋮	⋮	⋮	⋮	⋮
		1655		1
24	35396 [+354]		24	
		1679		2
25	37075 [+381]		26	
		1705		2
26	38780 [+406]		28	
		1733		2
27	40513 [+432]		30	
⋮	⋮	⋮	⋮	⋮
		2279		2
40	66500 [+730]		56	
		2335		6
41	68835 [+726]		62	
		2397		6
42	71232 [+714]		68	
		2465		6
⋮	⋮	⋮	⋮	⋮
79	298858 [−73139]			

Figure 4.3
Yixing's table of gnomon shadow lengths.

This obscure text is a compact representation of a numerical table, remarkably using third differences (and, elsewhere, fifth differences). Although second differences (the successive differences of the successive differences of the table's entries) were common in India and elsewhere, it is extremely rare to see an application of third differences—not only to represent the entries in the table but also actually to construct them. The text has been interpreted as follows (see figure 4.3[12]). We are told that the shadow length for one *du* is 1,379.

[12] The numbers in square brackets indicate the error in the entry in units of 1/10,000th *chi*. For instance, the entry at five *du* is given as 6915 [+17], so the correct value is 6,898.

The "differences" that begin with and are increased by one are the second differences of the eventual shadow length values (so that the third differences are all equal to one), starting at the beginning of the table and working up to the entry for 25 *du*. One fills in the entries from right to left. For instance, the initial first difference is 1,379; the initial second difference is one. Therefore, the next first difference must be 1,380, and so the shadow length for 2 *du* is 1,379 + 1,380 = 2,759. We can continue in this fashion until we have reached the entry for 25 *du*. At this point Yixing instructs us to change the third differences from one to two. We proceed until we reach the entry for 40 *du*. Now the third differences increase to six, and we continue until we reach the entry for 44 *du*. The table builds in this way up to the entry for 79 *du*, corresponding to just under 78°.

Just how Yixing arrived at this formulation in terms of third differences is not clear. The obvious hypothesis would be that he derived it from the sine table in the *Jiuzhi li*, interpolating within that table as needed. But such a table does not generate third differences with the pattern prescribed by Yixing: they generally change continuously, whereas Yixing requires abrupt changes at several places. Also, had a sine table been used, one would have expected small random fluctuations in the third differences due to rounding and other calculation errors, and there are none. Possibly Yixing calculated the table directly somehow, observed a pattern in the third differences, and then simplified the pattern to give a simple and memorable set of instructions for generating the table. This would lead to an error pattern like the one we find in the table.[13] The errors are rather large and would be worrisome if the table had been used as a foundation for trigonometric calculation. But as a collection of shadow lengths, the errors are small enough for the values to be useful. Indeed, even the very large errors in the entries at the end of the table are not meaningful in practice since these entries correspond to latitudes far to the north of where the table was to be used. In Yang cheng, the sun's noon zenith distance never exceeds a bit more than 58 *du*.

An important caveat: although we have been referring to the *du* as a form of angular measure, it has been pointed out that this may not be entirely accurate historically. The *du* never represented the inclination between two lines; it was used only as a unit of distance in the heavens.[14] We must take caution not to infer notions that may not be present in the historical texts.

The above episodes were not the only incidental contacts between Chinese astronomy and a foreign science infused with trigonometry. As early as

---

[13] [Cullen 1982, 26–30].
[14] [Cullen 1996, 91–92].

the tenth century, occasional connections were made also with the astronomy of medieval Islam.[15] In the late thirteenth century, Khublai Khan, the first emperor of the Yuan dynasty (AD 1271–1368), founded the Islamic Astronomical Bureau. It is unclear precisely what role this institute played, especially in its early years. However, about a century later we find a Chinese translation of an Islamic astronomical handbook (a *zīj*) entitled *Huihui li* ("Islamic Astronomical System"). Although this work is not identical to any extant *zījes*, it is closely related to the *Sanjufīnī Zīj*, a fourteenth-century Arabic work composed in Tibet.[16] The *Huihui li* is filled with mathematical astronomy based on trigonometric methods. However, no Chinese successors to the *Huihui li* have as yet been discovered. Generally, it appears that the practitioners of Chinese and Islamic astronomy seldom interacted; although they rubbed shoulders, they rarely rubbed off on each other.

▨ Indigenous Chinese Geometry

The episodes of transmission we have seen so far were the exception rather than the rule; in general, mathematics in premodern China remained independent. Even when new ideas arrived from foreign lands, differences in approach and in purpose rendered mutual interactions difficult. For instance, the geometry in Euclid's *Elements* is isolated from other disciplines or practical concerns. The goal is mathematical rather than physical: not to determine facts for use in practice but to explore and classify theorems as they develop from the axioms. Early Chinese geometry, on the other hand, never separated from practice in this manner. It contains problems that relate directly to state and agricultural concerns. In fact, there are not many passages that we might label as "theorems": when some mathematical situation arises for a second time in a new context, often the text deals with it as if for the first time. It is true that the practical scenarios in the texts were often artificial, possibly contrived to provide a space for the mathematics to grow. Nevertheless, the intimate connection between the physical world and the geometry was rarely broken.[17] In astronomy, much the same was true: the state employed astronomers primarily for the practical business of calendar reck-

---

[15] See [Martzloff 1997, 101–105].

[16] On the *Huihui li* in English see especially [Yabuuti 1997], [van Dalen/Yano 1998], [van Dalen 2002a], and [Shi 2014]. On the tables of planetary latitudes see [Yano 1999] and [van Dalen 1999]; for an analysis of the star table, see [van Dalen 2000]; for an analysis of the astronomical data, see [Chen 1987].

[17] For a survey of differences between Chinese and Western geometry, see [Martzloff 1997, 273–277].

oning. The geometrical models for the motions of the planets that had given birth to trigonometry in Greece and India did not exist in China; there was no need for them.

Chinese geometry is exemplified in the *Jiuzhang suanshu* ("Nine Chapters of the Mathematical Art"), a work that served as a standard in China much like Euclid's *Elements* in the West. Its character, however, could not be more different. The chapters deal with issues like mensuration of fields, commodity exchange (millet and rice), taxation, and surveying. Known mostly through an extensive commentary by Liu Hui in the third century AD, the *Jiuzhang suanshu* deals with problems related to geometry in the first book (areas, including Liu Hui's approximation of π), the fifth (volumes of solids), and the ninth (problems related to the Pythagorean Theorem). The last chapter comes closest to trigonometry, employing right-angled triangles to solve surveying problems.[18]

Liu Hui's commentary, however, went considerably further than the *Jiuzhang suanshu*'s original text. He added another set of nine sophisticated problems on surveying to the end of the ninth chapter, which were eventually separated from their parent work and given their own title, the *Haidao suanjing* ("Sea Island Computational Canon").[19] The title derives from the first of the nine problems, where Liu Hui demonstrates how to find the height of and distance to an inaccessible island using some simple observations. The other problems deal with the height of a distant tree, the distance to a walled city, the depth of a pool, and so on. Each problem requires the determination of some length that is inaccessible to direct measurement, and each problem uses a technique known in Liu Hui's time as *chong cha*, or "double difference."

Text 4.2
Liu Hui, Finding the Dimensions of an Inaccessible Walled City
(from the *Sea Island Mathematical Manual*)

Now, looking southward at a square [walled] city of unknown size, erect two poles 6 *zhang* [600 *cun*] apart in the east-west direction such that they are standing at eye level and are joined by a string. Assume that the eastern pole

---

[18] For a translation of the *Jiuzhang suanshu* into English, see [Shen/Crossley/Lun 2000]; see also [Dauben 2007, 228–288]. For a French translation, see [Chemla/Guo 2005]. On right-angled triangles in ancient Chinese mathematics in addition to other sources documented in this chapter, see [Lam/Shen 1984] and [Dauben 2007, 213–226].

[19] Translations of the *Haidao suanjing* may be found in various places. In English, see especially [Ang/Swetz 1986] and [Swetz 1992]. See also [Dauben 2007, 288–292].

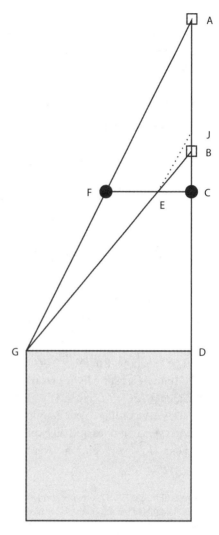

Figure 4.4
Liu Hui's determination of the
dimensions of a walled city. The
solid circles are sighting posts, and
measurements are taken at the points
indicated by empty squares.

is aligned with the southeastern and northeastern corners of the city. Move
northward 5 *bu* [300 *cun*] from the eastern pole and sight on the northwestern
corner of the city; the line of observation intersects the string at a point 2 *zhang*
2 *chi* 6½ *cun* [226.5 *cun*] from its eastern end. Once again, move backward
northward 13 *bu* 2 *chi* [800 *cun*] from the pole and sight on the northwestern
corner of the city; the corner coincides with the western pole. What is the
[length of the] side of the square city . . . ?

   The side of the squared city measures 3 *li* 43 ¾ *bu* [56,625 *cun*]. . . .

   Multiply the final distance from the pole by the observation measurement
obtained on the string and divide [the product] by the distance between poles.
What is thus obtained is the shadow difference. Subtracting the initial distance

from the pole [from the shadow difference], the remainder is the *fa*. Place the final distance from the pole [on the counting board] and subtract the initial distance from the pole. The remainder is multiplied by the observation measurement along the string, giving the *shi*. Dividing the *shi* by the *fa* yields the [length of the] side of the squared city.[20]

Explanation 1: In figure 4.4, we place poles attached by a string at C and F, separated by 600 *cun*.[21] We travel northward from C a distance of 300 *cun* to B and then sight the northwest corner of the city, obtaining $EC = 226.5$ *cun*. We then proceed further north until we sight the westernmost pole to coincide with the northwest corner of the city, arriving at A; we measure $AC = 800$ *cun*. Liu Hui does not indicate the reasoning behind his calculations, but a possible path is as follows. Let EJ be parallel to AF. Liu Hui first calculates the "shadow difference" $JC = AC \cdot EC/FC = 302$ *cun* (from similar triangles $\triangle ACF$ and $\triangle JCE$). Then $JB = JC - BC = 2$ *cun*, the *fa* (divisor). Next, calculate $AB \cdot EC = (AC—BC) \cdot EC = 113,250$ *cun*[2], the *shi* (dividend). Dividing the *shi* by the *fa* gives $GD = 56,625$ *cun*. To see why this is true we need two pairs of similar triangles: from $\triangle ABG$ and $\triangle JBE$ we have $BG/BE = AB/JB$; and from $\triangle BDG$ and $\triangle BCE$ we have $BG/BE = GD/EC$. Combining these two ratio equalities gives Liu Hui's result.[22]

The "shadow difference" JC might be thought of as a sort of a shadow of EC onto AC cast by light rays parallel to FA. In fact, the formula for JC is equivalent to $JC = EC \tan \angle AFC$. Does this line of reasoning constitute trigonometry? According to our definition (calculation of lengths from arcs/angles and vice versa), it does not; no explicit notion of the measure of an angle is to be found. However, opinions over the decades have varied.[23]

---

[20] [Swetz 1992, 21–22]. This text is abridged; the full problem also calls for the distance to the city. The units of measurement, in terms of the smallest unit, are as follows: 1 *li* = 18000 *cun*, 1 *zhang* = 100 *cun*, 1 *bu* = 60 *cun*, and 1 *chi* = 10 *cun*. For ease of use, all units in the explanations are translated to *cun*.

[21] The use of letters to represent vertices is not typical of Chinese geometry and is included to aid the reader.

[22] This interpretation is taken from [Ang/Swetz 1986, 109–111].

[23] Alexander Wylie took a liberal view of the term, ascribing the title of "trigonometry" to various problems involving geometry and calculation [Wylie 1897, 164–165] and "practical trigonometry" in this context [Wylie 1901, 114], but already [van Hee 1920] reacted to this terminology with caution. [Needham 1959, 109] states that ratios of sides in right-angled triangles were used but notes that the concept of angle was not present. [Libbrecht 1973, 122–123] agrees with Needham and encloses the word "trigonometry" in the title of his chapter on the subject in double quotation marks. [Ang/Swetz 1986, 111] refers to the use of ratios that might be interpreted as tangents as "prototrigonometry." One begins to wonder whether, in this context, these labels are at all useful.

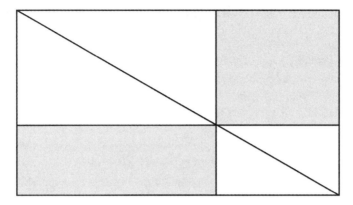

Figure 4.5
The out-in complementarity principle at work. The two shaded
rectangles are equal in area.

But is this really what is going on? Another interpretation that conforms
better to the rest of ancient Chinese mathematics has been posed[24] and has
been favored by the scholarly community. It relies on the "out-in complementarity principle," which in this case corresponds to the following statement
(see figure 4.5): if a rectangle is bisected by a diagonal and from some point
on the diagonal perpendiculars are drawn to the sides, then the areas of the
two rectangles not cut by the diagonal are equal. Adding the bottom right rectangle to both shaded areas, we may also conclude that the area below the
horizontal cross section is equal to the area to the right of the vertical cross
section.[25] It turns out that the "out-in" principle can replace the use of similar
triangles in geometric arguments.

> **Explanation 2:** In figure 4.6, the city, poles, and sighting locations are as be-
> fore; the various horizontal and vertical lines are defined as one would expect.
> By the "out-in" principle $\Box CX = \Box KZ$, so $EC \cdot CA = FC \cdot KA$, which gives us a
> newly defined "shadow difference" $KA = EC \cdot CA/FC$.
> Now, define M so that $CB = JM$. Then
>
> $$\Box CJ = \Box DJ - \Box DE = \Box JV - \Box EU = \Box JV - \Box JT = \Box MV.$$

[24] This hypothesis may be found in [Lam/Shen 1986], perhaps inspired by work in Chinese by
Wu Wenjun, and is adopted in [Swetz 1992]. Lam and Shen (p. 18) conclude that "if [these
methods] are accepted as the methods of derivation, then any suggestions that the *Haidao suanjing* is associated with trigonometry are without foundation."
[25] For a summary of the "out-in" technique see [Dauben 2007, 199–201].

Therefore $EC \cdot CK = GS \cdot MX = GS(KA-BC)$, and this gives a modern equivalent to Liu Hui's calculation as follows:

$$GD = GS + SD = \frac{EC \cdot CK}{KA - BC} + EC = \frac{EC(AC - BC)}{KA - BC}.\,^{26}$$

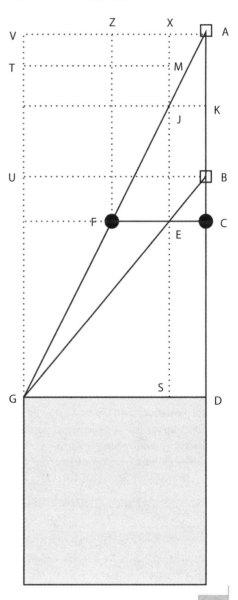

Figure 4.6
Another interpretation of Liu Hui's determination of the dimensions of a walled city.

---

[26] This interpretation is inspired by [Swetz 1992, 42–46].

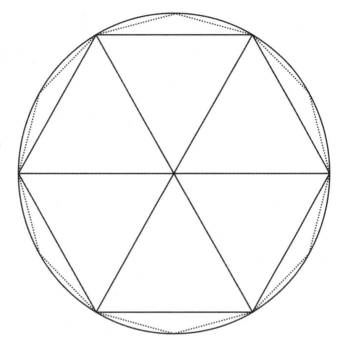

Figure 4.7
Determining π
by inscribing
regular polygons
within a circle.
Here we begin
with a hexagon;
doubling the
number of sides
gives a
dodecagon.

This understanding of Liu Hui's calculation does not use methods readable as trigonometry in any obvious way. Thus, we cannot assert the existence of trigonometry, under any definition, in this text.

Another example that places us in a gray area between geometry and trigonometry is the method of *geyuan* or circle division. Found in Liu Hui's commentary to the *Jiuzhang suanshu*, the method approximates a circle with a collection of thinner and thinner triangles. In figure 4.7, Liu Hui begins by inscribing a regular hexagon in a circle. He then doubles the sides, producing a regular dodecagon. Given the side length of the hexagon, he computes the area of the dodecagon. Liu Hui repeats the pattern: he doubles the sides again, and from the side length of the dodecagon he finds the area of the regular 24-gon. He is not explicit about how he finds the side lengths, but his method has been reconstructed.[27] Continuing in this way up to the area of a 192-gon and comparing the area with the square on the diameter, he obtains what we would call an approximation to π.[28]

[27] See [Shen/Crossley/Lun 2000, 87–117] and [Chemla/Guo 2005, 177–189] for English and French translations (respectively) and commentary. See also the reconstructions of the argument in [Li/Du 1987, 65–68] and [Martzloff 1997, 278–280].

[28] Approximating his fractional results, he concludes that the ratio of the circle to its circumscribed square is 314/400; that is, $\pi \approx 3.14$. This method is reminiscent of Archimedes's approximation of π but works with the areas of the regular polygons rather than their perimeters.

This may be argued to be a trigonometric calculation because the side lengths of the regular polygons relate directly to various sine values: the side of the hexagon is $2R \sin 30°$ (where R is the radius of the circle), the side of the dodecagon is $2R \sin 15°$, the side of the 24-gon is $2R \sin 7\frac{1}{2}°$, and so on. We find a similar calculation in the early fourteenth century in the work of Zhao Youqin, who begins with a square inscribed in the circle, repeatedly doubling the sides until he reaches the length of an inscribed 16,384-gon. This corresponds to calculating $2R \sin 45°$, $2R \sin 22\frac{1}{2}°$, and so on, all the way down to $2R \sin \frac{45°}{4096}$.[29] These calculations are similar to other approximations of π in various cultures, for instance, Archimedes, al-Kāshī, and Viète. Whether any of these calculations represent a form of trigonometry could be argued on both sides, but we are inclined to respond in the negative; the texts contain no systematic trigonometric intent. In any case, the question itself may be wrong-headed: it implies a disciplinary approach that does not match the character of the techniques being employed.

■ Indigenous Chinese Trigonometry

One might then wonder, given how many ways that a geometric prescription might be understood, whether one can find anything in indigenous Chinese geometry that can be clearly classified as trigonometry. The answer to this question depends to some extent on the reader's willingness to make analogies among mathematical activities in different contexts. Is the notion of "trigonometry" too rooted in the related cultures of Greece, India, Islam, and the West to have any relevance in China? These days, historians might be tempted to answer in the positive, but there is a remarkable episode in the astronomical work of Guo Shoujing that might pose a challenge.[30]

Guo came to prominence in the mid-thirteenth century as a hydraulic engineer, eventually working under Kublai Khan at the beginning of the Yuan dynasty. His attention was drawn to astronomy when he was asked to work on a reform of the existing calendar. This particular revival of interest in the heavens came along with the importation of some Islamic mathematical astronomy.[31] However, the Islamic influence was well contained and did not affect significantly those people who were working within the indigenous tradition, one of whom was Guo. His contribution to trigonometry may have

---

[29] [Chen 2010, 70], quoting a Chinese-language paper by Liu Dun.

[30] Guo Shoujing was part of a calendar reform project that involved several astronomers; the text that would have contained his derivations, the *Shoushi li cao* ("Workings of the Season-Granting Calendar"), is lost. It has been argued (see, for instance, [Sivin 2009, 152–153]) that Wang Xun may also be responsible for some of the mathematics.

[31] For surveys of the Islamic incursion of astronomy into China during the Yuan and Ming dynasties, see [Yabuuti 1987], [Yabuuti 1997], [van Dalen 2002b], and [Shi 2014].

been inspired tangentially by what he saw on the Islamic side of the astronomical fence. But his work is clearly within the Chinese context and does not require the presence of Islamic influence.

Guo's interest in the construction of astronomical instruments was part of the motive for his study of astronomy.[32] Although his writings are lost, his work has been reconstructed from various historical sources that had relied on them, and what we find is impressive. Without any preexisting trigonometric apparatus Guo seems to have been able to compute an array of astronomical quantities, including the conversion of ecliptic coordinates into equatorial (the declination and right ascension). Now, it is not entirely clear that what is reported of Guo's work is entirely his own or, rather, interpretations of commentators who brought their own trigonometric understandings to the text. It may not even be the case that Guo thought of the heavens as a spherical dome. Thus we must proceed with the caution that what follows might, or might not, be a fair reflection of Guo himself.

If indeed Guo is represented fairly, then we find an astronomer who needed to start from scratch. This meant that he had to invent a primitive trigonometric function, demonstrate a means of computing it, and use it to compile tables of the required astronomical functions. Rather than the chord or the sine, Guo settled on what we would call the ***versed sine*** as the foundation of his calculations.

In figure 4.8, Guo's goal is to compute the length of CM, given the length of $\widehat{AB} = 2\widehat{AM}$ and the circle's diameter d. Now, as we saw earlier in this chapter, arcs on the circle are measured not in degrees but in terms of the average distance traveled by the sun on the ecliptic in a day so that 365.25 of these *du* correspond to 360°. Thus, a table of the versed sine consists of values for every *du* from zero to 91, plus a value for 91.31 *du* corresponding to a 90° arc. These arcs are treated as lengths, and just as we saw in Indian trigonometry, the diameter is measured in the same units as the arcs so that $d = C / \pi = 365.25 / \pi$. Guo chooses the seemingly crude value $\pi \approx 3$ (a point to which we shall return), so $d = 121.75$ units.

To find the versed sine $v = CM$, Guo first finds two different expressions for the chord length $c = ACB$. Applying the Pythagorean Theorem to $\triangle AOC$ gives

$$\left(\frac{c}{2}\right)^2 = AO^2 - OC^2 = \left(\frac{d}{2}\right)^2 - \left(\frac{d}{2} - v\right)^2 = dv - v^2. \qquad (4.1)$$

The other expression for c was already available in the *Mengxi bitan* by Shen Kuo (AD 1086), a widely known work on a variety of topics in mathematics,

---

[32] The following description of Guo Shoujing's calculations is based largely on [Gauchet 1917] and [Martzloff 1997, 325–335]. See also [Sivin 2009] for a treatment of the astronomical reform of which Guo was a part.

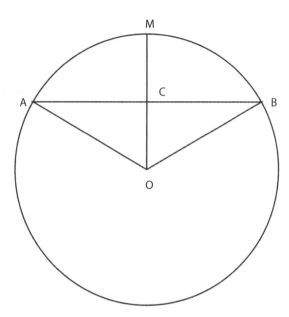

Figure 4.8
Guo Shoujing's versed
sine.

science, and music. In a sense it approximates an inverse trigonometric process, that is, finding the length of an arc from lengths of lines in a circle. In figure 4.8, Shen had asserted that

$$\theta = \widehat{AB} = c + \frac{2v^2}{d}. \tag{4.2}$$

Let $\phi = \widehat{AM} = \theta / 2$. Eliminating the common term c in (4.1) and (4.2) leaves Guo (after some algebra) with an equivalent to a quartic equation for v in terms of the arc φ and the diameter d:

$$v^4 + (d^2 - 2\phi d)v^2 - d^3 v + \phi^2 d^2 = 0. \tag{4.3}$$

So, the problem of finding the versed sine is reduced to that of solving a quartic equation, which does not seem like much progress. However, Chinese geometers and surveyors were accustomed to approximation, and they reduced many problems to polynomial equations. These were then solved using a numerical process equivalent to what we call today "Horner's method," which allows the calculation of as many digits of the unknown quantity v as the calculator has perseverance.[33]

It is natural to ask, but a bit tricky to answer, how accurate Guo's method for computing the versed sine is. The process relies on the approximation (4.2)

---

[33] The method works essentially as follows: estimate the solution v to a certain number of digits; call the estimate v_0. Substitute $v = v_0 + y$ into the polynomial, reducing the problem to the solution of a polynomial in terms of the remainder y. Repeat as desired.

φ	Guo's versed sine	Correct versed sine
10	0.826	0.899
20	3.347	3.568
30	7.637	7.929
40	13.689	13.852
50	21.315	21.165
60	30.133	29.649
70	39.682	39.055
80	49.572	49.105
90	59.563	59.503

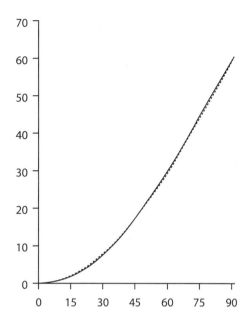

Figure 4.9
Guo Shoujing's calculation of the versed sine. The solid curve in the graph represents Guo's calculations; the dashed curve represents the correct values.

(which itself relies on an underlying approximation); the value of d is determined using a crude estimate of π, and the solution of the polynomial is found by an approximate method. So, to some extent the accuracy of the method depends on the diligence of the computer. If we assume that this diligence is effectively infinite (therefore assuming a precise solution of the quartic equation (4.3)) and we assume $d = 121.75$ (which comes from $\pi = 3$), we arrive at the values illustrated in figure 4.9. As one can see from the graph, the fit with the correct versed sines is surprisingly good.

One might wonder how quantities like declinations of points on the ecliptic may be found using nothing more than a versed sine. A plausible reconstruction works as follows. In figure 4.10 the sun is at C and the summer solstice is at D, both on the ecliptic (so that CD is the complement of the sun's longitude λ). AB is the equator, so $\widehat{DB} = \epsilon \approx 24°$, the obliquity of the ecliptic, and the desired declination is $\delta = \widehat{AC}$. Various perpendiculars are drawn in the figure as shown. On plane OBD, RB is the versed sine of ε, giving us OR (equivalent to the cosine of ε) and DR (equivalent to the sine of ε).[34] Likewise, on

---

[34] DR^2 is equal to the product of RB and $r + OR$. (see *Elements* VI.13), where r is the circle's radius.

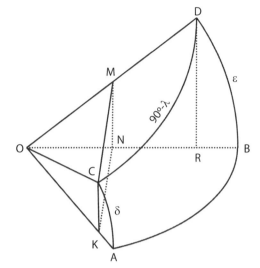

Figure 4.10
Guo Shoujing and the declination
of an arc of the ecliptic.

plane OCD, MD is the versed sine of $90° - \lambda$, giving us OM and CM. Triangles OMN and ODR are similar, and from this we can find ON and MN. Then $OK = \sqrt{ON^2 + KN^2} = \sqrt{ON^2 + CM^2}$, from which we obtain AK, the versed sine of δ. To get δ itself, we turn back to Shen's approximation (4.2) of the length of an arc, obtaining

$$\delta = \widehat{AC} \approx CK + \frac{AC^2}{d} = MN + \frac{AC^2}{d}. \tag{4.4}$$

One curiosity of Guo's determination of the declination is a mystery. Since the calculation depends on one's choice of the value of π, and the value chosen here is 3, one might expect significant errors. However, the errors interact in such a way that the use of $\pi = 3$ rather than the correct value actually *decreases* the maximum possible error by an order of magnitude.[35] It seems unlikely that this is simply a coincidence, although the thought process behind it remains obscure.

■ The Jesuits Arrive

Up to this time Chinese scientific culture had remained mostly separate from outside influences. This changed dramatically with the arrival of Jesuit missions, beginning in the late sixteenth century. The first settlement in Macau had little integrative effect; the European immigrants spoke only

---

[35] Toshio Sugimoto first drew attention to this fact in two Japanese-language publications in 1987; see [Martzloff 1997, 334–335] for a summary.

Portuguese. However, in 1579 Michele Ruggieri was brought in to help break the language barrier, and three years later he was joined by Matteo Ricci (1552–1610). Ricci's approach to mission work was to have a permanent effect. He attempted to forge cross-cultural connections by sharing his knowledge of European science and technology with the educated and by collaborating with leading Chinese intellectuals. In this way, substantial portions of European science (although by no means all of it), including mathematics, astronomy, mechanics, architecture, and geography, began to enter China. Over time, the missionaries began to play significant roles in government and administration.[36]

The mathematics of the Ming dynasty at this time, perhaps reacting to an increase in trade, interacted more with mercantile concerns than had the earlier Song dynasty. Some of the most prominent older Chinese texts, including parts of the *Jiuzhang suanshu*, were not readily available when the Jesuits made their appearance.[37] Nevertheless, when Ricci and his scientific colleague Xu Guangqi (1562–1633)[38] presented a translation of the first six books of Euclid's *Elements* (the *Jihe yuanben*, based on a Latin commentary by Ricci's former teacher Christopher Clavius) in 1607, it entered a mathematical culture that was in several ways incommensurable with its message.[39] The axiomatic-deductive structure did not resonate; the local audience tended to retain only the theorems that supported their computational needs. On the other hand, European mathematical astronomy was more easily adapted to its new context, but it too differed from the corresponding Chinese science: it sought an ultimate model of the motions of the heavenly bodies while the Chinese worked with variable, continually improving models.[40] As more books were translated, different scholars took different approaches to the influx. Some attempted to transform the new material to reconcile it with their indigenous mathematics; others tried to demonstrate the compatibility of the

---

[36] The literature on the Jesuit mission to China is extensive, and we shall not attempt to summarize it here. See [Brockey 2007] and [Mungello 2009, 1–80] for good starting points and [Hart 2013], [Elman 2005], [Chu 2003], and [Udias 1994] in dealing specifically with scientific interactions. [Jami 1992] is a study of the differences between the entrance of Western science through the Jesuits and the Westernization that took place later, beginning in the middle of the nineteenth century. [Jami 2009] surveys Jesuit mathematics in particular.

[37] On the state of Chinese mathematics upon the Jesuits' arrival, see [Engelfriet 1998, 98–102].

[38] On Xu Guangqi's life, work, and context see [Hart 2013, 195–256] and the collection [Jami/Engelfriet/Blue 2001].

[39] On the *Jihe yuanben* and its reception, see [Engelfriet 1998], [Martzloff 1997, 112–118], and [Li/Du 1987, 191–196]. [Hashimoto/Jami 2001, 264–267] describes the interactions between Ricci and Xu Guangqi as the translation was being completed. [Huang 2005] is a study of how linguistic barriers between the Jesuits and Chinese scientists could be overcome as a result of local coordination.

[40] See [Martzloff 1993–94].

two systems. We shall see examples of both in the following pages. As more of the primary literature becomes accessible to us, it is becoming clearer that the period from the seventeenth to the middle of the nineteenth centuries was a complicated narrative of appropriation and naturalization of foreign knowledge rather than a straightforward account of transmission.

Even before he completed his translation of the *Elements* with Xu Guangqi, Ricci had begun to consider how he might contribute Western astronomical resources to reforming the calendar. This was an ongoing concern; the existing calendar (the *Datong li*) often was unable to predict eclipses accurately. However, the Bureau of Astronomy refused to consider any changes, and efforts made to improve eclipse predictions within the existing structure were unsuccessful. In 1628, the beginning of the Chongzhen's reign, Xu Guangqi used Western methods to predict an eclipse successfully. This led him (and his successor, Li Tianjing [1579–1659]) to be appointed to supervise a project called the *Chongzhen lishu* ("Chongzhen Treatises on Calendrical Astronomy"), independent of the Bureau of Astronomy.[41] This large compendium, presented to the throne between 1630 and 1635, was assisted by a vast translation effort of Western mathematical and astronomical texts performed by missionaries such as Giacomo Rho (1593–1638) and Johann Adam Schall von Bell (1592–1666) as well as Chinese collaborators. The resulting eclipse predictions were superior to the *Datong li*.

▨ Trigonometry in the *Chongzhen lishu*

The *Chongzhen lishu* contains four main mathematical works: the *Chousuan*, on Napier's rods for calculation; the *Bilgui jie*, on Galileo's sector compass; the *Dace* ("Grand Measure," 1631); and the *Celiang quanyi* ("Complete Principles of Measure," 1635).[42] The *Dace*, compiled with Jesuits Johann Schreck (1576–1630) and Johann Adam Schall von Bell, is the first systematic account of plane trigonometry in China. The *Celiang quanyi*, by Giacomo Rho, contains within it the first Chinese treatise on spherical trigonometry with other works involving measurement of the circle and computational geometry. There had been briefer encounters with trigonometry before this; for instance, Sabatino de Ursis (1575–1620) had produced a table of shadow lengths, essentially a table of tangents.[43] But the *Dace* and the *Celiang quanyi* are really the texts where European-style trigonometry was announced.

---

[41] The story of the victory of European astronomy is told in [Elman 2005, 63–106] and [Lü 2007].

[42] See a summary in [Martzloff 1997, 23].

[43] In *Biaodu shuo* ("On Table and Measure"); see [Chen 2015, 495].

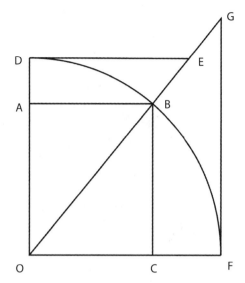

Figure 4.11
Defining the eight trigonometric lines
in the *Dace*.

The *Dace* was based on relatively recent sources from the European era of textbooks in trigonometry, especially Pitiscus's *Trigonometriae* and Simon Stevin's *Memoires Mathématiques*, but also on works by Clavius, Regiomontanus, and Magini.[44] All six of the now-standard trigonometric functions are defined in a single diagram, plus two more—the versed sine and the versed cosine. In figure 4.11, a diagram that was to have a long tradition in China, $\overset{\frown}{BD} = \angle BOD$ are the given arc and angle.[45] As in the European trigonometry of the time, the fundamental quantities are *lengths* of line segments (rather than today's *ratios*) and therefore depend on the value of the circle's radius R. The quantities and their names are:

AB	Sine	*zhengxian*
BC	Cosine	*yüxian*
DE	Tangent	*zhengqie*
OE	Secant	*zhengge*
OG	Cosecant	*yüge*
FG	Cotangent	*yüqie*
AD	Versed sine	*zhengshi*
CF	Versed cosine	*yüshi*

---

[44] [Iannaccone 1998, 703]; see also [Martzloff 1997, 23], both quoting a Chinese paper by Bai Shangshu.

[45] Chinese geometry used the concept of an arc but not often that of an angle; this would have been a novelty. Even the standard Western practice of referring to line segments and arcs by letter names of the vertices at the endpoints would have been unfamiliar.

These types of diagrams date at least as far back as tenth century Islam[46] and may be found in a number of European texts of the time.[47]

The *Dace* contains a description of methods of computing trigonometric tables, the *liuzong sanyao erjianfa* ("six essentials, three important methods, and two simple methods"), following the standard European approach to computing sines dating back eventually to Ptolemy.[48] The "six essentials" refer to six regular polygons inscribed in a circle (triangle, square, pentagon, hexagon, decagon, 15-gon), whose side lengths may be determined from propositions in the 1607 translation of the *Elements*. From these side lengths, several sines may be computed easily. The "three important methods" are the Pythagorean Theorem (for finding the sine of the complement an arc for which the sine is given), a sine double-angle formula, and a sine half-angle formula. One of the "two simple methods" is a statement of the sine sum and difference laws; the other is

$$\operatorname{Sin}\theta = \operatorname{Sin}(60° + \theta) - \operatorname{Sin}(60° - \theta). \tag{4.5}$$

All but this latter formula (which was known in Europe) have ancient roots.

For the first time in China, the *Dace* encountered the problem that had afflicted European and before them Islamic and Greek mathematicians: given these tools, the only sines of arcs that are computable are of the form $3m/2^{no}$. The *Dace* table (with $R = 10,000,000$) was to be subdivided into minutes of arc, but Sin 1′ is not a computable value. The authors' solution begins with the computed values $\operatorname{Sin} 22'30'' \left(= \operatorname{Sin} \frac{3°}{8}\right) = 65,449$ and $\operatorname{Sin} 11'15'' \left(= \operatorname{Sin} \frac{3°}{16}\right) = 32,724.5$. Since both these arcs and these sines (to the given level of rounding) have a ratio of precisely 2:1, one might assume that arcs and sines for all values this small are also in proportion to each other. Hence

$$\operatorname{Sin} 10' \approx \frac{10'}{22'30''} \cdot \operatorname{Sin} 22'30'' = 29,088; \tag{4.6}$$

and dividing by ten, we get sin 1′ ≈ 2,909. This approach had been common in Europe, for instance, in the works of Levi ben Gerson, Georg Rheticus, and Bartholomew Pitiscus.[49] From these values, the remaining sines could be computed easily (if tediously) from the given tools. The values of the other

---

[46] See Abū'l-Wafā"s diagram in [Van Brummelen 2009, 155–156].

[47] See for instance [Pitiscus 1600, 34] or [Stevin 1608b, vol. 1, *Cosmographiae*, 1].

[48] On construction methods for trigonometric tables in the *Dace*, see [Chen 2015, 497–501] and [Li/Du 1987, 205–207].

[49] For descriptions of Levi ben Gerson's and Rheticus's methods, see [Van Brummelen 2009, 231–232 and 278–279]. Pitiscus carries out a calculation very similar to the method in the *Dace* in [Pitiscus 1600, 52].

trigonometric functions then could be determined from the sine values in the usual ways.

However, in practice (just as in Europe) the entries in the tables often were simply borrowed or modified from other sources. This is true of the 100-page *Geyuan baxianbiao* ("Tables of Eight Lines Dividing the Circle," 1631) in the *Chongzhen lishu*, compiled by the Jesuits Schreck, Schall von Bell, and Giacomo Rho.[50] Although the "eight lines" in the title refer to the eight trigonometric quantities defined above, these tables contain only the six now-standard quantities, omitting the versed sine and the versed cosine.[51] The entries are taken from Pitiscus's 1613 tables[52] and use the same layout: the sine, tangent, and secant are three columns read from top to bottom while the cosine, cotangent, and cosecant are the same three columns, read from bottom to top.

The use of degrees to measure angles and arcs was a hurdle for Chinese astronomers accustomed to using the *du* as their unit of arc measure (365.25 $du \approx 360°$). The Jesuit compilers noted that a table converting between *du* and degrees would be helpful.[53] In 1644, when appointed to the Bureau of Astronomy at the beginning of the Qing dynasty, Schall von Bell controversially replaced *du* with degrees and also converted time measurements to European units.[54] In theory these conversions present only a trivial obstacle. But in practice, the mere necessity of unit conversion may have been a significant barrier in the integration of the two systems: it signals difference and the possibility of incompatibility.

The last work we shall consider from the *Chongzhen lishu* is Giacomo Rho's *Celiang quanyi* ("Complete Principles of Measurement"). In addition to material deriving from Archimedes's measurement of the circle, an approximation of π accurate to 19 decimal places, and other geometry,[55] this book contains a complete treatise on spherical trigonometry. Fundamental to quantifying motions on the celestial sphere, this subject could serve as part of the mathematical basis of calendar production. The contents of Rho's book appear to be based on European trigonometric texts; for instance, the theorem we shall examine below comes from Magini's *Trigonometricae sphaericorum*.[56] The material on right spherical triangles is derived by constructing

---

[50] On this table and the activities of the Jesuits who compiled it; see [Iannaccone 1998].

[51] Nevertheless, the values of these functions can be found readily from the table, simply by subtracting the cosine and sine respectively from *R*.

[52] [Pitiscus 1613a]; see [Iannaccone 1998, 703].

[53] [Chen 2015, 501].

[54] [Elman 2005, 103].

[55] See [Martzloff 1997, 23].

[56] [Chen 2010, 76], quoting a Chinese-language article by Bai Shangshu. Our coverage is taken from [Chen 2010, 73–76]. See also [Magini 1609, 18v].

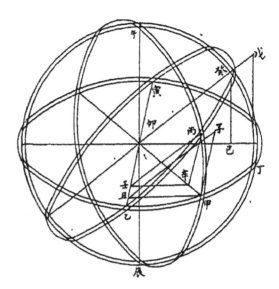

Figure 4.12
A spherical diagram in the
Celiang quanyi (1635).

similar right-angled planar triangles within the sphere and establishing equalities of ratios (*lü*) among them,[57] a common procedure in European trigonometric texts of the time.

Figure 4.12 shows a diagram from the *Celiang quanyi* that is used to produce a number of theorems.[58] We shall use the simplified diagram of figure 4.13 to sample one of them. In right-angled spherical triangle *ABC*, both $\widehat{BJ}$ and $\widehat{BD}$ are 90°, so $\angle B = \widehat{DJ}$. Then, from the eight-line diagram defining the trigonometric quantities (figure 4.11), we know that $DE = \text{Tan } B$. Similarly, $b = \widehat{AC}$, so $AK = \text{Tan } b$;[59] and $LA = \text{Sin } \widehat{AB} = \text{Sin } c$. Now, planar triangle *AKL* is similar to triangle *DEN*, so

$$\frac{ND}{DE} = \frac{LA}{AK}, \text{ or } \frac{R}{\text{Tan } B} = \frac{\text{Sin } c}{\text{Tan } b}, \tag{4.7}$$

which is one of the ten standard identities of a spherical right triangle.

▨ Logarithms in China

The *Chongzhen lishu* introduced the fundamentals of European trigonometry to Chinese scholars. However, it made no reference to the computational technology that was beginning to sweep across Europe, namely, logarithms.

---

[57] The use of *lü* goes as far back as the *Jiuzhang suanshu*; see [Chemla/Guo 2005, 956–959] or [Shen/Crossley/Lun 1999, 50]. The expression of the trigonometric properties of spherical right-angled triangles as equalities of ratios had been standard since the medieval period.

[58] Image from [Chen 2010, 74].

[59] Here, as usual, the lowercase letters represent the sides opposite the angles with corresponding uppercase letters.

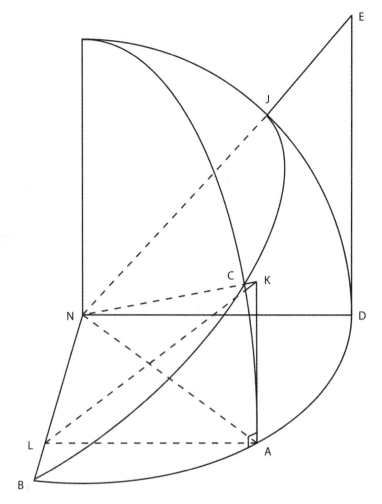

Figure 4.13
The simplified diagram. Dashed lines are within the sphere. *NK* is not drawn
in the diagram in the manuscript.

This omission would be rectified a couple of decades later. Nikolaus Smogulecki (1610–1656), a Polish Jesuit, spent the last decade of his life as a missionary in several sites in China. From 1651 to 1653 he worked with Xue Fengzuo (1600–1680) in Nanjing, and the pair produced a series of mathematical
and astronomical works.[60] The result of their collaboration was *Tianbu
zhenyuan* ("True Principles of the Pacing of the Heavens," 1653), part of Xue
Fengzuo's collection *Lixue huitong* ("An Integration of Calendrical Studies,"

---

[60] On the story of Smogulecki's and Xue Fengzuo's collaborations, see [Shi 2007] and [Jami
2004, especially 90–91].

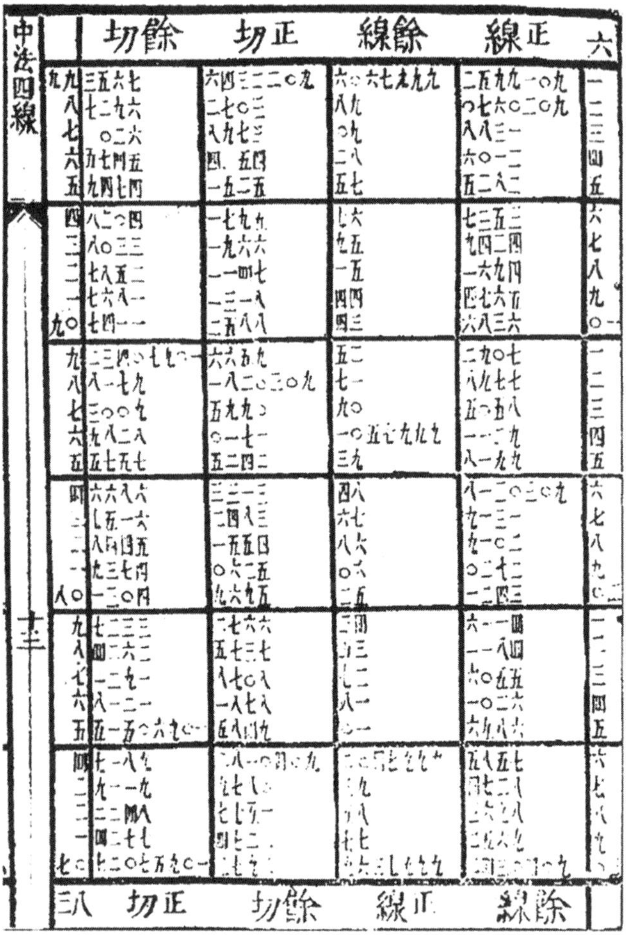

Figure 4.14
A page from the "Table of Four Lines."

1662–1665). Within its pages are found European astrology[61] and a criticism of the astronomical system in the *Chongzhen lishu* based on Tycho Brahe. This adaptation of a book by the Copernican astronomer Philippe van Lansberge (1561–1632) kept its cosmological features deliberately hidden. Astronomically, the *Tianbu zhenyuan* was a dissident work, undermining confidence in the earlier Jesuit texts. However, its reception in the Chinese audience was muted either due to a lack of readership or because the nature of the underlying astronomical model made little difference to the practically motivated astronomers of the time.[62]

[61] See [Standaert 2001].

[62] On the *Tianbu zhenyuan* see [Shi 2007]. For the method of presentation of Lansberge's astronomy, see p. 64 (quoting Shi's earlier Chinese-language article) and pp. 103–106.

There is a treatise on trigonometric calculations within the *Tianbu zhenyuan*,[63] but logarithms first appear in another treatise, the *Bili duishu biao* ("Tables of Proportional Logarithms"). The words used to describe logarithms, *bili shu* ("proportional numbers") and *jia shu* ("false numbers"), derive from similar terms being used in Europe, such as "artificial numbers." A table of base 10 logarithms for arguments from one to 10,000 is included, but there is no direct reference to trigonometry.[64] Another of the texts in the *Lixue huitong* is of more interest to us: the *Bili sixian biao* ("Tables of the Proportions of the Four Trigonometric Lines"; see figure 4.14).[65] The "Four Lines" are the sine, cosine, tangent, and cotangent; presumably it was thought that values of the other four lines (secant, cosecant, versed sine, and versed cosine) could be safely ignored since they may be found easily from the others. The values given in the table are $10 + \log_{10} \sin \theta$, $10 + \log_{10} \cos \theta$, and so on, matching the quantities tabulated in Vlacq's *Arithmetica Logarithmorum* (1628) and Briggs's *Trigonometria Britannica* (1633).

Of some interest is the fact that Xue Fengzuo subdivided degrees not into minutes but into centesimal units of 0.01°. Besides the convenience of a purely decimal system of fractional degrees, this decision also harks back to the traditional Chinese system where the *du* is subdivided into 100 *fen*. Xue was well aware of this connection:

> However, when I was engaged in the project of emendation and integration recently, I found that, whereas the Chinese method was too scattered and simple, the old [trigonometric] method was in the sexagesimal system, both being incompatible with each other. Therefore, I integrated them by converting the sexagesimal system in both the eight trigonometric lines and related books into the centesimal system, so that the old and new, the Chinese and Western can be unified into one system, which might become a ladder for this study [i.e., astronomy].[66]

Ironically, this quotation rules out Briggs's *Trigonometria Britannica* as Xue's source: one of Briggs's contributions was precisely to subdivide the degree into hundredths (see chapter 2). Indeed, traces of the use of linear

---

[63] The *Suan sanjiao fa* ("Methods of Calculating Triangles"); see [Shi 2007, 208] and [Li/Du 1987, 205–206].

[64] See [Shi 2007, 68], [Roegel 2011a], and [Li/Du 1987, 208]. This text and the *Bili sixian biao* may originally have been part of the *Tianbu zhenyuan*; see [Shi 2007, 78]. The entries in the tables may derive from Vlacq's *Arithmetica logarithmorum* [Vlacq 1628]; see [Roegel 2011a, 4].

[65] On the *Bili sixian biao* see [Chen 2015, 502–504] and [Roegel 2011b]. Image from [Chen 2015, 503].

[66] From the *Zhongfa sixian* ("Four Trigonometric Lines in the Chinese Method"), quoted in [Shi 2007, 111].

interpolation from a sexagesimal table can be found in Xue's entries, so Vlacq's tables remain as a potential origin.[67] The *fen* unit would endure; a couple of decades later the relatively obscure Li Zijin (1622–1701) abbreviated the *Geyuan baxian biao* (which had used units of 1′) in his *Tianhu xiangxian biao* ("Table of Celestial Arcs in a Quadrant," before 1673) by taking every sixth entry of the original table accompanying the *Dace* and renaming the tabular increments as not 6′ but 10 *fen*.[68] However, the *fen* was not widely accepted, and somewhat later Jiang Yong (1681–1762) would complain that Xue was "changing the convention and looking for troubles."[69]

Xue's motive for introducing logarithms recalls European thinking:

> The logarithms of the eight [trigonometric] lines are subtle and wonderful, easy and simple, being free of the pain of multiplication and division. But these are not any new and special things. It is just a matter of choice between easiness and simplicity, fineness and clumsiness, which might be done slowly.[70]

(As we shall see, not all Chinese astronomers agreed, at least initially.) Logarithms were used as one might expect; for instance, when applying the planar law of sines the reader is instructed to calculate according to

$$\log b = \log a + \log \sin B - \log \sin A.^{[71]} \qquad (4.8)$$

The *Lixue huitong* contains one other document of interest, the *Zhenxian bu* ("Section of Sine"), in which we find another description of how to compute a sine table. The method is mostly conventional, applying the standard geometric methods such as the Pythagorean Theorem, the sine addition law, and the sine half-angle theorem. To find sin 1° Xue applies something like a weighted average to his previously calculated values of $\sin\frac{3°}{4}$ and $\sin\frac{3°}{2}$:

$$\frac{2}{3}\sin\frac{3°}{4} + \frac{1}{3}\sin\frac{3°}{2} < \sin 1° < \frac{4}{3}\sin\frac{3°}{4}. \qquad (4.9)$$

For $R = 10{,}000{,}000$ this gives bounds of 174,520 and 174,528, from which Xue selects sin 1° ≈ 174,524. This material is taken from Stevin's *Hypomnemata*

[67] See the analysis in [Roegel 2011b, 5–7]
[68] See [Chen 2015, 507–509]. Chen argues that the adoption of the *fen* did not really help to embed trigonometry within traditional Chinese astronomical computations; it "only gives an illusion of integration of the Chinese and Western methods."
[69] [Shi 2007, 99].
[70] [Shi 2007, 89].
[71] [Li/Du 1987, 209].

mathematica.[72] However, the *Zhenxian bu* likely did not play a significant role in the construction of Xue's own table: this text does not mention logarithms, and in any case the smallest entry in Xue's table is Sin 0.01°, not Sin 1°.[73]

The Kangxi Period and Mei Wending

In 1662, the Kangxi emperor took the throne as the second emperor of the Qing dynasty at age seven. Lasting until his death in 1722, his reign was the longest in Chinese history and one of the most stable. When Kangxi took the throne, Western science was seen not so much as scholarly knowledge but rather as a collection of methods that could be put to the use of the state. This placed mathematics and astronomy in the service especially of eclipse prediction and calendar production. Nevertheless, Kangxi had a strong interest in the mathematical sciences, including the Western imports, and even taught them himself. Kangxi's continuing engagement with Western science also had a political dimension: as a Manchu, his interactions with the Han scholarly community had overtones that sometimes erupted into power struggles.[74] In 1668, a test of predictions of gnomon shadow lengths was won by Jesuit Ferdinand Verbiest (1623–1688) over his Chinese rivals, helping to return Western methods to the Bureau of Astronomy.[75] However, near the end of Kangxi's reign he shifted back against the Jesuits, partly for political reasons. Indeed, he recast the history of the subject in keeping with this opinion:

> Astronomy originated in China and was transmitted to the Far West. Westerners received it, did measurements without fail, revised it year after year without end, and therefore obtained precision in differences. It is not that they have different procedures.[76]

He also ascribed to Chinese antiquity the birth of trigonometry:

> In general the test for this [accuracy in surveying] is nothing but triangles. Although the word "triangle" (*sanjiaoxing*) was not used formerly, calculation methods in past dynasties must have had some foundations. For example, the base and altitude (*gougu*) method is pretty close to triangles. In fact this method must have been

---

[72] [Stevin 1608b, 14].

[73] For a detailed description of the contents of the *Zhenxian bu*, see [Chen 2015, 504–507].

[74] For the role of the Kangxi Emperor in the interactions between Western and Chinese science, [Jami 2012] is the essential source, but see also (among others) [Han 2014], [Jami 2007], [Jami/Han 2003], and [Bai 1995].

[75] See [Golvers 1993, 60–65] and [Jami 2012, 59–62].

[76] [Jami 2012, 249].

transmitted from antiquity. It has not been found in books, therefore we do not know when it started.[77]

These comments resonate with some of the challenges we encountered earlier in this chapter in applying the term "trigonometry" to certain ancient Chinese geometrical procedures. In any case, whatever its origin, Kangxi clearly valued trigonometry as the mathematical foundation not only of surveying but also of astronomy:

> Reckon the multitude of angles in squares and circles, classify the measurements of latitude and longitude, then astronomical methods prevail for one thousand years: why reject them? With triangles, the subtleties in the multitude of angles in squares and circles are easy to understand. If one rejects them and searches by another way, one is bound to end in confusion, and astronomy cannot succeed.[78]

The most significant mathematician/astronomer during Kangxi's reign was Mei Wending (1633–1721). From a learned family, Mei was one of several scholars who never worked for the Qing dynasty established by the invading Manchus. His approach to scholarship at a time when European science was competing with the traditional Chinese system was to treat knowledge as a universal entity and to combine the best of all available work regardless of its origins. His early books emphasized the importance of meticulous calendar scholarship, including European products, at a time when some Jesuits were being condemned to death (including Schall, who was eventually freed). But Mei also defended the importance of Chinese mathematics, extending several traditional topics. A prolific author, Mei wrote at least 80 treatises on topics such as arithmetic, linear algebra, geometry, geodesy, and of course calendrical calculations.

A meeting with the Kangxi emperor around 1705, at which it is said that the two men discussed mathematics for three days, enhanced Mei Wending's reputation immeasurably. At this meeting Mei presented to the emperor his *Sanjiaofa juyao* ("Essentials of Trigonometry"), a typical introduction to the Western subject—although Mei retained the traditional terms *gougu* (base and altitude) when referring to right-angled triangles. Mei's inclination to harmonize Western and Chinese mathematics was compatible with the emperor's belief that Western methods had originated in China; both men went away impressed with the other.[79] The meeting helped to trigger a sequence of events that led to the birth of a college of mathematics, the *Suanxue guan* (Office of

---

[77] [Jami 2012, 245].

[78] [Jami 2012, 249–250].

[79] See [Jami 2012, 251–253] and [Han 1997, 24–28]. [Jami 1994] explores the subtleties of Mei's position on this issue in detail.

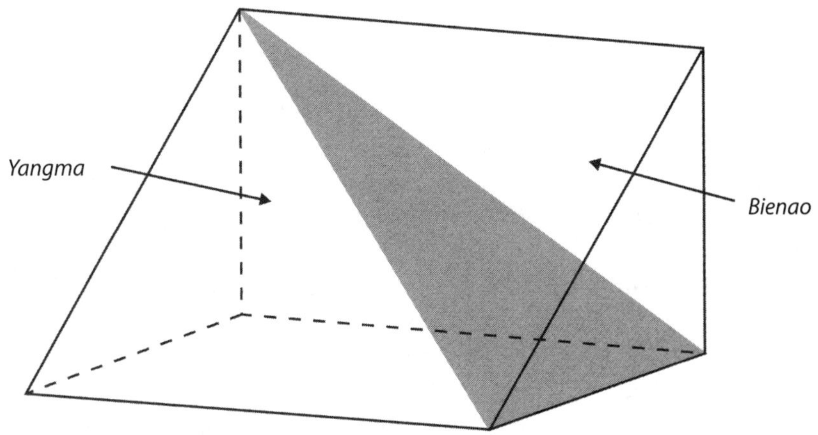

Figure 4.15
A *qiandu* divided into a *yangma* and a *bienao*.

Mathematics), in 1713. Mei's grandson Mei Juecheng (1681–1763) eventually would become chief compiler of the *Yuzhi shuli jingyun* ("The Essence of Numbers and Principles Imperially Composed," 1722), a 5,000-page mathematical encyclopedia organized by the *Suanxue guan*, surveying the mathematical sciences in a synthesis that reflected many approaches.[80]

Mei Wending's position with respect to the harmony between European and Chinese mathematics is exemplified especially in his approach to spherical trigonometry.[81] Among Mei's textbooks we find a pair dealing with planar (the *Sanjiaofa juyao*, renamed as *Pingsanjiao juyao*) and spherical trigonometry (*Husanjiao juyao*, 1684). In the latter we find an approach similar to that taken by Giacomo Rho in the 1631 *Celiang quanyi*, which we saw earlier in this chapter. Both Rho and Mei use families of similar right-angled triangles constructed within the sphere to generate their results.

Much more interesting is Mei's return to the subject a couple of decades later in his *Qiandu celiang* ("Measured with Prisms," 1701). The *qiandu*, a solid formed by cutting a rectangular prism diagonally in half, had had a long tradition in Chinese mathematics; in Liu Hui's commentary on the *Jiuzhang suanshu*, a prescription for finding its volume had already been given.[82] As part of his analysis, Liu had divided the *qiandu* into two parts according to the gray triangular sectioning in figure 4.15: a *yangma* (the pyramid with rectan-

---

[80] On the life and work of Mei Wending, see [Jami 2014], [Engelfriet 1998, 405–431], [Li/Du 1987, 212–216] and [Martzloff 1997, 25–29]. See [Martzloff 1981a] for Mei's transformation of Euclidean geometry, and [Martzloff 1981b] for a survey of some of Mei's mathematics.

[81] The following discussion relies on [Chen 2010, 76–84]; see also [Martzloff 1981b, 284–290].

[82] See [Shen/Crossley/Lun 1999, 267–269] and [Chemla/Guo 2005, 428–429]; see also [Guo/Tian 2015, 250–252].

gular base and right triangular sides) and a *bienao* (the tetrahedron with right triangular sides), equal to half the volume of the *yangma*.

One might wonder what such polyhedra have to do with spherical geometry. Mei Wending recognized how they might be used and employed them in a method he called *shishu* ("concrete number"). In figure 4.16 (simplified from Mei's original diagram), ABC is a spherical triangle in a sphere of radius $R = OB = OF = OD$. Arcs $c = \widehat{BA}$ and $a = \widehat{BC}$ are extended outward to lengths of 90° to D and F respectively so that $\angle B = \widehat{DF}$ and OFB is a quadrant. Draw tangent FH upward to intersect OD at H, then extend $\triangle OHF$ forward so that $\triangle BGE = \triangle OHF$, forming a *qiandu*. Extend OA to M and draw a vertical to it from C, forming K. Finally, cut the *qiandu* across from both A and C parallel to the front face to form two more triangles so that $\triangle BGE = \triangle LXZ = \triangle IWY = \triangle OHF$. The triangles formed within them by dropping perpendiculars to base $OBEF$ are all similar within all four of these triangles. Mei proceeds to "unfold" tetrahedron $OBMN$ into its **development**,[83] resulting in figure 4.17 and revealing several collections of similar triangles.

From all these similar triangles, Mei could deal with the spherical right triangle just as well as his Western counterparts. Sometimes he uses the development, but sometimes he doesn't need to. For instance, from the definitions of the eight lines (see figure 4.11) we know that $EG = FH = \text{Tan}\,\widehat{FD} = \text{Tan}\,B$, $CK = \text{Tan}\,\widehat{CA} = \text{Tan}\,b$, and $CL = \text{Sin}\,\widehat{CB} = \text{Sin}\,a$. Therefore

$$\frac{EG}{BE} = \frac{CK}{CL}, \text{ or } \frac{\text{Tan}\,B}{R} = \frac{\text{Tan}\,b}{\text{Sin}\,a}. \tag{4.10}$$

This is one of the ten fundamental identities of a right-angled spherical triangle.

Mei turns next to the *yangma* constructed from figure 4.16 by extending OM and HG to an intersection point U and extending ON and FE to an intersection point V (figure 4.18). He names the resulting square pyramid *fangzhi yi* from a now-lost work by Guo Shoujing. This pyramid, with base $UVFH$ and apex O, has faces containing various families of similar triangles as before. Mei unfolds the pyramid into its development, as we saw with the *qiandu*, and determines these families.

How does Mei see this work as a synthesis of Western and Chinese mathematics? He associates the tetrahedron $OBMN$ and its similar triangles with Giacomo Rho's *Celiang quanyi*; indeed, there is a strong similarity here with the European work. Next, Mei associates the *yangma* with the indigenous

---

[83] From descriptive geometry, the **development** of a solid is its unfolding onto a plane.

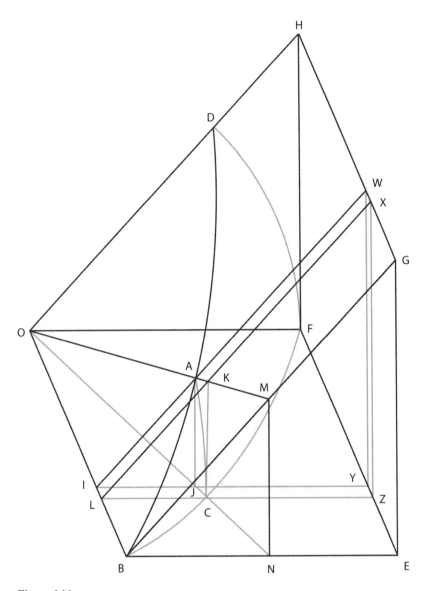

Figure 4.16
Mei Wending's *qiandu* leading to relations in spherical trigonometry.

work of Guo Shoujing. The two figures are indeed related: the tetrahedron corresponds to $\triangle ABC$ while the *yangma* corresponds to spherical quadrilateral $ACFD$. Since $\overset{\frown}{BA}$ and $\overset{\frown}{AD}$ are complements of each other, and likewise $\overset{\frown}{BC}$ and $\overset{\frown}{CF}$, Mei calls the solids "complementary." In this way, he felt he had

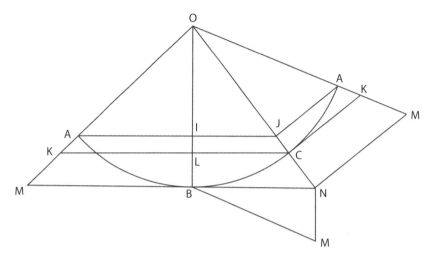

Figure 4.17
The development of tetrahedron *OBMN*.

shown that the Western and Chinese approaches were complementary as well. He concludes:

> [If the] method is worth utilizing, [it should be adopted] no matter [whether it is applied] in the East or West; the mathematical principles are supposed to be made clear; why [should one] differentiate [whether they are] new or old. . . . Get rid of the prejudice of Chinese versus Western (methods). Fairly seek the principles. . . . Gather the merits from all and seek the integration.[84]

Elsewhere, in a similar spirit, he writes:

> That . . . the opinions of the two schools [Chinese and Western] stand wide apart, is also the fault of those who study it. I, however, only strive to gain understanding by means of the Way of learning. If what I cannot grasp is grasped by someone else, then what difference do ancient or modern, east or west make?[85]

Did Mei Wending concern himself with the business of constructing trigonometric tables? This question is harder to answer than one might guess. In *Lisuan quanshu* ("Complete Writings on Mathematics and Calendric Astronomy," 1723), Mei's work that appeared shortly after his death, the table of contents indicates that a table would be published later. Another section, the

---

[84] Quoted in [Chen 2010, 84].
[85] Quoted in [Engelfriet 1998, 431].

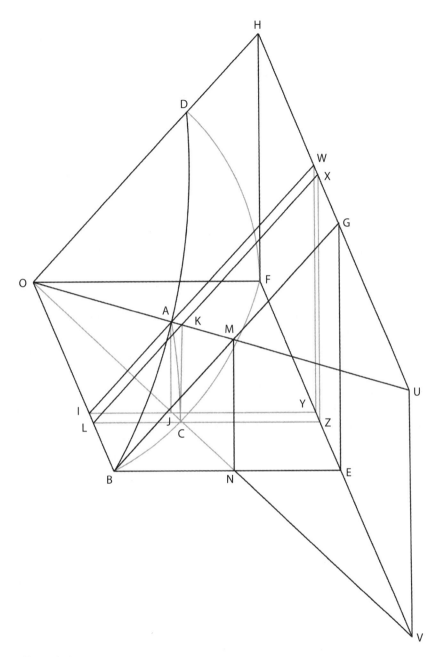

Figure 4.18
Mei Wending's extension of figure 4.16.

Jie baxian geyuan zhi gen ("Explicating the Origins of [How] Eight Lines Divides the Circle"), seems to have been written by Yang Zuomei, although it might have been based on a manuscript by Mei.[86] While its explanations are unprecedented in their thoroughness, its methods are similar to its predecessors.[87] We do have evidence of what Mei thought of some of the advances he had witnessed over his life. Commenting on Xue Fengzuo's logarithmic tables, he says:

> [Xue] converted the sixty minutes [of the arc] into one hundred minutes . . . , which is really convenient to use. However, he conducted calculations with logarithms, which in my opinion is not as proper a method as the direct use of multiplication and division.[88]

He must have changed his mind eventually, for in the *Bilishu jie* ("Explanation of Proportional Numbers") he calls logarithms "a wonder of the arts," particularly when extracting higher order roots.[89] Even so, logarithms seem to have played little role until the end of Kangxi's reign.

It was only a few months before Kangxi died that the *Suanxue guan* released the monumental *Yuzhi shuli jingyun*. One of a number of editorial projects in different disciplines, it was ten years in the making and covered almost all known mathematics.[90] It included basic arithmetic and geometry, ratios, systems of linear equations, root extraction, problems in the Chinese *gougu* (base and altitude) tradition, mensuration, algebra, logarithms, and the sector compass. The latter section demonstrates how to add markings for pure logarithms to the instrument as well as logarithms of trigonometric functions. As a foundation of mathematical education and research, this massive work remained influential for almost two centuries until and even during the second wave of Western learning in the nineteenth century.

Gathered together at the end of the work are a number of large tables: of trigonometric functions, logarithms, and logarithms of trigonometric functions. These tables were partly responsible for restoring interest in logarithms, which had been introduced more than half a century earlier but had not seen much use.[91] The trigonometric tables are monumental: displaying all six

[86] [Chen 2015, 510], quoting a 2003 Japanese-language dissertation by Tatsuhiko Kobayashi.

[87] For a description, see [Chen 2015, 510–512].

[88] [Shi 2007, 96–97].

[89] [Shi 2007, 98–99].

[90] The political and administrative story behind the construction of the *Yuzhi shuli jingyun* is told in [Jami 2012, 267–277], [Han 2014, 1220–1222], and [Han 2015, 79–82]; the work itself and its context are described in [Jami 2012, 315–384] and [Martzloff 1997, 163–166].

[91] The descriptions of logarithms in this work are translated from Henry Briggs. See [Martzloff 1997, 165], quoting a Chinese-language paper by Han.

functions with $R = 10,000,000$ and computed to every ten seconds of arc, their magnitude competes with the great European computational achievements of Rheticus and Pitiscus.[92] The enhancement of the grid from $1'$ to $10''$ of arc, requiring more than 10,000 entries for each of the six functions, seems to have been accomplished mostly by interpolation. However, for tangents and secants of large arcs, which are hard to compute due to numerical instability, the compilers explicitly report that they changed their methods of calculations.[93]

Much earlier in the *Yuzhi shuli jingyun*, within the section on the division of the circle (*geyuan*), we find the typical topics related to the definitions of the eight lines and the construction of tables. One aspect of its contents, however, is far from typical. The author begins by objecting to the *Dace*'s approximation of sines of values below $45'$. Recall that in that treatise, it had been assumed that the sines of small arcs are proportional to the arcs themselves (see (4.6)). The *Yuzhi shuli jingyun* proposes to avoid this unnecessary assumption by introducing a pair of related algorithms. The first, called *yishi guichu* ("the method of augmenting dividend and dividing numbers"), allows the reader to solve the following problem: given four quantities in continued proportion, $\dfrac{a}{b} = \dfrac{b}{c} = \dfrac{c}{d}$, where $a = 100{,}000$ and $a+d=3b$, solve for b, c, and d.[94] This reduces to

$$b^3 + (100{,}000)^3 = 3 \cdot (100{,}000)^2\, b, \qquad\qquad (4.11)$$

a depressed cubic equation. The author finds a way to use the structure of these equations to represent the sine triple-angle formula, also a depressed cubic. Thus, the problem of finding Sin $\theta/3$ given the value of Sin θ reduces to the problem of solving the relevant depressed cubic equation. There already had been a very long tradition in China of solving equivalents to polynomial equations by extending the long division algorithm, effectively solving these equations iteratively.[95] Hence the author was able to trisect the angle using a combination of algebra and iteration, just as Jāmshīd al-Kāshī had done three centuries earlier in Samarqand[96] and as François Viète had done two

---

[92] However, these tables are not as accurate as those in the *Opus palatinum*. The tables of logarithms of trigonometric functions may be based on Adrian Vlacq; see [Roegel 2011f, 5–7].

[93] [Chen 2015, 512–514].

[94] The second problem replaces $a+d=3b$. with $a+d=2b+c$.

[95] One might think of the cubic equation $\sin 3° = 3 \sin 1° - 4 \sin^3 1°$ (where $\sin 1°$ is the unknown) as a variation of long division by writing $\sin 1° = \dfrac{\sin 3°}{3} + \dfrac{4}{3}\sin^3 1°$. Then $\sin 1°$ is the result of the division algorithm applied to $\sin 3°$ and 3, modified appropriately to take account of the $\frac{4}{3}\sin^3 1°$ term.

[96] See [Van Brummelen 2009, 146–149].

centuries earlier in France.[97] Mei Wending, with his view of the harmony of mathematics in different cultures, would have approved.

▨ Dai Zhen: Philology Encounters Mathematics

The state-sponsored scholarship of the last years of the Kangxi reign was to affect much of the rest of the eighteenth century. While the Manchus were often brutal in their suppression of academic activities that led to actions undermining their authority, they consistently supported efforts that did not have political implications. One of the consequences was the rise to prominence of *kaozheng* (evidential research). Practitioners of this discipline turned their attention to the historical classical texts, paying close attention to issues of textual interpretation and questions of philology. Their careful work could on occasion undermine the authenticity of certain texts and challenge neo-Confucian perspectives, but it did not lead to any potential threats to the regime in power. Scholarly research became more structured: large, coordinated efforts were organized with the benefit of governmental support; academic associations were formed; and research institutions were created.

Unlike textual criticism or philology today, the practice of *kaozheng* included as a fundamental part of its makeup an emphasis on empirical scientific analysis, especially mathematics, astronomy, and geography. These subjects were crucial to the Confucian tradition: one could not properly understand the texts without knowledge of them. *Kaozheng* scholars sought detailed and precise research on the basis of evidence drawn from impartial sources; mathematics and astronomy thus fit naturally within the mission "to search truth from facts" (*shishi qiushi*).[98] Combined with the continued governmental interest in calendar production, *kaozheng* promoted mathematical astronomy and helped it to flourish. In addition, the emphasis on classical texts and the impetus given to claims of the ancient Chinese origin of mathematics and astronomy that had been promulgated by the Kangxi emperor led to a climate that welcomed efforts to unify current mathematical astronomy with classical approaches.[99]

Such an effort was to come from one of the most eminent of the *kaozheng* classicists. Dai Zhen (1724–1777) actually made his first contribution in mathematics with an explanation of Napier's rods, but he was primarily interested in philosophical issues. His *Mengzi ziyi shuzheng* ("Evidential Study of

---

[97] See chapter 1. On the sine triple-angle formula in the *Yuzhi shuli jingyun*, see [Chen 2015, 515–517].

[98] [Elman 2004, 42–43].

[99] On the history and content of the *kaozheng* movement see [Elman 2001] and [Elman 2005, 225–280].

the Meaning and Terms of the *Mencius*"), a meticulous study of a collection of sayings of the ancient Confucian scholar, included an attack on Song dualism. In the Song system, *qi* is the physical stuff out of which the universe is made and is responsible for passions and evil, whereas *li* is the spiritual, rational, and moral and is the source of good.[100] Contrary to this view, Dai Zhen took a materialist position. He denied the value of *li*, arguing that *qi* is sufficient to account for the universe. Further, he felt that *li* was used as a tool to propagate power imbalances between the strong and the weak and was therefore a negative influence on society. He based his position partly on an analysis of the ancient meaning of *li*, which originally denoted the texture or fiber of things.[101]

Dai's attention to terminology and meaning, exemplified here, also was applied to great effect in his approach to mathematics. His *Gougu geyuan ji* ("Records of Base-Altitude and Circle Division"), a controversial work in trigonometry, went beyond textual analysis to textual reconstruction with historiographic motives.[102] It is divided into three parts: the first part is on the foundations of plane trigonometry, the second on right spherical triangles, and the third on general spherical triangles. We shall concentrate on the first two parts. From the beginning, we realize we are in for something different. Rather than the usual definitions in terms of lines and planar triangles, Dai starts with arcs, chords, and the versed sine. He takes his lead not from any contemporary accounts of trigonometry but from the thirteenth-century astronomer Guo Shoujing, whose work we saw earlier in this chapter. The key here for Dai is that Guo predates the arrival of the Jesuits. Dai goes on to divide the circle not into the standard 360° but into 96 parts (*xian*). This time he takes inspiration from the *Jiuzhang suanshu*'s division of the circle into 96 parts in its approximation of π.[103]

The connections to indigenous traditions continue in the definitions of the trigonometric quantities themselves. This time Dai turns to the *Zhoubi suanjing* ("Computational Canon of the Gnomon of Zhou," 100 AD), one of the ancient classics of astronomy and mathematics. Dai refers to this text's use of the *ju*, an L-shaped rectangular region now called a trysquare.[104] The names chosen by Dai to represent the eight lines are different from the

---

[100] The character *li* here is different from the character referring to a calendrical system.

[101] On Dai Zhen and his philosophical positions, see (among others) [Chen/Freeman 1990], [Hummel 1943–1944, vol. 2, 695–700], and [Elman 2001, 18–21].

[102] Our account of this text relies on [Chen 2010, 84–104].

[103] Indeed, one of Dai Zhen's most important works was a reconstruction of the *Jiuzhang suanshu*. See [Chemla/Guo 2005, 74–79]. One can find in Dai's writings traces of several units he planned to use and discarded, including 360°, 384 *xian*, and 100 *du*; see [Chen 2011].

[104] [Cullen 1996, 77–80]. Although the operations used in this text might be explained in modern terms by the application of similar right triangles, Cullen is careful to avoid imposing this view on the *Zhoubi suanjing*'s practice. See [Cullen 1996] for a study of the entire work.

standard names we saw earlier: appealing to the *Zhoubi suanjing*, they have the ring of ancient authority:

Tangent	*jufen*
Secant	*jingyinshu*
Cotangent	*cijufen*
Cosecant	*ciyinshu*
Sine	*neijufen*
Cosine	*cineijufen*
Versed sine	*shi* or *zhengshi*
Versed cosine	*cihubei zhi shi*

When Dai goes on to the mathematics itself, he uses the same two tools shared by all trigonometers at the early stages of the theory: the Pythagorean (*gougu*) Theorem and similar right triangles. Dai discusses the former in the context of the *Zhoubi suanjing*. He addresses the latter by referring to the ancient practice of *yicheng tongchu* to solve for an unknown when its ratio to some other quantity is known to be equal to another known proportion, noting that Guo Shoujing had used this method in his thirteenth-century astronomical work. He then builds the basic theory of plane trigonometry, showing how to compute sine and cosine tables, presenting algorithms for solving triangles, and using the laws of sines and tangents.

In part 2 Dai turns to spherical right triangles, and here he chooses to follow Mei Wending. Recall that to solve problems on the sphere, Mei had constructed families of similar planar right triangles. This he had accomplished by extending a spherical right triangle or quadrilateral into a solid region and then constructing right triangles on the faces of these solids. Dai begins with a set of constructions reminiscent of Mei's solids. In figure 4.19, $\triangle ABC$ is the original right triangle; A is a pole of $\overset{\frown}{EDN}$. He calls quadrilateral *BCED* the *fangzhi yi* ("straight rectangular device"), $\triangle ABC$ the *ciwei yi* ("auxiliary latitude device"), and $\triangle BDN$ the *cijing yi* ("auxiliary longitude device"). For each device he applies a method leading to a development of the corresponding solid as Mei had done (for example) in figure 4.17. As before, this achieves the identification of families of similar right triangles. By bringing the three figures together,[105] he can link the three separate triangle families into an organized system.

---

[105] Incidentally, this forms a Menelaus configuration, used in ancient Greece to solve problems in spherical astronomy. See [Van Brummelen 2009, 56–63].

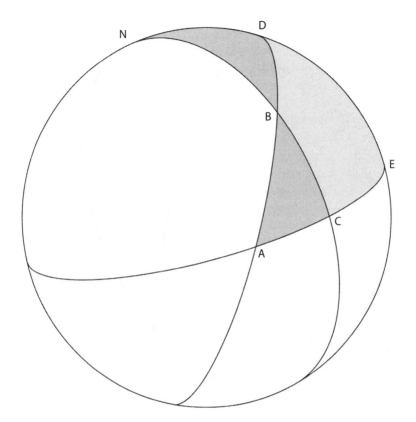

Figure 4.19
Dai's devices. The light gray region is the straight rectangular device; the two shaded triangles are the auxiliary devices.

But Dai is not done yet. Next, he proceeds to build a *zong yi* ("comprehensive device"), an approach that systematizes everything there is to be known about right-angled configurations on the sphere. In figure 4.20 (an elaboration of figure 4.19), we add one more great circle *PGK*, the equator to *B*'s pole. In addition to the collection of devices *ACEDNB*, he identifies four analogous collections: *BCLKGA*, *AKJTPG*, *GTRFNP*, and *BDIFPN*.[106] Each collection consists of two right triangles and a quadrilateral; each right triangle is used in two collections. By combining the relations that can be obtained from the similar triangles in all these collections, Dai has provided the as-

---

[106] Points *L* and *J*, completing their respective quadrilaterals, are on the back side of the sphere behind the quadrilaterals and are not visible on the diagram. Points *R* and *I* are also on the back of the sphere but are visible.

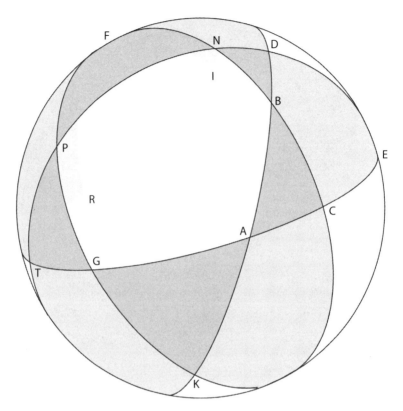

Figure 4.20
Dai's comprehensive device.

tronomer with a systematic method to solve any problem, regardless of the location of the triangle or quadrilateral.

In fact, Dai's arrangement parallels the method that had been used in Europe 150 years earlier to systematize the theory of spherical right triangles. In figure 4.20, the configuration consisting of the five right triangles and the pentagon joined by the five hypotenuses is precisely what came to be known in Europe as the *pentagramma mirificum* (see chapter 2). It was this diagram that John Napier used to derive the ten standard identities for right spherical triangles and to identify the symmetries among them. Each of Dai's collections of three devices is a pair of triangles from the *pentagramma*, along with the quadrilateral that joins them. Thus, just as Napier had achieved a grand synthesis of the various relations in spherical right triangles, so had Dai Zhen. It was a remarkable accomplishment.

Over the years, Napier's encapsulation became the basis of mnemonics used by European students for centuries to memorize the identities of a right spherical triangle, but the mathematical basis of his system was gradually forgotten. Dai Zhen's structure suffered a different fate. A surprisingly large number of criticisms of his work appeared in print; it has been suggested that this may be because it was widely circulated thanks to its being appended to the *Suanjing shishu* ("Ten Computational Canons").[107] The complaints are predictable: Dai chose to invent an entirely different terminology, and his approach was seen as incomprehensible, especially for those familiar with the more common Jesuit-inspired trigonometric style. Presumably as a result of the work's opacity, it generated few followers. Dai also was chastised by some colleagues for not acknowledging his debt to the Jesuit tradition; although he was not the only one to believe that trigonometry had its origins in indigenous Chinese mathematics, his attempt to forge an explicit pathway was seen as revisionist history. Finally, claims by several contemporaries that Dai's trigonometry could already be found in the work of Mei Wending indicate that Dai's self-contained systematic theory may not have been appreciated fully by his readers.[108]

▨ Infinite Series

One of Mei Wending's successors was his own grandson Mei Juecheng.[109] The younger man's talents appear to have been recognized quickly, for he was appointed as an Imperial Student, and for a decade he served as chief editor of the vast mathematical compendium *Yuzhi shuli jingyun* (1722). He remained influential for many years through the Yongzheng and Qianlong courts. Although emperors succeeding Kangxi did not share his zeal for scholarly pursuits in mathematics and astronomy, Mei Juecheng and his colleagues nevertheless were able to serve at high levels in the administration. He was involved in updating and writing several mathematical works, including a rediscovery of Song dynasty work on polynomials which he interpreted in terms of European cossic algebra. Generally, he took the attitude (more

---

[107] [Chen 2010, 102]. The *Suanjing shishu* is a collection of ten classic Chinese mathematical texts, seven of which had been rediscovered by Dai himself. Published shortly after Dai's death, the series adds to the ten classics his own work on trigonometry. On the *Suanjing shishu*, see [Martzloff 1997, 123–141] and [Li/Du 1987, 225–230].

[108] See [Chen 2010, 102–103] for a reappraisal of Dai's contribution, emphasizing that his approach is "a quintessence of a self-contained system of trigonometry."

[109] On Mei Juecheng see [Elman 2005, 156–160, 176–182], [Martzloff 1997, 353–357], [Jami 2012, 306–307, 377–383], and [Han 2015, 83–86].

stringent than his grandfather) that the mathematics and astronomy passed on to them through the Jesuits was originally Chinese, and he even displayed an antipathy toward Western astronomers.

This attitude lends an ironic flavor to one of Mei Juecheng's last contributions. Around 1762 he produced the *Meishi congshu jiyao* ("Selected Essentials of Mei's Collection"), an edited collection of his grandfather Mei Wending's work. One of the two chapters that he added to this work was the *Chishui yizhen* ("Pearls Recovered from the Red River"). Among its topics we find the following passage.

Text 4.3
Mei Juecheng, On Calculating the Circumference of a Circle from Its Diameter
(from the *Meishi congshu jiyao*)

Given [a] diameter, 20 *yi,* find [the length] of the circumference. The more numerous the number of digits of the diameter are, the more precise the tall-number [i.e., the decimals] are. Here we take ten digits as an example. Method: diameter multiplied by 3. Result: 60 *yi* as first number taken as dividend. Multiply it by the first multiplier [i.e., "1"] ["1" multiplied by this number: nothing changes]. Divide it by the first divisor [24]. Result: 25,000,000 as [the] second number again taken as dividend. Multiply it by the second multiplier. . . . Multiplications and divisions stop when units are obtained.[110]

Explanation: The word *yi* represents the number 100,000,000, so the diameter d of the circle is chosen to be 2,000,000,000. With such a large diameter, calculations can be performed with whole numbers; the procedure is meant to be repeated until the result is found to the closest whole number. The paragraph describes a progression of terms to be added. The first term is $3d$; the second term is $3d$ multiplied by the "first multiplier" and divided by the "first divisor," 24. Thus far, then, we have $3d + 3d \cdot \frac{1}{24} = 6,000,000,000 + 250,000,000 = 6,250,000,000.$ We are instructed to continue by multiplying the second multiplier by the second divisor and so on. The resulting sequence is

$$\text{Circumference} = 3d\left(1 + \frac{1^2}{3! \cdot 4} + \frac{1^2 \cdot 3^2}{5! \cdot 4^2} + \frac{1^2 \cdot 3^2 \cdot 5^2}{7! \cdot 4^3} + \cdots\right). \quad (4.12)$$

Dividing both sides of this equation by d gives an infinite series for π which had been derived by Isaac Newton from his series for the arc sine.[111]

[110] [Martzloff 1997, 353–355].
[111] [Jami 1988a, 45].

In the same work we also find infinite series for the sine,

$$\text{Sin}\,\theta = \theta - \frac{\theta^3}{3!\,R^2} + \frac{\theta^5}{5!\,R^4} - \frac{\theta^7}{7!\,R^6} + \cdots; \qquad (4.13)$$

and the versed sine,

$$\text{Vers}\,\theta = \frac{\theta^2}{2!\,R} - \frac{\theta^4}{4!\,R^3} + \frac{\theta^6}{6!\,R^5} - \frac{\theta^8}{8!\,R^7} + \cdots. \qquad (4.14)$$

These correspond to the Taylor series for sine and cosine respectively.[112] In each case Mei Juecheng simply asserts the series without any sort of justification, citing the Jesuit scholar Du Demei. This is a reference to Pierre Jartoux (1669–1720), whose presence in China for the last two decades of his life would have given him plenty of opportunities to pass on this knowledge. Since Mei says nothing more, we do not know whether Jartoux justified these results or what other results he might have presented.[113]

This tantalizing set of formulas, with their obvious advantages for calculating sine values, must have caught the eyes of Chinese mathematical astronomers. Mei Juecheng himself was impressed:

> The old circle division methods [are] . . . most refined [and] most accurate . . . but extracting the square root of a number of tens of digits, it cannot be done in less than tens of days. Now [we] can set up multiplication and division numbers from a regular hexagon inscribed in a circle to find them. They can be obtained instantly and are not different from the values obtained from repeated uses of right triangles. Therefore, they are called quick methods [*jiefa*].[114]

Nevertheless, it seems that Mei Juecheng did not pursue the matter any further, and it remained to one of his colleagues to pick up this thread.

Minggatu (1692–1763), a Mongolian student at Kangxi's court, had worked as an assistant editor with Mei Juecheng on the compilation of the 1722 *Yuzhi shuli jingyun*. Most of his work was within the Imperial Board of Astronomy, over which he presided starting in 1759. In addition to his astronomical work he took part in a topographic survey; mathematics seems to have been just one of several interests. Perhaps due to a lack of family connections, he was not considered to be a *chouren* until after his work was

---

[112] Recall that trigonometric quantities like the sine and versed sine were considered at this time to be lengths in a circle of radius R, not ratios.

[113] See [Li/Du 1987, 234] and [Martzloff 1997, 353–355]. On Jartoux, see [Pfister 1934, vol. 2, 584–586].

[114] [Chen 2015, 522–523].

published posthumously in 1839.[115] Nevertheless, Jartoux's infinite series captivated him for 30 years until his death was near. Minggatu's student Chen Jixin reports his words:

> Indeed, nothing like [Jartoux's series] exists in ancient or modern mathematics. I would have liked my colleagues to have been able to use them; unfortunately, Jartoux's text contains recipes without justifications. Therefore, I did not wish to divulge them, for fear of providing those who might find them with the "golden needle" without the secret to the way of using it. I have built up arguments over many years without succeeding in completing the task I gave myself. Thus, I would like my work to be continued.[116]

True to his teacher's wish, Chen Jixin completed the demonstrations in 1774 in *Geyuan milü jiefa* ("Quick Methods for Trigonometry and for Determining the Precise Ratio of the Circle"). Although it was not published until over half a century later (1839), it circulated in an abridged manuscript form and influenced a number of scholars. In it we find a total of nine series: the original three supplied by Jartoux (for π, the sine, and the versed sine), one for the chord of a given arc, another for the versed sine as a function of twice the given arc, and four for the inverses of these functions.[117]

Although presumably Jartoux's series had been derived originally with the aid of calculus, Minggatu had no such advantages.[118] His method of proof relies on a geometric argument containing an infinite regress reminiscent of the derivations of similar series found in fourteenth- and fifteenth-century Kerala. We examine Minggatu's derivation of the series for the chord of a given arc, beginning with a lemma. In the isosceles triangle ABC in figure 4.21, draw CD so that $CD = CB$; then $\triangle ABC$ is similar to $\triangle CDB$. Next draw DE so that $DE = DB$; then $\triangle CDB$ is similar to $\triangle DBE$. We may repeat this process as many times as we like to produce a series of isosceles triangles; since they are all similar, we have a series of ratios in continued proportion (*lian bili*):

$$\frac{AC}{CB} = \frac{CB}{DB} = \frac{DB}{DE} = \cdots.$$

---

[115] On Minggatu's life and work, see [Jami 1990, 36–40], [Jami 1991, 105–106], and [Martzloff 1997, 355–360].

[116] Quoted in [Martzloff 1997, 358].

[117] All nine of the series became identified with Jartoux, but it is probable that Minggatu derived the last six on his own; see [Jami 1990, 44–45].

[118] Thanks to the Jesuits there were various calculus texts available, but they were not translated from Latin, and there is no evidence that they were studied at this time.

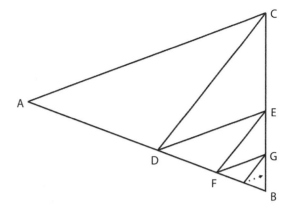

Figure 4.21
Iterated isosceles triangles.

It is possible that Minggatu was inspired in this construction by Mei Wend-ing; the latter performs precisely this iterative construction (*dijia fa*) in his work on the golden ratio.[119]

Text 4.4
Minggatu, On Calculating the Chord of a Given Arc
(from the *Geyuan milü jiefa*)

The method: Takes the length of an arc as the first item. Use the radius as the first *lü* of continued proportion and the [length of the] arc as the second *lü* of continued proportion to find the third *lü* of continued proportion. Next, mul-tiply the first item with the third *lü* and divide [the product] by the first *lü* to get the fourth *lü*. Divide [the fourth *lü*] by 4, by 2, and then by 3 to get the second item, which is supposed to be subtracted. Write it separately. Next, take the second item, multiply [it] with the third *lü* and divide [the product] by the first *lü* to get the sixth *lü*. Divide [it] by 4, again by 4, and then by 5 to get the third item, which should be written under the first item. Next, take the third item, multiply [it] with the third *lü*, and divide [the product] by the first *lü* to get the eighth *lü*. Divide [it] by 4, by 6, and then by 7 to get the fourth item, which is supposed to be subtracted. Write [the fourth item] under the second item. Add the first and third items; add the second and fourth items. Subtract one sum from the other to get the value of the chord [of the arc].[120]

Explanation: Let θ be the given arc in a circle of radius R; the goal is to find the length of the chord c. The first four *lü* are quantities in continued pro-portion. The first four *lü* are R, θ, $\dfrac{\theta^2}{R}$, and $\dfrac{\theta^3}{R^2}$. The sixth *lü* is equal to

[119] See [Engelfriet 1998, 418–420].
[120] [Chen 2015, 524–525].

$\dfrac{\theta^5}{4\cdot3!\cdot R^4}$; the eighth *lü* is equal to $\dfrac{\theta^7}{4^2\cdot5!\cdot R^6}$. The "items" are the successive terms in the series. The odd-numbered items and even-numbered items are summed separately, then the former sum is subtracted from the latter to produce the chord. The resulting calculation is equivalent to

$$c = \theta - \frac{\theta^3}{4\cdot3!\cdot R^2} + \frac{\theta^5}{4^2\cdot5!\cdot R^4} - \frac{\theta^7}{4^3\cdot7!\cdot R^6}. \tag{4.15}$$

Now that we have the expression in modern notation, we recognize it as the beginning of a power series for c in terms of θ. However, the lack of symbolism and the use of the language of continued proportion should warn us that the modern expression conveys a meaning and context somewhat different from that of a Taylor series, even if the text is mathematically isomorphic.

The above text simply asserts the series; Minggatu has yet to show us how it is derived. Space precludes an account of the entire argument, so we present a simplified version of one part of it.[121] The goal (see figure 4.22, simplified from the original) is to find an expression for the chord $c=BD$ of a given arc $\theta = \widehat{BCD}$ in a circle of radius $R=AB=AC=AD$. Minggatu begins by dividing the arc into some number of equal segments; in our figure, that number is two. Here and elsewhere, he consciously embeds his method in a Chinese historical context: in this case the practice of *geyuan* or circle division. Apparently, the attempt to nativize his work was not entirely successful; some traditionalists later boycotted it due to its Western influences.[122] The first step is to express c in terms of the chord of half the arc, $c_2 = BC = CD$. $\widehat{BC}$ is bisected at E; E is joined to A, B, and C. Construct $BG = BC$ and $DH = DC$; join CG and CH. Then $\triangle BGC = \triangle DHC$, and both are similar to $\triangle ABE$.[123] Now, we may think of $\triangle CGH$ as deriving from $\triangle BCG$ in the process illustrated in figure 4.21 so it is also similar to the others. Draw $BF = BE$; we have now applied the same process to $\triangle ABE$, giving us yet another similar triangle $\triangle BEF$. Thus, we have the collection of ratios

$$\frac{AB}{BE} = \frac{BE}{EF} = \frac{BC}{CG} = \frac{CG}{GH},$$

---

[121] Our narrative of the simplified form of Minggatu's argument is based on [Li/Du 1987, 234–239] and [Horng 1991, 150–153]. To see a full account, see [Jami 1988b] and [Jami 1990].

[122] [Engelfriet 1998, 447].

[123] Let $x = \angle BAE$; then since $\triangle ABC$ is isosceles, $\angle ACB = 90° - x$. So $\angle DCB = 180° - 2x$, and since $\triangle DCB$ is isosceles, $\angle DBC = \angle BDC = x$.

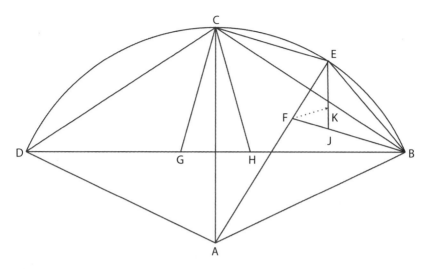

Figure 4.22
Minggatu's derivation of c from c_2 (simplified).

which (among other relations) gives

$$GH = EF \cdot \frac{BC}{R}.$$ (4.16)

Then

$$c = BD = BG + DH - GH = 2BC - GH,$$

which, applying (4.16), gives

$$c = 2BC - EF \cdot \frac{BC}{R} = 2c_2 - \frac{c_2}{R} \cdot EF.$$ (4.17)

Now we have c in terms of c_2 and EF. Minggatu iterates: let $EJ = EF$; by the process in figure 4.21 we have yet another similar triangle, ΔEFJ. Using the continued proportions, we can replace EF in (4.17) with an expression involving FJ. Then we let $FK = FJ$ and so on. Eventually, as the similar triangles become smaller, the pattern in the expression for c becomes clear:

$$c = 2c_2 - \frac{c_2{}^3}{4 \cdot R^2} - \frac{c_2{}^5}{4 \cdot 16 \cdot R^4} - 2\frac{c_2{}^7}{4 \cdot 16^2 \cdot R^6} - \cdots.$$ (4.18)

Curiously, using this method when n is even, one gets an infinite series for c in terms of c_n; but when n is odd the result is only a finite series.[124] For instance,

$$c = 5c_5 - \frac{c_5^3}{R^2} + \frac{c_5^5}{R^4}. \tag{4.19}$$

From (4.18) and (4.19), Minggatu is able to build a formula for c in terms of c_{10}; simply use the general result (4.19) to represent c_2 in terms of c_{10} (since it is true for any given chord c, it is true for c_2) and then substitute into (4.18). In this manner he builds formulas for c in terms of ever-smaller chords, eventually producing a formula for c in terms of $c_{10,000}$:

$$c = 10,000c_{10,000} - 166,666,665,000\frac{c_{10,000}^3}{4 \cdot R^2}$$
$$+ 3,333,333,000,000,003,000\frac{c_{10,000}^5}{4 \cdot 16 \cdot R^4}. \tag{4.20}$$

The first term is essentially equal to θ, so

$$c = \theta - 0.166666665\frac{\theta^3}{4 \cdot R^2} + 0.03333333\frac{\theta^5}{4 \cdot 16 \cdot R^4}. \tag{4.21}$$

The coefficients in this series are obviously close to the simple fractions 1/6 and 1/30.[125] Substituting and continuing the pattern, we conclude:

$$c = \theta - \frac{\theta^3}{4 \cdot 3! \cdot R^2} + \frac{\theta^5}{4^2 \cdot 5! \cdot R^4} - \cdots \tag{4.22}$$

identical to (4.15) above.

Minggatu then clearly developed power series for trigonometric functions, but (as in the case of the Keralite power series) there are shades of meaning here. Neither Minggatu nor his colleagues were thinking of functions in the modern sense. One might make an analogy between his process and long division: after each iterative step, Minggatu has a quantity that is closer to what is sought than what he had before. The way that Minggatu lays out his continued proportions on the page, in fact, suggests something like this. Finally, these series were developed as quick methods for determining trigonometric quantities rather than analytical tools as in calculus. For in-

[124] See [Mikami 1913, 146–147] for a sketch of the argument for trisecting the chord.
[125] Here we simplify the argument slightly. In the full treatment, Minggatu examines the coefficients of θ^n in the collection of formulas he has derived for different numbers of subdivisions of chords, and he notices that these coefficients are converging to obvious values as the number of subdivisions increases.

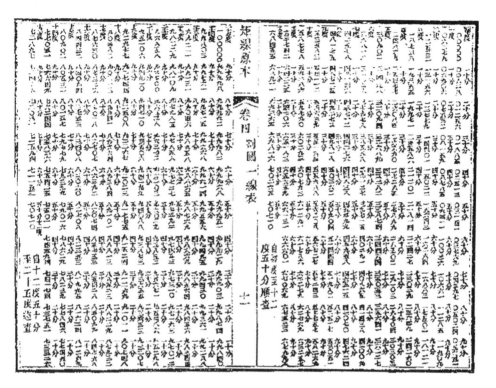

Figure 4.23
An Qingqiao's "Table of One Line."

stance, there is no discussion of issues of convergence, nor is there any need for one.[126] Thus the context for Minggatu's series is perhaps closer to trigonometry than it is to Western-style analysis.

Whether Minggatu's series were actually used to compute tables in practice is unclear. As far as can be told, there seems to have been little interest in computing new tables at this time. One possible exception is An Qingqiao's *Yi xianbiao* ("Table of One Line," 1817–1818), a sine table computed around the time that Minggatu's manuscript was circulating.[127] This unusual table (see figure 4.23) divides the circle not into 360° or even into 365¼ *du* but rather into 100 units, also called *du*. These are subdivided into 100 *fen*, and each *fen* into 100 *miao*. With increments of 10 *fen* from zero to 25 *du* (a right angle), the table

---

[126] In the first chapter of the book, Minggatu discusses the use of trigonometric difference formulas in association with the values of these functions for the arcs 30°, 45°, 60°, and 90° [Jami 1988b, 313]. This speeds the series' convergence, but whether the series converge at all does not arise.

[127] For a discussion of An Qingqiao's table (which uses $R = 100,000$), see [Chen 2015, 517–521]. The treatise also includes a partial tangent table. Image from [Chen 2015, 518].

has only 250 entries. Although this grid leads to large errors when interpolating, An thought of the table's compactness and its simplicity as assets.[128]

The division of the right angle into 25 parts has a significant implication for the table's construction. To derive the sine values, the usual geometric procedures (sine addition and subtraction formulas, sine half-angle formula) are insufficient. This is also true for a right angle divided into 90°; however, we have seen several times before that the use of a sine $\frac{1}{3}$-angle formula—which requires the solution (precise or approximate) of a cubic equation—is sufficient to fill in the rest of the table in this case.[129] For An's *du* units a sine $\frac{1}{3}$-angle formula does not suffice; instead, he needs a sine $\frac{1}{5}$-angle formula, which requires the solution of an equivalent to a quintic equation. An may have relied on some of Minggatu's work, but all that can be said with confidence about the table's values is that An computed independently of any previous tables—perhaps for the first time in China.[130]

Minggatu's contributions did provoke interest. This was partly due to some characterizations of his work not as *xi fa* ("Western method") but as *xin fa* ("new method").[131] Dong Youcheng (1791–1823) was among the first. In 1819 he saw the manuscript of the first chapter of Minggatu's book, containing the nine series but no formulas. Using an entirely different geometric method, he reconstructed proofs of all nine series in *Geyuan lianbili shu tujie*.[132] He related his derivations to the Chinese practice of "methods of piles" (*duoji shu*), which deals with recursive formulas for (finite) sums of powers of integers, Pascal's triangle, and so on. Two years later he was able to access Minggatu's proofs and must have realized that he had come to the same conclusions in a different way. Dong also extended Minggatu's results, giving series for Crd $m\theta$, Crd $\dfrac{\theta}{m}$, Vers $m\theta$, and Vers $\dfrac{\theta}{m}$ where not θ but Sin θ and Vers θ are given.[133]

Before long the trigonometric series, thus far restricted to sines, chords, versed sines and their inverses, were extended to other functions. Xu Youren (1800–1860) dealt with the tangent and its inverse in his *Ceyuan milü* ("Measuring Circle and Its Precise *lü*") perhaps even before the publication of Min-

---

[128] [Chen 2015, 531–532].

[129] We have seen methods like this in the works of Jāmshīd al-Kāshī ([Van Brummelen 2009, 146–149]), François Viète (see chapter 1), and the author of the *Yuzhi shuli jingyun* (earlier in this chapter).

[130] [Chen 2015, 520–521, 525].

[131] See [Horng 1991, 33–38]. According to Luo Shilin, who helped bring Minggatu's work to print in 1839, it is similar to ancient methods of interpolation and "cannot be regarded as the Western method and thereby ignored."

[132] For a summary of Dong Youcheng's method, see [Horng 1991, 153–158].

[133] On Dong Youcheng see [Bréard 2013, 26–28], [Li/Du 1987, 239–240] and [Chen 2015, 526–527].

ggatu's treatise in 1839. In another of Xu's works, he extends Dong's contribution by writing $\tan m\theta$ as an infinite series in terms of $\sin \theta$:[134]

$$\tan m\theta = m\sin\theta + \frac{2m^2+1}{3!\cdot R^2}\sin^3\theta + \frac{16m^4+20m^2+9}{5!\cdot R^4}\sin^5\theta$$
$$+\frac{272m^6+560m^4+518m^2+225}{7!\cdot R^6}\sin^7\theta$$
$$+\frac{7,936m^8+22,848m^6+31,584m^4+25,832m^2+11,025}{9!\cdot R^8}\sin^9\theta+\cdots.$$

$$(4.23)$$

Several other mathematicians expanded the theory of infinite series in other directions. For instance, Li Shanlan (1811–1882) included algorithms for the secant and its inverse in *Hushi qimi* ("Opening the Secrets of Arcs and Sagitta," 1845); and Dai Xu (1806–1860) found series for the cotangent, the cosecant, and their inverses in *Waiqie milü* ("Tangent Outside the Circle and the Precise *lü*," 1852). However, the series seem to have remained as theoretical entities. Although both Xu Youren and Dai Xu wrote on constructing tables with them, in practice series calculations did not take over.[135] Even so, series were a powerful source of mathematical inspiration, and in the hands of the most dominant figure in nineteenth-century Chinese mathematics, they were to provide more surprises.

Li Shanlan's interest in mathematics started at an early age. It is said that he discovered and worked through the problems of the *Jiuzhang suanshu* at age eight. In his mid-teens he mastered Ricci and Xu Guangqi's translation of the *Elements*; the *Ceyuan haijing* (a thirteenth-century algebraic text); and perhaps most impressively, the *Gougu geyuan ji*, Dai Zhen's nearly impenetrable text on trigonometry. An early work demonstrating a new method for computing logarithms brought him to the attention of British missionary Alexander Wylie. Following in the footsteps of Xu Guangqi and Ricci, the two men produced a collection of translations of Western mathematical and scientific texts, including Loomis's *Elements of Analytical Geometry and of the Differential and Integral Calculus* in the late 1850s—although they made no explicit connection between the contents of these books and the Chinese work on series. These translations helped to trigger a process of interaction—sometimes called the second wave of European mathematics in China—between indigenous and Western mathematics over the next few decades. Nevertheless, while Li devoted much of his career to topics inspired by European subjects, he was also a powerful practitioner of native methods and often applied them in creative ways in novel contexts. Li also played a crucial role in professionalizing mathematical

---

[134] [Martzloff 1997, 189–190], quoting a Chinese-language article by Yan Dunjie; the treatise's name is *Geyuan baxian zhuishu*.
[135] [Chen 2015, 528–529].

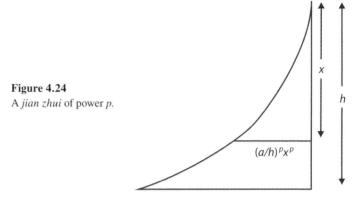

Figure 4.24
A *jian zhui* of power *p*.

$(a/h)^p x^p$

x

h

practice in China, positioning his colleagues with the institutional support for success as they entered the twentieth century.[136]

Li's early work on series is a fascinating example of his ability to adapt traditional methods to new contexts. He embedded his study of series not within *geyuan* as Minggatu had done but rather within the "method of piles" mentioned above. In geometry, areas of figures and volumes of solids were found by approximating them with collections of smaller slices (often rectangular prisms or collections of cubes). They were then added together—a practice that, in the case of an infinite sum, has some parallels with integration. One of the early theorems in Li's *Fangyuan chanyou* ("Explanation of the Square and the Circle") gives the area of a *jian zhui*, portrayed in figure 4.24. The height of the figure is *h*; the horizontal distance across the figure at any vertical distance x from the top is $\left(\dfrac{a}{h}\right)^p x^p$. Li gives only the area formula, but his method has been reconstructed in a way that sounds very much like the limiting process used in a definite integral. The region is divided into many thin horizontal layers. As the number of layers "tends to be infinite," each layer "becomes thin enough to be like a piece of paper." When *n* is finite, the sums of the layers may be evaluated using known results about sums of powers. When *n* is permitted to be infinite, Li has a value for the area that is mathematically equivalent to the statement[137]

$$\int_0^h \left(\frac{a}{h}\right)^p x^p \, dx = \frac{a^p h}{p+1}. \tag{4.24}$$

---

[136] On Li Shanlan's life and work see especially [Horng 1991], [Wang 1996], and also [Martzloff 1997, 173–176]. On the translation of Loomis's calculus see [Horng 1991, 330–353].
[137] See a full treatment in [Horng 1991, 125–133].

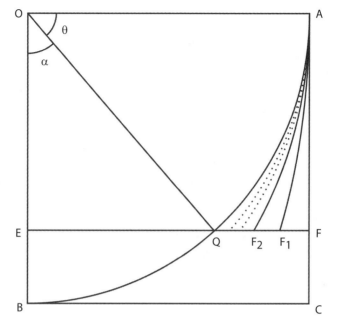

Figure 4.25
Deriving Li
Shanlan's series
for the arc sine.

Armed with this result we may follow Li's path to his infinite series; here we follow him to a series for the arc sine. In figure 4.25 we have a quadrant $OAQB$ of a unit circle. Let $\alpha = \angle BOQ$, and draw a horizontal line EQF through Q. Let $y = AF = \cos \alpha$; then

$$QF = \text{covers } \alpha = 1 - \sqrt{1 - y^2} = \frac{1}{2} y^2 + \frac{1}{2 \cdot 4} y^4 + \frac{1 \cdot 3}{2 \cdot 4 \cdot 6} y^6$$

$$+ \frac{1 \cdot 3 \cdot 5}{2 \cdot 4 \cdot 6 \cdot 8} y^8 + \cdots. \tag{4.25}$$

Li does not explain how he arrived at this series, but we are fortunate to have the word of his brother Li Xinmei. In the expression $1 - \sqrt{1 - y^2}$, let y^2 be the very small number 10^{-8}. Then

$$1 - \sqrt{1 - y^2}$$
$$= 0.000000005000000012500000062500000390625002734375 \cdots.$$

The pattern in this sequence of decimal digits is clear:

$$1 - \sqrt{1 - y^2} = \frac{1}{2} \cdot 10^{-8} + \frac{1}{2 \cdot 4} \cdot 10^{-16} + \frac{1 \cdot 3}{2 \cdot 4 \cdot 6} \cdot 10^{-24} + \frac{1 \cdot 3 \cdot 5}{2 \cdot 4 \cdot 6 \cdot 8} \cdot 10^{-32} + \cdots.$$

So in general it must be true that

$$1 - \sqrt{1 - y^2} = \frac{1}{2}y^2 + \frac{1}{2 \cdot 4}y^4 + \frac{1 \cdot 3}{2 \cdot 4 \cdot 6}y^6 + \frac{1 \cdot 3 \cdot 5}{2 \cdot 4 \cdot 6 \cdot 8}y^8 + \cdots. \quad (4.26)$$

Li Xinmei tries this again for $y^2 = 2 \cdot 10^{-8}$, arriving at the same series and thereby verifying its correctness. Of course, by modern Western standards this would not be acceptable apart from as a heuristic verification, but we are in neither a modern nor a Western context.[138]

Returning to figure 4.26, Li now calculates the area AFQ as a collection of *jian zhui* figures. The first is AFF_1, where $FF_1 = \frac{1}{2}y^2$; the second is AF_1F_2, where $F_1F_2 = \frac{1}{24}y^4$; and so on. Although all the *jian zhui* figures after the first have a curved right edge, the area formula (4.24) still holds. Also, we know that the stacks of *jian zhui* figures fill up the area to the right of $\overset{\frown}{AQ}$ by the series (4.26). Therefore, applying Li's equivalent of (4.24) to each of the *jian zhui* figures in turn, we have

$$AFQ = \frac{1}{2} \cdot \frac{y^3}{3} + \frac{1}{2 \cdot 4} \cdot \frac{y^5}{5} + \frac{1 \cdot 3}{2 \cdot 4 \cdot 6} \cdot \frac{y^7}{7} + \frac{1 \cdot 3 \cdot 5}{2 \cdot 4 \cdot 6 \cdot 8} \cdot \frac{y^9}{9} + \cdots. \quad (4.27)$$

Now, the area of sector AOQ is equal to $\theta/2$, where $\theta = \angle AOQ = 90° - \alpha$. But it is also equal to rectangle $AOEF$, minus triangle OQE, minus our summed *jian zhui* AFQ. This gives us

$$\frac{\theta}{2} = y - \frac{1}{2}y\sqrt{1 - y^2} - AFQ,$$

so

$$\theta = y\left(2 - \sqrt{1 - y^2}\right) - 2AFQ.$$

Substitute series (4.26) for $1 - \sqrt{1 - y^2}$ and series (4.27) for AFQ. We then get

$$\theta = y\left(1 + \frac{1}{2}y^2 + \frac{1}{2 \cdot 4}y^4 + \frac{1 \cdot 3}{2 \cdot 4 \cdot 6}y^6 + \frac{1 \cdot 3 \cdot 5}{2 \cdot 4 \cdot 6 \cdot 8}y^8 + \cdots\right)$$
$$- 2\left(\frac{1}{2} \cdot \frac{y^3}{3} + \frac{1}{2 \cdot 4} \cdot \frac{y^5}{5} + \frac{1 \cdot 3}{2 \cdot 4 \cdot 6} \cdot \frac{y^7}{7} + \frac{1 \cdot 3 \cdot 5}{2 \cdot 4 \cdot 6 \cdot 8} \cdot \frac{y^9}{9} + \cdots\right). \quad (4.28)$$

---

[138] [Horng 1991, 134–141].

Combining terms, we have

$$\theta = y + \frac{1}{2 \cdot 3} y^3 + \frac{1 \cdot 3}{2 \cdot 4 \cdot 5} y^5 + \frac{1 \cdot 3 \cdot 5}{2 \cdot 4 \cdot 6 \cdot 7} y^7 + \cdots. \tag{4.29}$$

But $y = \sin \theta$, so Li has a series for the arc sine:[139]

$$\theta = \sin \theta + \frac{1}{2 \cdot 3} \sin^3 \theta + \frac{1 \cdot 3}{2 \cdot 4 \cdot 5} \sin^5 \theta + \frac{1 \cdot 3 \cdot 5}{2 \cdot 4 \cdot 6 \cdot 7} \sin^7 \theta + \cdots. \tag{4.30}$$

Minggatu's and Li Shanlan's series derivations, both brilliant, nevertheless reveal a sharp distinction in character. Both men arrived at their series using an iterative process. But Minggatu worked geometrically, embedding smaller and smaller isosceles triangles within his diagram. One might draw an analogy between it and Archimedes's approximations to π obtained by embedding regular n-gons within a circle. Li Shanlan's method, however, is algebraic. Even his *jian zhui*, admittedly geometric shapes, are defined with the algebraic freedom of permitting the length across a *jian zhui* to depend on a power of x.[140] The geometry of the specific situation plays a much smaller role, suggesting that the method might be applied in other contexts. Given time to develop, this freedom might have led to a systematic theory similar to some of the work of Fermat and even Isaac Newton. Some of Li's series were derived using an extension of the process of long division, just as Newton had done.

The shift in emphasis here from geometry to algebra raises the interesting question whether what we have seen here of Li's work can even be described as trigonometric. Certainly, Li's goal was not specifically related to questions that trigonometry was asking at the time. In this respect he was not doing trigonometry, at least in the way that his predecessors had done. However, one cannot resist an analogy to what had happened to trigonometry in Europe: as calculus and analysis began to revolutionize mathematical practice, trigonometry itself was subjected gradually to a reinvention in algebraic terms. One wonders whether Li's combination of trigonometric functions and infinite series might have provoked a similar movement in China.

But, given the lack of evidence, one does not wonder for very long. We will never know whether Li Shanlan's work would have developed into a Chinese equivalent to calculus, and a corresponding re-visioning of trigonometry. His translations of European mathematical texts were a part of the sec-

[139] [Horng 1991, 159–161].
[140] [Horng 1991, 165].

ond flood of Western learning, advancing the field considerably but disrupting indigenous approaches. The influx was provoked in part by the two Opium Wars (1839–1842 and 1857–1860) pitting the United Kingdom and France against the Qing dynasty. The resulting weakening of the Qing caused the collapse of the closed-door policy to the West, and China was subsequently opened for trade with Europe and America. The military superiority of the outnumbered European forces convinced many in China that the West had knowledge worth knowing. The European texts that entered China, including Li Shanlan's translations, helped eventually to incorporate China into a global, Western-inspired science and mathematics power. Trigonometry participated in this process. In 1877 John Hymers's *Treatise on Plane and Spherical Trigonometry* appeared in China, defining trigonometric quantities as we do now—as ratios of sides in a triangle rather than as lengths of line segments.[141] Hymers's text includes a chapter on the relation between trigonometric and exponential functions through complex numbers and provides a direct treatment of trigonometric quantities as functions in the modern sense. After two centuries of near separation, the assimilation of trigonometry into the Western perspective had begun.

[141] [Hymers 1858]; see [Chen 2015, 532].

5 ⚞ Europe After Euler

Revolution does not happen overnight. While analysis was clearly the way of the future in Europe, trigonometry had ancient roots that continued to provide nourishment in traditional directions. After Euler, research continued in two different styles—some of it injected with analytical approaches, some of it remaining unremittingly geometric. The same was true in mathematics education where a fierce debate over the proper manner to engage students in the subject flared up in textbooks.

The geometric side continued well into the nineteenth century in a manner that one might describe as "normal science." A number of new theorems, especially in spherical trigonometry, were added to the collection, but major breakthroughs were not found. One does find a continuation of the trend that saw trigonometry embedded more and more deeply in disciplines beyond astronomy, particularly in geodesy and navigation. While analysis had something to do with this, it is fairer to say that science and various practical disciplines were simply beginning to use more of the tools that mathematics provided and that traditional approaches found applications just as well. The increased visibility of trigonometry outside of mathematics drove forward various advances especially in the computational side of the subject, dealing for instance with the efficient generation of improved numerical tables and studies of the propagation of error.

Some of the most critical advances in trigonometry after Euler also took place in conjunction with advances in other areas of both pure and applied mathematics. We saw in chapter 3 the birth of infinite polynomial series representing trigonometric quantities. Already in Euler's work sines and cosines were starting to be used as the building blocks of another kind of series, and this reached fruition with Fourier's studies of the transfer of heat through a medium. The discovery of non-Euclidean geometries merged with the ancient study of spherical astronomy to consider the question of doing trigonometry on surfaces other than the flat plane. This was motivated partly by interest in foundational mathematical questions. There had already been interest in working out the nature of "spheroidal" trigonometry due to the discovery that the earth is not a perfect sphere.

This chapter covers developments that are more or less post-Euler, from the late eighteenth century onward. Our treatment is somewhat episodic. As trigonometry gradually became a supporter of research in other areas and part of the collection of subjects in a general mathematics education, developments occurred more and more frequently to respond to various outside pressures.

A collection of short stories is thus more appropriate here than a continuous narrative.

▨ Normal Science: Gap Filling in Spherical Trigonometry

After Euler, some of mathematical life went on much as before. There were fewer opportunities for dramatic improvements in trigonometry, but there were still openings to exploit. Many of these opportunities were in spherical trigonometry where interesting theoretical questions had yet to be answered that may not open up new horizons but could yet be meaningful. These included, for instance, thorough treatments of cases where the triangle's solution is ambiguous (such as angle-angle-side triangles); new identities that grouped triangle quantities in novel ways; and new methods for calculating a triangle's spherical excess (and hence its area) without knowledge of its angles, such as L'Huilier's Theorem.[1] Since spherical trigonometry is not as familiar to a modern audience as it once was, we shall not attempt to cover all of the many developments that took place. However, the topic was still of considerable interest to the historical actors of the time, and we would do them a disservice to ignore it entirely.

Many of the new theorems were the result of attempts to solidify, generalize, or extend existing knowledge into wider domains of applicability or new contexts. Some included considering the validity of theorems where the arcs and angles take on values greater than 90°. Geometer and astronomer August Ferdinand Möbius took this further in an 1846 paper, in which he treated spherical trigonometry with his barycentric coordinate system and dealt with arcs and angles up to 360°.[2] Others sought for symmetries in the formulas of the subject, either to unify the structure of an increasingly complicated collection of identities or to present a system to users of spherical trigonometry that could be remembered easily in practice. One typical effort was by Alexandre Guy Pingré (1711–1796). Originally a priest who fell under suspicion of Jansenism, Pingré's career continued in astronomy and naval geography. His most important works were a two-volume study of comets and a history of seventeenth-century astronomy, but he also wrote a nautical almanac and

---

[1] $\tan\dfrac{E}{4} = \sqrt{\tan\dfrac{s}{2}\tan\dfrac{s-a}{2}\tan\dfrac{s-b}{2}\tan\dfrac{s-c}{2}}$, where the spherical excess E is the excess above 180° of the angle sum of a triangle and s is the triangle's semiperimeter. Another result worth noting, from spherical geometry but sometimes proved with trigonometry, is Lexell's Theorem: given a spherical triangle with a known base and area, the locus of possible locations for the third vertex is a certain circle on the sphere. The result would eventually become important in the development of hyperbolic geometry. See [Papadopoulos 2014] and [Atzema 2017].

[2] [Möbius 1846]; see [von Braunmühl 1900/1903, vol. 2, 186–189] for a discussion. For a description of Möbius's barycentric coordinates, see [Gray 1993, 81–85].

participated in observations of the transit of Venus. His poor eyesight shifted his focus from observational practice to calculations, and it is from this that he proposed an extension to spherical trigonometry, "La trigonométrie sphérique réduite à quatre analogies."[3]

Text 5.1
Pingré, Extending Napier's Rules to Oblique Spherical Triangles

The proofs of the analogies of spherical trigonometry are easily conceived, but they are not retained in the same way, the multitude of these analogies fatigues the memory; it takes a practice that is constantly followed to ensure that we do not confuse them: consequently we prefer take the analogies in the books, and this help coming to naught, we try in vain to remind ourselves of principles too long neglected.

I speak of this embarrassment by my own experience. . . . I finally saw a succinct trigonometric treatise; I found there all the thirteen analogies of right spherical triangles reduced to only two.[4] . . . Convinced of the usefulness of these two rules, I believed that it may be possible to reduce the theory of oblique triangles similarly to a small number of rules, and I believe I have succeeded. . . .

Third general rule: In every spherical triangle, the sines of analogous segments are proportional to the tangents of the adjacent parts.

Fourth general rule: In every spherical triangle, the cosines of analogous segments are proportional to the sines of the opposite parts.[5]

Figure 5.1
Pingré's representation of Napier's rules. An appearance of "co" with an angle or side instructs the user to apply the complement of that magnitude, rather than the magnitude itself.

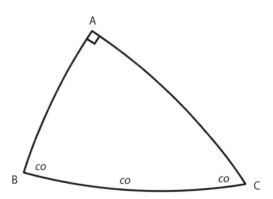

Explanation: (See figure 5.1.) Pingré is referring to Napier's rules of circular parts for right spherical triangles, which we discussed in chapter 2. In Pingré's

---

[3] [Pingré 1756].

[4] Pingré refers here to John Keill's *Introductio ad veram physicam* (1702), but he likely confused it with Keill's *Trigonometriae planae et sphaericae: Elementa*. See [Keill 1726].

[5] [Pingré 1756, 301–303], translated from the French.

diagram and terminology, the right angle is at *A*, and the five circular parts, traveling around the triangle, are *AC*, co-*C*, co-*BC*, co-*B*, and *AB*. ("Co-" refers to the complement of the quantity.) Napier's rules state that (i) the sine of any of these parts is equal to the product of the tangents of the two parts adjacent to it,[6] and (ii) the sine of any part is equal to the products of the co-sines of the two parts that are not adjacent to it. For instance, (i) cos *B*=cot *BC* tan *AB* and (ii) sin *AB*=sin *BC* sin *C*.

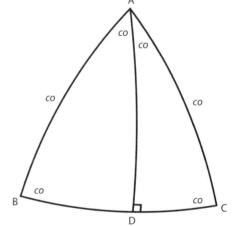

Figure 5.2
Pingré's third and fourth rules.

Pingré adds third and fourth rules to the list that apply to oblique triangles and gives two examples. Suppose (figure 5.2) that we are given *AB*, *BC*, and angle *B*; the goal is to find angle *C*. Drop a perpendicular from *A* onto *BC*, defining *D*.[7] We are told first to apply Napier's rule (i) to the left triangle, which gives cos *B*=tan *BD* cot *AB*. This gives us *BD*, and since we know *BC*, we also have *CD*. Pingré's "analogous segments" are either the two arcs *BD* and *DC* or the two angles *B* and *C*, depending on how one wishes to apply the rule. In this case he chooses the former, and his rule gives

$$\frac{\sin BD}{\sin CD} = \frac{\tan(\text{co-}B)}{\tan(\text{co-}C)} = \frac{\cot B}{\cot C}. \tag{5.1}$$

This allows him to solve for the sought angle *C*. (Later in the article he proves this result.)

Pingré's fourth rule works similarly. In the same diagram, given *AB*, ∠*B* and ∠*C*, his goal is to find ∠*A*. He begins as before by using the first of Na-

[6] The five parts are in a cycle, so *AB* and *AC* are considered to be adjacent. Pingré's work is given for a base circle of radius *R*; we express his results here in terms of the unit circle.
[7] Pingré also gives instructions for dealing with cases where the perpendicular does not lie within *BC*.

pier's rules on triangle ABD to get cos $AB = $ cot $\angle BAD$ cot $\angle B$, giving him $\angle BAD$. Then, applying the fourth rule, he has

$$\frac{\sin \angle BAD}{\sin \angle CAD} = \frac{\cos \angle B}{\cos \angle C},$$

(5.2)

which gives him $\angle CAD$. He adds $\angle BAD$, giving the desired $\angle A$.

Pingré's method, ingenious as it is, does not completely succeed in capturing all situations: as he admits, it does not provide a method for solving triangles when all the sides or all the angles are known.

Pingré's work was typical of its time in several ways. As the number of identities began to proliferate, astronomers needed compact and easily remembered tools to navigate them. The Laws of Sines and Cosines along with the rules for right triangles were already sufficient for most tasks, but as Pingré says, a more streamlined approach would ease computational labors. This kind of work was not really transformative as it proceeded in the traditional geometric style of its predecessors dating back centuries. There was some argument about the merits of analytical methods in spherical trigonometry, but mathematical style is a personal choice rarely altered by debate.

Proponents of analysis had no qualms about subsuming trigonometry within their purview. One of its leading figures, who we saw in chapter 3 in his youth collaborating with Euler, was Joseph-Louis Lagrange. Turning his attention to spherical trigonometry at the end of the eighteenth century, he did his utmost to wrestle the subject away from geometry as much as he thought it was possible to accomplish. After noting that Euler's 1782 paper (which we saw in chapter 3) had left him requiring a mere three geometrical derivations on which to build the entire subject, he says:

> But can one not simplify this system more, by reducing it to a single fundamental equation? This reduction would serve to perfect the analytical theory of spherical triangles; for, in analysis, perfection consists in using only the fewest possible principles, and in removing from these principles all the truths that they can contain, by the force of analysis alone.[8]

He is true to his word. His system relies on just a single theorem proved using geometry: the spherical Law of Cosines.[9]

---

[8] [Lagrange 1799, 280].

[9] Lagrange notes that Jean Paul de Gua de Malves had done the same thing in 1783 but that his work "contains calculations so complicated that they seem more proper to show the inconvenience of his method than to adopt it." [Chemla 1989, 355] points out that both he and De Gua

One collection of new theorems did gain traction, and again the result came from astronomy rather than mathematics via a solution by M. Henri to a problem of calculating parallax effects of the planets. Henri's solution involved the use of formulas closely related to Napier's analogies, which we saw in chapter 2.[10] The article was followed by an 1809 commentary by Jean Baptiste Joseph Delambre generally praising Henri's solution but pointing out a problem in astronomical practice.[11] The method required the use of the term $\sin\frac{1}{2}(A - B)$ in a denominator, but in the situation where A and B are close, this results in a denominator with a small value, leading to a loss of accuracy in calculation. Fortunately Delambre had an alternative at hand. In modern form, **Delambre's analogies** (where A, B, and C are the angles of a spherical triangle and a, b, and c are the opposite sides respectively) are as follows:

$$\frac{\sin\frac{1}{2}(A+B)}{\cos\frac{1}{2}C} = \frac{\cos\frac{1}{2}(a-b)}{\cos\frac{1}{2}c}$$
$$\frac{\sin\frac{1}{2}(A-B)}{\cos\frac{1}{2}C} = \frac{\sin\frac{1}{2}(a-b)}{\sin\frac{1}{2}c}$$
$$\frac{\cos\frac{1}{2}(A+B)}{\sin\frac{1}{2}C} = \frac{\cos\frac{1}{2}(a+b)}{\cos\frac{1}{2}c}$$
$$\frac{\cos\frac{1}{2}(A-B)}{\sin\frac{1}{2}C} = \frac{\sin\frac{1}{2}(a+b)}{\sin\frac{1}{2}c}. \tag{5.3}$$

(Although two of these identities contain the same sorts of terms that had caused problems for Henri's methods, there is always a way to bypass those terms in practice.) Delambre stated the identities without proof, although he hinted that he had known them for a while.

Over the following two years, the analogies appeared twice more: in Gauss's 1809 *Theoria motus corporum coelestium*, again without proof;[12] and in a November 1808 article by German astronomer Karl Mollweide (known best for a map projection), containing the first proof.[13] This tangled case of near-simultaneous discovery by three different authors was made even thornier by the fact that Delambre's statement of the theorems was published

---

"take for granted that the variables which these formulas imply correspond to precise geometrical entities to which one can therefore apply permutations." See also [Chemla/Pahaut 1988, 166–171]. An ironic twist to this tale is Friedrich Theodor von Schubert's effort to base spherical trigonometry solely on Menelaus's Theorem, thereby both advancing the subject and returning it to its ancient roots [von Schubert 1796].

[10] [Henri 1809].
[11] [Delambre 1809].
[12] [Gauss 1809, 51].
[13] [Mollweide 1808].

in an issue of the French astronomical journal *Connaissance des Temps*, which dated their issues two years after publication. So Delambre's announcement was thought to date from 1809, whereas it actually appeared in 1807. As a result the theorems were known as Gauss's analogies in textbooks until the situation was cleared up later in the century.[14] Delambre's analogies became a staple of spherical trigonometry textbooks not because of their numerical stability but rather because they contain all six elements of a triangle. Thus they can be used to check whether a purported solution of a triangle contains any errors.

Two of Delambre's analogies are spherical equivalents of **Mollweide's formulas** of plane trigonometry, which once appeared in textbooks but today are increasingly rare:

$$\frac{a-b}{c} = \frac{\sin\frac{1}{2}(A-B)}{\cos\frac{1}{2}C}$$
$$\frac{a+b}{c} = \frac{\cos\frac{1}{2}(A-B)}{\sin\frac{1}{2}C}.$$

(5.4)

Indeed, these latter formulas also appear in Mollweide's 1808 article proving Delambre's analogies. However, it remains a puzzle how Mollweide's names became associated with them: even in his own article he cites Antonio Cagnoli's classic 1786 textbook, in which the results appear.[15] Mollweide's formulas may also be found in various earlier works, starting not only with Isaac Newton's 1707 *Arithmetica universalis* but also in the writings of Anthony Thacker, Friedrich Wilhelm von Oppel, Thomas Simpson, William Emerson, Antoine René Mauduit, Basil Nikitin, and Prochor Souvoroff.[16]

Delambre's analogies were manipulated into one last set of important identities by German textbook writer Friedrich Reidt in his 1872 *Sammlung von Aufgaben und Beispielen der Trigonometrie und Stereometrie* (1872).[17] **Reidt's analogies**, arrived at with the sine and cosine sum-to-product formulas and some algebraic prestidigitation, made triangles easier to solve. In an

---

[14] Isaac Todhunter, author of the leading spherical trigonometry textbook of the second half of the nineteenth century, resolved the controversy in [Todhunter 1873b].

[15] See [Cagnoli 1786, 117–118].

[16] [Newton 1707, 121–122]. The convoluted history of the Mollweide formulas is outlined carefully in [von Braunmühl 1901b] and in [Wu 2007]. It has been suggested that the story goes back to the originator of the Law of Tangents (a closely related formula), Thomas Fincke [Schönbeck 2004]. At least once, it has been given the impressive name "Newton-Oppel-Mauduit-Simpson-Mollweide-Gauss formula."

[17] [Reidt 1872, 212–213], in a different form. The identities are also in Serret's *Traité de Trigonometrie* but not until the fifth edition [Serret 1875, 156–157]. Thanks are due to my student Kailyn Pritchard, who untangled the mathematics of Reidt's analogies.

oblique triangle, let $s = \frac{1}{4}(A + a)$, $s' = \frac{1}{4}(B + b)$, and $s'' = \frac{1}{4}(C + c)$; also let $d = \frac{1}{4}(A - a)$, $d' = \frac{1}{4}(B - b)$, and $d'' = \frac{1}{4}(C - c)$. Then

$$\tan^2(45° - s'') = \cot(s - s')\,\tan(s + s')\,\tan(d - d')\,\tan(d + d'),$$

$$\tan^2(45° - d'') = \tan(s - s')\,\tan(s + s')\,\cot(d - d')\,\tan(d + d'),$$

$$\tan^2 d'' = \tan(45° - s + d')\,\tan(45° - s + d')$$
$$\tan(45° - d - s')\,\tan(45° + d - s'), \text{ and}$$

$$\tan^2 s'' = \tan(45° - s - d')\,\tan(45° + s - d')$$
$$\tan(45° - d - s')\,\tan(45° - d + s'). \tag{5.5}$$

The benefits of these new theorems over Napier's and Delambre's analogies are twofold: they deal more smoothly with the situation where a, b, and A are known (an ambiguous case with two different triangles that satisfy the conditions), and they reduce the required number of logarithm lookups when calculating with logarithms (as all practitioners did at the time; see figure 5.3).[18] With today's modern computational tools, Reidt's analogies have fallen into disuse even among those few who still practice spherical trigonometry.

Working largely without mechanical aids, astronomers, geographers and students at this time were still strongly motivated to find the simplest pencil-and-paper solutions. This was especially true in spherical trigonometry where the problems could get complicated and the formulas frequently required that one multiply or divide irrational quantities. Tables of logarithms flourished in this environment and were in ready supply throughout the nineteenth century. Formulas that employ both addition/subtraction and multiplication/division, such as the planar and spherical Laws of Cosines, are not amenable to calculation with logarithms: while $\log(ab) = \log a + \log b$, there is no simple formula for $\log(a + b)$. Thus, although by 1800 every triangle could be solved with existing tools in principle, there were still reasons to search for new identities like Delambre's and Reidt's analogies (see figure 5.4). Often it was not the mathematical community that drove the search for logarithm-friendly alternatives but rather those who could benefit directly from such discoveries: astronomers, scientists, and textbook writers.

Some of these new formulas are on offer in Johann Heinrich Lambert's extensive article "Comments on and additions to trigonometry." This innocuous title obscures what is actually a comprehensive approach to solving plane and spherical triangles, including dozens of identities that are hardly known today. For instance,

$$(\sin y + \sin z)(\sin y - \sin z) = \sin(y + z)\,\sin(y - z), \tag{5.6}$$

---

[18] See [Casey 1889, 41–42, 59–60] and [Todhunter/Leathem 1901, 40–41, 78–79].

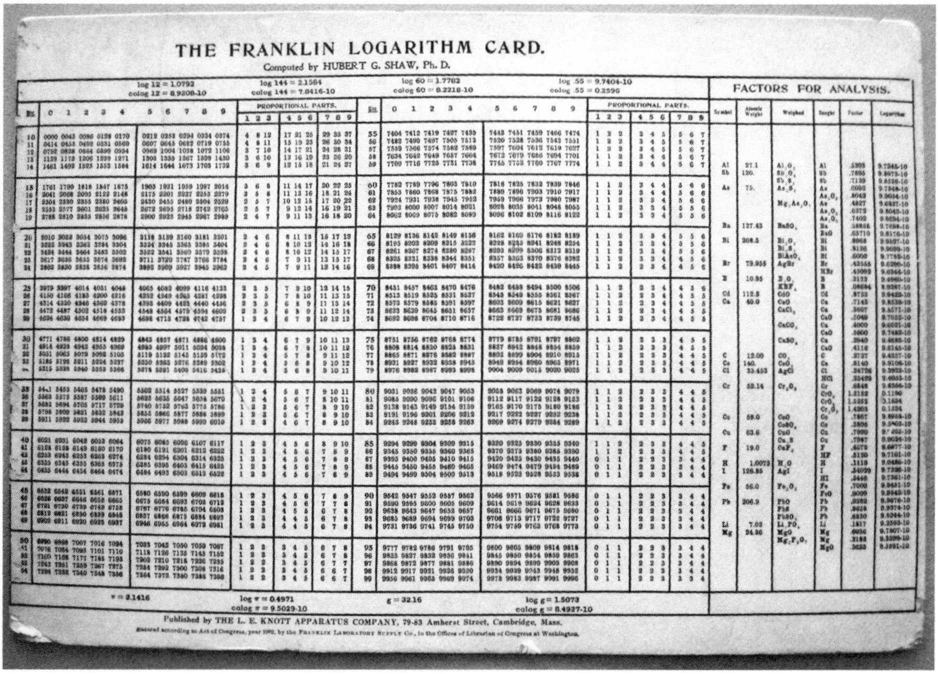

Figure 5.3
A typical table of logarithms for use in schools (1902). Credit: The author.

which may be verified by expanding the product on the left and expanding using the sine sum and difference laws on the right.[19]

The quintessential analyst, Lambert applies this identity to find an alternative to the spherical Law of Cosines without reference to a single diagram. Beginning with

$$\cos a = \cos b \cos c + \sin b \sin c \cos A, \tag{5.7}$$

we substitute $\cos A = 1 - 2\sin^2 A/2$. After some manipulation and the application of the cosine difference law, we have

$$\frac{\cos a}{2\sin b \sin c} = \frac{\cos(b-c)}{2\sin b \sin c} - \sin^2 \frac{A}{2}.$$

Set φ so that

$$\sin^2 \varphi / 2 = \frac{\cos(b-c)}{2\sin b \sin c}. \tag{5.8}$$

[19] [Lambert 1792, 389].

§. 37.] Allg. sphaer. Dreieck. Fund. Aufgaben. 213

$$\operatorname{tang} d^2 = \operatorname{tang}(s' - s'') \,.\, \operatorname{tang}(d' - d'') \,.\, \operatorname{cotg}(s' + s'') \,.\, \operatorname{tang}(d' + d''),$$
$$\operatorname{tang} s^2 = \operatorname{tang}(s' - s'') \,.\, \operatorname{tang}(d' - d'') \,.\, \operatorname{tang}(s' + s'') \,.\, \operatorname{cotg}(d' + d'').$$

1938.	$b = 70^0\,40'$	$\log \sin b = 9{,}97479$	$\beta_1 = 69^0\,34'\,30''$
	$c = 40.\ 20$	$\log \sin c = 9{,}81106$	$\beta_2 = 110.\ 25.\ 30$
	$\gamma = 40.\ 0$	$0{,}16373$	Vergl. 1939.
		$\log \sin \gamma = 9{,}80807$	
		$\log \sin \beta = 9{,}97180$	

$4s' =$	$140^0\,14'\,30''$
$4d' =$	$-1.\ 5.\ 30$
$4s'' =$	$80.\ 20.$
$4d'' =$	$-0.\ 20.$
$s' =$	$35.\ 3.\ 37{,}5$
$d' =$	$-0.\ 16.\ 22{,}5$
$s'' =$	$20.\ 5.$
$d'' =$	$-0.\ 5.$
$s' - s'' =$	$14.\ 58.\ 37{,}5$
$d' - d'' =$	$-0.\ 11.\ 22{,}5$
$s' + s'' =$	$55.\ 8.\ 37{,}5$
$d' + d'' =$	$-0.\ 21.\ 22{,}5$

$\log \operatorname{tang}(s' - s'')$	$= 9{,}42736$	$45^0 - s$	$= 0^0\,36'\,7'',4$
$\log \operatorname{tang}(d' - d'')$	$= 7{,}51968\,n$	$45^0 - d$	$= 40.\ 21.\ 2{,}4$
$\log \operatorname{tang}(s' + s'')$	$= 0{,}15710$	$90^0 - 2s$	$= 1.\ 12.\ 14{,}8$
$\log \operatorname{tang}(d' + d'')$	$= 7{,}79364\,n$	$90^0 - 2d$	$= 80.\ 42.\ 4{,}8$
$\log \operatorname{tang}(45 - s)^2$	$= 6{,}04306$	$2s$	$= 88.\ 47.\ 45{,}2$
$\log \operatorname{tang}(45 - d)^2$	$= 9{,}85842$	$2d$	$= 9.\ 17.\ 55{,}2$
$\log \operatorname{tang} d^2$	$= 4{,}58358$	α_1	$= 98.\ 5.\ 40$
$\log \operatorname{tang} s^2$	$= 9{,}31050$	a_1	$= 79.\ 29.\ 50$
	$8{,}02153$	oder d	$= 0.\ 6.\ 43{,}8$
	$9{,}92921$	s	$= 24.\ 19.\ 42{,}5$
	$7{,}29179$	$2d$	$= 0.\ 13.\ 27{,}6$
	$9{,}65525$	$2s$	$= 48.\ 39.\ 25{,}0$
		α_2	$= 48.\ 52.\ 53$
		a_2	$= 48.\ 25.\ 57.$

Figure 5.4

From Reidt's *Sammlung von Aufgaben und Beispielen aus der Trigonometrie und Stereometrie* (1872). Two of his analogies appear at the top of the page. The rest is the solution of a spherical triangle using his analogies and logarithms.

Then

$$\frac{\cos a}{2\sin b \sin c} = \sin^2 \frac{\varphi}{2} - \sin^2 \frac{A}{2} = \left(\sin \frac{\varphi}{2} + \sin \frac{a}{2}\right)\left(\sin \frac{\varphi}{2} - \sin \frac{a}{2}\right).$$

Applying (5.6) above, we have

$$\frac{\cos a}{2\sin b \sin c} = \sin\left(\frac{\varphi + A}{2}\right)\sin\left(\frac{\varphi - A}{2}\right),$$

or

$$\cos a = 2\sin b \sin c \sin\left(\frac{\varphi + A}{2}\right)\sin\left(\frac{\varphi - A}{2}\right). \tag{5.9}$$

This bit of wizardry illustrates the distance that computers were willing to go to avoid multiplying numbers by hand. The use of (5.9) requires that one must first find φ by (5.8), and (5.8) and (5.9) require multiplications, divisions, and a square root. Now apply logarithms to both formulas: products become sums; divisions become differences; and the square root becomes a halving.[20] No multiplications or divisions are needed.

▪ Symmetry and Unity

Lambert's paper exemplifies the dramatic expansion of the list of trigonometric formulas at this time. The identities that he labels with numbers, not including the equations within his own derivations, amount to over 60, and new identities beyond these were being discovered continually. We have seen examples of two different kinds of efforts to cut the mountain down to size: Pingré, with the practical goal of producing an easily memorized system for the benefit of clients, and Lagrange, with the philosophical goal of enclosing the subject within analysis.

At the same time, symmetries in many of the subsets of identities (especially in spherical trigonometry) were signals of an underlying unity that could be lost in the algebraic weeds. In chapter 2 we saw already that John Napier had tied together the ten formulas of right-angled spherical triangles in the *pentagramma mirificum*. But earlier in chapter 1, we also saw the discovery of the polar triangle by Viète, Snell, and van Lansberge.[21] The three sides of

---

[20] [Lambert 1792, 415–419]. In the case where $\cos(b-c)/(2\sin b \sin c) > 1$, then according to (5.8), φ does not exist. In these situations Lambert provides another process involving tangents rather than sines.

[21] Given a spherical triangle ABC with sides a, b, and c, construct another triangle by taking as its vertices a pole of each of the three sides. Every great circle divides the sphere into two hemi-

the polar triangle turn out to be the supplements of the angles in the original triangle, and the three angles turn out to be the supplements of the original sides. It can also be shown that the polar triangle of the polar triangle is the original triangle; therefore, the mapping between the original and the polar triangle is a duality relation.

Since $\sin(180° - \theta) = \sin \theta$ and $\cos(180° - \theta) = -\cos \theta$, this relation allows any identity to be transformed simply by switching the upper case letters (the angles) to lower case letters (the sides) and vice versa and by changing the signs of the cosine terms. For instance, the Law of Cosines

$$\cos c = \cos a \cos b + \sin a \sin b \cos C \qquad (5.10)$$

becomes the Law of Cosines for Angles:

$$-\cos C = \cos A \cos B - \sin A \sin B \cos c. \qquad (5.11)$$

These facts were well known at the time of Euler, although the term "duality" was not yet used. However, in the early nineteenth century duality relations were also appearing in other mathematical disciplines, particularly projective geometry. One of those to take notice was Joseph Diaz Gergonne (1771–1859). Originally a military man, he settled into mathematics in Nîmes and was influenced by Gaspard Monge, who had advanced both differential and descriptive geometry. One of his biggest contributions was institutional: he established the *Annales de Mathématiques Pure et Appliqués* (1810–1832), often called the *Annales de Gergonne* and a predecessor of Crelle's famous *Journal*. Gergonne's journal was a meeting place for work in projective and algebraic geometry, and polarity and duality became major topics; it was in this journal that these words were introduced. They were applied to the study of conic sections, certain curves and surfaces, polyhedra, projective geometry, and spherical trigonometry, among other subjects.

Now, one can approach duality in different ways. In chapter 3, we saw that Euler had used dual formulas to assert the existence of the polar triangle rather than using the polar triangle to assert the existence of dual formulas. Duality can apply not only to geometrical diagrams and theorems but also to proofs, a fact that Euler had exploited. Suppose one has two theorems in a dual relationship; it is often possible to produce proofs of the two theorems that are themselves dual to each other. Of course, if all one cares about is producing new theorems, it is unnecessary to prove the

---

spheres, each of which contains one of the great circle's poles. In all three cases (a, b, and c), choose the pole in the hemisphere that contains the original triangle.

two theorems in this way. However, Gergonne was interested in revealing the inherent symmetry even in the logic of spherical trigonometry and other disciplines. Gergonne demonstrated that both analytic and geometric proofs could be dual to each other. He did this by presenting the dual proofs side by side in two columns so that the reader could observe the parallels between them. He considered duality to be not so much a product of the new analysis but a unifying structure that could be applied across subdisciplines.[22]

Structures like duality were to find a home within abstract algebra over the following decades, and spherical trigonometry interacted with this new discipline once more as the nineteenth century drew to a close. In 1893 the German mathematician Eduard Study, known for his work in invariant theory, published the treatise *Spherical Trigonometry, Orthogonal Substitutions, and Elliptic Functions*.[23] His work, relying heavily on the tools of group theory, was soon succeeded by Grace Chisholm Young's PhD dissertation, *Algebraic-Group Theoretic Investigations of Spherical Trigonometry*.[24] Young's volume, done in Göttingen under Felix Klein's supervision, was the first doctorate in mathematics by a woman completed with an examination in Germany.[25]

▨ The Return of Stereographic Projection

The stereographic projection of a sphere onto a plane is not inherently linked with trigonometry, but the two subjects have been associated with each other for many centuries, mostly through their connections with astronomy. The astrolabe is a physical representation of a stereographic projection of the celestial sphere onto a metal plate. Its advantages are that it maps circles onto circles and lines and that it preserves angles through the mapping; this makes the astrolabe easy to build and to analyze. Stereographic projection has

---

[22] Gergonne's main article on duality in spherical trigonometry is a presentation of work by Sorlin, probably rewritten by Gergonne [Sorlin/Gergonne 1825]. On duality in Gergonne's work, see [Gergonne 1824] and [Gergonne 1826]; the latter contains several two-column presentations of dual geometric arguments. These works are analyzed, along with related works by Euler and other predecessors, in [Chemla/Pahaut 1988], [Chemla 1989], and [Chemla 1994]. Also on the topic of unity in early nineteenth-century geometry, see [Chemla 1998] on Lazare Carnot's study of generality in geometry using the planar version of Menelaus's Theorem.

[23] [Study 1893].

[24] [Chisholm 1895].

[25] Sophia Kovalevskaya's doctorate, granted in 1874, did not involve course work or an examination.

already appeared several times in this book: a number of authors adopted it to help solve spherical triangles, and Lambert used it to examine the *pentagramma mirificum* behind Napier's rules.

In the nineteenth century, stereographic projection became a method not only to solve triangles but also to solve problems in spherical astronomy.[26] The most creative use, however, was a new method that went beyond projecting triangles: projecting theorems. It began with *Des Méthodes en Géométrie* (1855), a textbook by French geometer Paul Serret.[27] He opens his book with a lament that modern students (especially at the École Polytechnique) are unaware of the rich history of mathematics, a complaint with which we are inclined to agree. His goal is to bring together the achievements of the past with those of the present, to synthesize geometry as it had been practiced up to his time. Serret's own words would prove to be ironic: his revolutionary method forged a completely novel path to the theorems of spherical trigonometry but was forgotten almost immediately. It would be rediscovered under another name half a century later.

Recall that stereographic projection maps points on a sphere onto the plane passing through the sphere's equator as follows. On figure 5.5 (a simplified version of Serret's diagram), join the arbitrary point B on the sphere and the south pole S with a straight line. The location B' where the line crosses the plane is the projection of B. From this figure, one sees immediately that the north pole A on the sphere is projected onto A', the center of the sphere.

Given any spherical triangle ABC, orient it so that vertex A is at the north pole. Then two of the sides, $\widehat{AB}$ and $\widehat{AC}$, project onto line segments $A'B'$ and $A'C'$ while the third side $\widehat{BC}$ projects onto arc $\widehat{B'C'}$ on the plane. Serret draws tangents to $\widehat{B'C'}$ from endpoints B' and C', which defines an intersection point t and leads to the configuration of figure 5.6. Since angles are preserved through the mapping, the angles at B' and C' are still equal to angles B and C. Serret then proves that if we connect B' and C' with a straight line, the angles in the resulting plane triangle are A, $B - E$, and $C - E$, where E is the spherical triangle's spherical excess (half the excess of the angle sum

---

[26] One example of this is [Mollweide 1809], which provides an elegant derivation of a solution to a problem discussed in [Gauss 1808] dealing with altitudes of stars for the purpose of navigation. Within it we find a derivation of one of Napier's analogies in an astronomical context.

[27] [Serret 1855, 30–44]; not to be confused with Joseph Alfred Serret, another nineteenth-century French mathematician. An account of Serret's methods also may be found in [Casey 1889, 121–127].

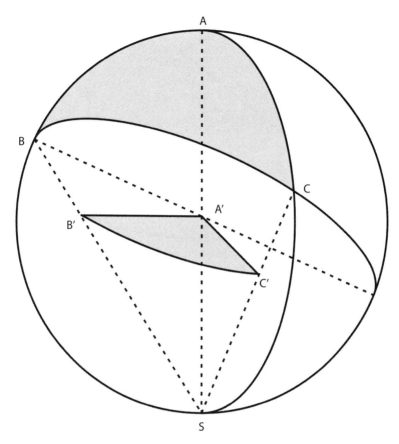

Figure 5.5
Serret's stereographic projection of spherical triangle ABC onto the planar figure $A'B'C'$.

above 180°).[28] He also derives expressions for the lengths of the three sides of the plane triangle:

$$c' = A'B' = \tan\frac{c}{2}, \; b' = A'C' = \tan\frac{b}{2}, \text{ and } a' = B'C' = \frac{\sin\frac{a}{2}}{\cos\frac{b}{2}\cos\frac{c}{2}}. \quad (5.12)$$

Serret is now in a position to derive a number of *spherical* identities simply by applying *planar* identities to the triangle in figure 5.6 and substituting these

---

[28] We can see this by noting that the two bottom parts of the angles at B' and C' must sum to $2E$. However, triangle $B'C't$ is isosceles by its construction, so the bottom parts of the angles must both be equal to E.

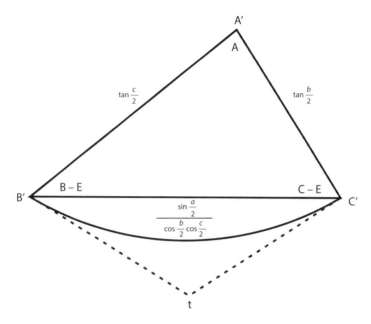

Figure 5.6
Serret's plane triangle. Multiply each of the three sides by $\cos\frac{b}{2}\cos\frac{c}{2}$, and we have Cesàro's triangle of elements.

values for the sides and angles. He begins by applying the planar Law of Cosines:

$$a'^2 = b'^2 + c'^2 - 2b'c'\cos A.$$

Substituting the values of (5.12) into this equation and applying the sine half-angle formula to the terms in front of $\cos A$, he gets

$$\sin^2\frac{a}{2} = \sin^2\frac{b}{2}\cos^2\frac{c}{2} + \sin^2\frac{c}{2}\cos^2\frac{b}{2} - \frac{\sin b \sin c}{2}\cos A.$$

From here, applying some more algebraic manipulation (replacing $\cos^2$ terms with $1 - \sin^2$ terms), he arrives at

$$1 - 2\sin^2\frac{a}{2} = \left(1 - 2\sin^2\frac{b}{2}\right)\left(1 - 2\sin^2\frac{c}{2}\right) + \sin b \, \sin c \, \cos A.$$

Finally, applying the cosine half-angle formula, he has the spherical Law of Cosines:

$$\cos a = \cos b \cos c + \sin b \sin c \cos A. \tag{5.13}$$

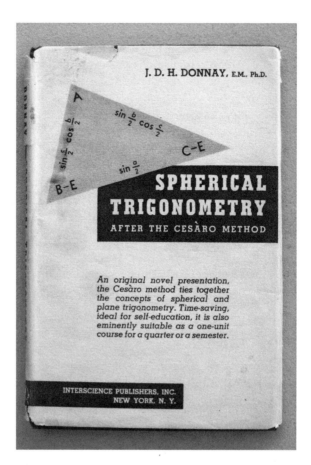

Figure 5.7
Donnay's textbook outlining the Cesàro method of spherical trigonometry. Cesàro's triangle of elements is displayed prominently on the cover.

This same technique transforms any planar identity, not just the Law of Cosines, into a spherical one. But the identity that comes out of this algebraic process is not always what one might expect. For instance, the planar Law of Sines produces two of Delambre's four analogies.

Serret's method appeared within his general textbook surveying a variety of methods in pure geometry and may not have been noticed by the astronomers who were practicing spherical trigonometry at the time. In any case, astronomers might not have had much use for a new technique to prove theorems that they already knew, and it faded into obscurity. Fifty years later, in apparent ignorance of Serret's earlier breakthrough, the method was rediscovered by Giuseppe Cesàro (1849–1939), a crystallographer and older brother of mathematician Ernesto, for whom the term "Cesàro summability" is named. He follows almost the same path that Serret had trodden by referring to the triangle of figure 5.6 as the "triangle of elements."[29] He then extends

[29] Cesàro multiplies the three sides by $\cos b/2 \cos c/2$ to simplify the expressions.

the method by defining three new planar triangles: (i) the "derived triangle" corresponding to the projection of a colunar triangle to the original spherical triangle,[30] (ii) the triangle of elements of the polar triangle of the original spherical triangle, and (iii) the derived triangle of the polar triangle. This gives Cesàro the ability to demonstrate a wide variety of spherical identities by passing planar identities on any one of the four planar triangles through the projection back onto the sphere.

Cesàro's two 1905 articles on the subject received a fate similar to Serret's, at least upon their publication.[31] Late in life Cesàro shared his method with J. D. H. Donnay, who used it in his classes. After Cesàro's death, Donnay published the method in a 1945 textbook, proposing that the entire subject of spherical trigonometry be based on stereographic projection in this way (see figure 5.7).[32] The book met with some interest but also resistance, depending on whether one believed that the purpose of teaching spherical trigonometry was for theoretical insight or for practical applications.[33] In any case, at this time the question in the mathematics education community was becoming not how to teach spherical trigonometry but whether it should be taught at all. With its almost complete disappearance from the curriculum a decade later, the debate became moot.

▨ Surveying and Legendre's Theorem

Starting around the middle of the eighteenth century, political, economic, and scientific interests converged with some of the largest geodetical operations ever conducted. In the 1730s and 1740s, scientific expeditions were sent to Lappland (as near to the North Pole as possible) and to Peru (as near to the equator as possible) to determine whether the sphere of the earth bulges at the equator, at the poles, or neither. The survey's result was in favor of a bulge at the equator. The survey of the meridional arc through Paris by Delambre and Mechain, conducted in the last decade of the eighteenth century, determined the length of the meter—defined to be precisely 1/10,000,000 of the distance from the North Pole to the equator, through Paris. This time the goal was mostly political: by establishing a universal system of weights and measures, that part of the human experience was taken out of the hands of power brokers who might manipulate it to their advantage and placed in the domain

---

[30] A *colunar triangle* is obtained by extending two of the spherical triangle's sides until they meet at the point on the sphere antipodal to their intersection.

[31] [Cesàro 1905a], [Cesàro 1905b]. Stereographic projection has a substantial history in geology; for a survey see [Howarth 1996], especially pp. 504–510.

[32] [Donnay 1945].

[33] Two reviews representing opposing perspectives are [Brown 1946] and [Craig 1946].

Figure 5.8
The "trig station," or survey marker, atop University of Canterbury Mount John Observatory (Lake Tekapo, New Zealand), the southernmost optical research facility in the world.

of science. As Condorcet put it, the metric system was to be for all people, for all time. The Great Trigonometrical Survey of India, begun in 1802 and performed over the better part of the nineteenth century, helped to consolidate the British East India Company's holdings by correcting significant errors in available cartographic information. Along the way, it determined the altitudes of the mountains in the Himalayas; Mount Everest is named after the survey's second leader. The Principal Triangulation of Great Britain and Ireland over the first half of the nineteenth century originated with military requirements; the British army needed more accurate maps to track down dissenters and to plan for defense of the island. These various motivations, coming from places of power and wealth, had the financial resources to do land surveying on a much grander scale than had taken place before.

The success of each mission was not assured at the outset; the goal in each case was unprecedented accuracy while at the same time spanning unprecedented distances. The standard method is known as ***triangulation***. One sets up a series of locations (sometimes called "trig stations") in an interconnected grid of triangles (figure 5.8). Imagine that we measure with great precision the length of one side of one of these triangles. If we can measure accurately the angles of the triangle at either end of the line, then we can calculate the other angle and the two remaining distances with the same accuracy. We can then move to an adjacent triangle and measure its angles and calculate its lengths. We continue, stepping from one triangle to the next, until all distances have been determined.

However, this method works only if the triangles are small enough that the curvature of the earth does not need to be taken into account. On the large scale of the new projects, one could not simply ignore the fact that the earth

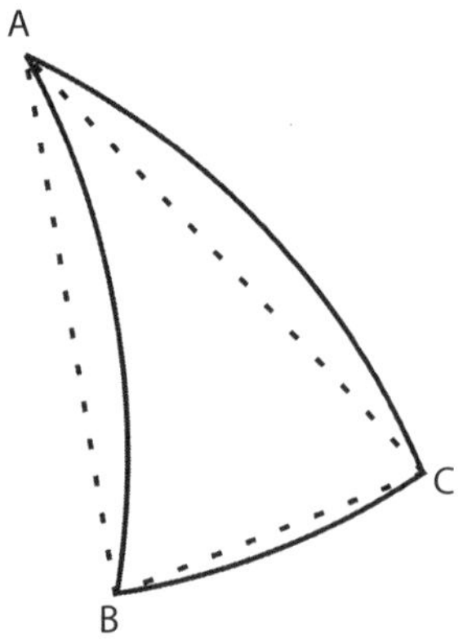

Figure 5.9
A spherical triangle and its chordal triangle.

is round. Three methods were used in practice to deal with this challenge. One is simply to treat all the triangles in the grid as spherical triangles and to use the standard spherical identities to compute the missing angles and lengths. In theory, this is precise. But in practice it is rather cumbersome and leads to calculation difficulties in cases where the triangle's sides are small relative to the sphere, as they are here. The second was the method of ***chordal triangles***. If one knows the lengths of two sides b and c of the spherical triangle and the angle A between them, one may calculate approximately the small difference θ between A and the corresponding planar angle in the triangle connecting the three vertices with straight lines (figure 5.9). Then we may bypass more complicated spherical identities in favor of simpler planar equivalents. The formula,

$$\theta \approx \tan\frac{A}{2}\sin^2\frac{b+c}{4} - \cot\frac{A}{2}\sin^2\frac{b-c}{4}, \tag{5.14}$$

is obtained by applying the spherical Law of Cosines to a smaller sphere centered at A. The approximation comes from applying the cosine difference law to the quantity $\cos(A - \theta)$ and replacing $\cos\theta$ and $\sin\theta$ with 1 and θ respectively.[34]

The third method, the most popular of the three, was to use ***Legendre's Theorem***. The result is stated simply: given a sufficiently small spherical triangle with angles A, B, and C and a plane triangle with side lengths equal to the arcs of the spherical triangle and angles A', B' and C', then it is approximately true that

$$A = A' + E/3,\ B = B' + E/3,\ \text{and}\ C = C' + E/3, \tag{5.15}$$

---

[34] For the full derivation, see [Todhunter/Leathem 1901, 171–172].

where E is the spherical excess[35] of the spherical triangle. This fact greatly eased the surveyors' computational burden, making their task practicable: for instance, to determine the angles in their triangulations, they merely used the much simpler formulas of plane trigonometry and arrived at A, B, and C by apportioning the spherical excess equally among them.[36] Legendre's theorem was not necessarily more accurate than the method of chordal triangles, but it was much easier to execute.

The theorem was first announced by Adrien-Marie Legendre in 1789. Since he was involved in a triangulation survey to determine the relative positions of the Greenwich and Paris observatories, he reported that he used this method successfully while surveying between the French and English coasts and intended to continue using it. A proof of the theorem came later in 1798 as part of the report on the French meridional arc.[37]

Some initial reactions from the scholarly community were skeptical. Textbook writer Abraham Kästner did not believe that the method would lead to any benefit in practice,[38] and Johann Karl Burckhardt was concerned that a sequence of approximations might cause problems.[39] However, as the method started to prove its worth, several mathematicians extended the theorem to improve its accuracy, thereby allowing it to be used on larger spherical triangles, or even within spheroids.[40] One of early improvements was due to Karl Buzengeiger, who in 1818 preserved more terms in the series approximations that had led to Legendre's result. This gave him the following:[41]

$$A = A' + \frac{E}{3} + \frac{E}{180}(b^2 + c^2 - 2a^2),$$

$$B = B' + \frac{E}{3} + \frac{E}{180}(a^2 + c^2 - 2b^2),$$

$$C = C' + \frac{E}{3} + \frac{E}{180}(a^2 + b^2 - 2c^2). \tag{5.16}$$

The real proof of the theorem's worth would be found not within mathematics but literally in the field. In fact it may have found its first application almost half a century before, by Charles Marie de la Condamine in the expedition to

---

[35] The amount by which the angle sum of the spherical triangle exceeds 180°.

[36] The spherical excess may be approximated by using the area of the plane triangle.

[37] For the original paper see [Legendre 1789, 358–359]; for the article containing the proof, see [Legendre 1798, 12–14]. On Legendre's Theorem, see also [Frischauf 1916] and [Nádeník 2005]. [Hauer 1938] is a detailed survey of the history of the theorem.

[38] [Kästner 1791, 453–458]; see also [Hauer 1938, 650].

[39] [Burckhardt 1799].

[40] See especially [Gauss 1828, 47–50].

[41] [Buzengeiger 1818]; see also [Hauer 1938, 584–585]. Even further precision was obtained in [Grunert 1847], [Nell 1874], and [Helmert 1880/1884, vol. 1, 92–93]; see also [Nádeník 2005, 43].

Peru (1735–1745), within the project to determine the shape of the earth.[42] In the French survey of the meridional arc through Paris, the team (including Legendre) used all three methods (spherical trigonometry, the method of chordal triangles, and Legendre's Theorem). Finally, the Principal Triangulation of Great Britain and Ireland applied the method of chordal triangles early in its tenure. However, eventually Legendre's Theorem took over, and after that it became a near-universal practice through the mid-twentieth century.[43]

▦ Trigonometry in Navigation

If trigonometry was given a prominent role due to the political importance of surveying in the eighteenth and nineteenth centuries, the same and perhaps more could be said with respect to navigation. The global empires controlled by several European nations during these times were made possible and continuously maintained through seafaring routes. Far from a perfectly honed science, navigation was a dangerous business where mistakes could (and did) prove fatal. While mathematical solutions to the astronomical problems that navigators needed to solve were readily available, we shall see that mathematicians' and seafarers' definitions of an acceptable solution were quite different. As a result, even in the domains of navigation dominated by mathematics, practical needs drove important innovations long after the theory had been completely established.[44]

Determining one's latitude at sea is relatively easy in principle: simply measure the altitude of the celestial North Pole above the horizon. In practice it isn't so simple: Polaris, the North Star, is not precisely at the North Pole, and it is difficult to design instruments that measure accurately while at sea. Various methods were adopted that involved measuring altitudes not of Polaris but of the sun or certain stars above the horizon. The **double altitude** problem, that of finding one's local latitude from a pair of observations of the sun or a star, occupied navigators throughout this period and involved trigonometric work and table construction.

However, finding one's *longitude* at sea was considerably more vexed: it involved finding one's difference in longitude from an arbitrarily chosen location on the earth's surface (today, it is the Greenwich prime meridian) that

---

[42] [Wolf 1890/1892, vol. 2, 184].

[43] On the French survey see [Delambre 1810, 7]; on the British survey see [Clarke 1858, 244–246]; on the Indian survey see [Everest 1847, clviii]. See also a summary in [Todhunter/Leathem 1901, 188].

[44] For general sources on the history of nautical astronomy in addition to the references in this section, see for instance [E. G. R. Taylor 1957], [Cotter 1968], [Hattendorf 2007] (especially the articles "Nautical astronomy and celestial navigation" and "Western navigation"), and [Bennett 2017].

has no special astronomical significance. In the late eighteenth and early nineteenth centuries, two collections of methods were being proposed to determine longitude: one group using chronometers, and the other using lunar distances. They were put in competition with each other through several European governmental agencies. Prominent among them was the English Board of Longitude, which awarded a substantial prize for the demonstration of a method that was "practicable and useful at sea." The ideas behind both methods are relatively simple. The chronometer methods require the use of a timepiece able to keep accurate time for the period of an ocean voyage. One sets the clock to Greenwich time at the beginning of the journey. At sea, one observes the local time and compares it to Greenwich as recorded on the clock. The difference between these two times reflects the difference in longitude: since the earth rotates 360° in 24 hours, each hour of time difference corresponds to a difference of 15° in longitude.

The lunar distances methods rely on the rapid movement of the moon through the heavens, about ½° per hour (or about its own diameter every hour), which is an order of magnitude faster than any other heavenly body. Suppose that one has accurate tables that predict the moon's position as a function of time at Greenwich. On board, one observes the moon's position (in particular, its distance to the sun or some prominent star) and adjusts for refraction and parallax—that is, the moon's position is recalculated to where it would be seen from the center of the earth. This process was known as ***clearing the distance***. Then, using the lunar position tables, one finds the time in Greenwich; and as before, the difference between that time and the local time gives the longitude.

These methods are much simpler in theory than they are in practice. The various factors involved made the process extremely arduous, even after appropriate measurements had been taken. For instance, one of the earliest uses of a lunar distance method (Edmund Halley's, in this case) was reported to take six hours to complete the calculations.[45]

Methods by chronometer are simpler, but they took a long time to be adopted. The major drawback was the lack of availability of reliable enough timepieces. British clockmaker John Harrison famously succeeded at the longitude challenge through his mechanical expertise,[46] but adequate

---

[45] The reporter was Robert Waddington, assistant to Nevil Maskelyne on the 1761 voyage to St. Helena for a transit of Venus; see [Bennett 2019, 74]. This article describes how the mathematicians' and shipboard navigators' expertise combined to produce successful results. See also [Croarken 2003], which describes Maskelyne's methods in his later construction of the *Nautical Almanac*. Finally, see [Forbes 1970] on Tobias Mayer's table constructions using the method of lunar distances.

[46] The story of John Harrison and his struggles with the Board of Longitude has been popularized in [Sobel 1995] and converted into a television miniseries.

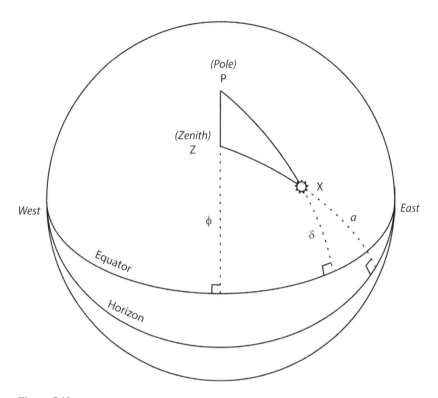

Figure 5.10
The astronomical or *PZX* triangle.

chronometers in good enough supply were not readily available until the middle of the nineteenth century. However, as always the technology improved, and by the beginning of the twentieth century chronometer methods eventually won out.

Chronometer methods are not entirely free from the problems of observation; one needs to accurately look up in order to determine the current local time. This may be done by measuring the sun's (or some other star's) altitude as long as one knows its declination and one's local latitude. The process of converting this observation (a *sight*) to the needed navigational information is known as **sight reduction**. The calculations rely on the **astronomical triangle** (often called the **PZX triangle**), which has been part of astronomy and navigation since premodern times.[47] In figure 5.10, it is morning and the sun, having risen in the east, is traveling westward to the left and upward across the celestial sphere. The local latitude φ is equal to the distance

---

[47] We discussed its use in medieval India in [Van Brummelen 2009, 128–129].

from the zenith Z to the equator; the sun's declination δ, the distance from the sun to the equator, may be found from the time of year. The sun's altitude a above the horizon is measured by an observer. The sun is carried by the rotation of the celestial sphere around the pole P so that the angle $\angle XPZ$—the **hour angle**—changes at a uniform rate. If we can find its value, then we simply need to divide 360° by 24 hours to get the number of hours before local noon.[48]

Determining $\angle P$ was child's play for eighteenth-century mathematicians. The three sides of the PZX triangle are known immediately: $PZ = 90° - \varphi$, $PX = 90° - \delta$, and $ZX = 90° - a$. From this the spherical Law of Cosines gives

$$\cos P = \frac{\cos ZX - \cos PZ \cos PX}{\sin PZ \sin PX} = \frac{\sin a - \sin\varphi\sin\delta}{\cos\varphi\cos\delta}. \tag{5.17}$$

The major problem with the Law of Cosines was not the correctness of this solution but the fact that it does not play well with logarithms. Although one can take the logarithm of the right side of this equation and convert the division into a difference, one is left with the logarithm of a difference in the numerator, for which there is no simple formula. This meant an extra time cost to the navigator and, more seriously, an increased probability of a mistake.[49]

A surplus of alternatives to the spherical Law of Cosines were proposed to get around this and other practical problems of calculation over the course of the nineteenth century, and some were adopted widely. Some of them took inspiration from the past, capitalizing on the **versed sine**, a lesser-known trigonometric function that dates back to medieval India:[50]

$$\text{vers } \theta = 1 - \cos \theta \tag{5.18}$$

The versed sine has a major advantage to navigators: it never takes on negative values, so one never needs to consider whether the value of a versed sine must be added or subtracted, thus eliminating a frequent source of error. The versed sine had been fading away in favor of sines, tangents, and secants in mathematical texts. But it had not disappeared, and it survived especially well in textbooks related to navigation. One of its more well-known appearances was the 90-page-long table in Jonas Moore's 1681 *A New Systeme of the Mathematicks*, a textbook written for the Royal Mathematical School designed to teach the techniques of navigation.[51]

---

[48] A small adjustment depending on the time of year, the **equation of time**, must be made.

[49] [Cotter 1974b, 540].

[50] See [Van Brummelen 2009, 96].

[51] See the list of trigonometric tables in [Glaisher et al. 1873, 46–47]. On *A New Systeme of the Mathematicks* see [Moore 1681] (the table of versed sines, provoked by John Collins, is on pp. 261–351); also see the biography [Willmoth 1993].

The dawn of the nineteenth century saw the introduction of a new function, seemingly a minor modification of the versed sine, that was to become dominant in navigational practice. In 1805 the Spanish scientist Josef de Mendoza y Ríos published a magisterial 670-page collection of tables for use in nautical astronomy. Within it one finds a table entitled "logarithmic versed sines." However, in an explanatory note Mendoza y Ríos states that they are, in fact, logarithms of *half* the versed sines.[52] The same year, James Andrew, an educator who trained military and engineering students, published his own *Astronomical and Nautical Tables*. Almost half the book is taken up by a table of "Squares of Natural Semi-Chords," which he uses frequently for solving astronomical problems relevant to navigation. What Andrew meant by the "semichord" was $\sin(\theta/2)$, so the function tabulated is $\sin^2(\theta/2)$. But, by the cosine double-angle formula, this is equal to

$$\sin^2\frac{\theta}{2} = \frac{1}{2}(1 - \cos\theta) = \frac{1}{2}\text{vers}\ \theta. \tag{5.19}$$

This quantity eventually would be dubbed the ***haversine*** by James Inman in 1835, a short form of "half the versed sine," abbreviated "hav."[53]

Text 5.2
James Andrew, Solving the PZX Triangle Using Haversines

PROBLEM VIII. *The Sun's Altitude and Declination being given, together with the Latitude of the Place of Observation, to find the apparent time.*

Take the difference of the squares of the natural semichords of the sum and difference of the sun's distance from the elevated pole and the co-latitude[54] of the place, as a divisor; and the difference of the squares of the natural semichords of the complement of the sun's given altitude and meridian zenith distance, with seven cyphers annexed, as a dividend; the quotient, after division, will be the square of the natural semichord of the apparent time, or distance from apparent noon, as required.

Example: July 8th, 1795, in latitude 34°55′ N. when the Sun's declination was 22°23′31″ N. his altitude was found to be 36°59′46″: What was then the apparent time of observation?

---

[52] The tables may be found in [Mendoza y Ríos 1805, 388–477]; the explanatory note is on pp. 8–9 in the section of explanations of the tables at the end of the work. It is sometimes stated incorrectly that haversines are first found in Mendoza y Ríos's 1801 tables.

[53] [Inman 1835]. On the haversine, see also [Goodwin 1910].

[54] The text reads "co-altitude," but this is a typographical error.

Co-decl.	67	36	29								
Co-lat. -	55	5	0								
Sum	122	41	29	S.S.C.	7700569	co-alt.	53	0	14	S.S.C.	1991196
Diff.		12	31	29	S.S.C.	118987	less z. d.	12	31	29	S.S.C.	118987
				Diff.	7581582					Diff.	1872209

$$\frac{1872209 \times 10000000}{7581582} = \frac{\text{(S.S.C.)}}{2469417} = 59^\circ\,35'\,40'' = 3^h\,58^m\,22\tfrac{2}{3}^s$$

Or $20^h\,1'\,37\tfrac{1}{3}''$ Apparent time, as before.[55]

Explanation: (See figure 5.10.) The given quantities are $\delta = 22°23'31''$, so $PX = 90°$ $- \delta = 67°36'29''$; the observer's latitude $\varphi = 34°55'$, so $PZ = 90° - \varphi = 55°5'$; and the sun's observed altitude $a = 36°59'46''$, so $ZX = 90° - a = 53°0'14''$. Haversines are given in units of a base circle of 10,000,000 and called "SSC" (square of the semichord) in the calculations. In modern terms, Andrew's calculation applies the formula

$$\text{hav } P = \frac{\text{hav } \bar{a} - \text{hav}(\varphi \sim \delta)}{\text{hav}\left(180° - (\varphi \sim \delta)\right) - \text{hav}(\varphi \sim \delta)}$$

$$= \frac{0.1991196 - 0.0118987}{0.7700569 - 0.0118987} = 0.2469417, \tag{5.20}$$

where $\sim$ represents the positive difference between the two quantities. Using his haversine table, Andrew gets $P = 59°35'40''$; dividing by 15, he finds local time to be $3^h58^m22\tfrac{2}{3}^s$ before noon.

There is no derivation of the result nor even a formal expression of the equation. Rather, addressing the needs of his navigator readers, Andrew provides only a concise description of the required calculations and gives an example. It is possible to show that this formula is equivalent to the spherical Law of Cosines by substituting the definition of the haversine and applying trigonometric identities; we leave this to the interested reader.

Andrew's use of the haversine was one of several attempts to streamline navigational calculations a little more efficiently than their predecessors. The theorem that eventually triumphed at the turn of the twentieth century was the ***cosine-haversine formula*** or simply the ***haversine formula***:

$$\text{hav } P = \frac{\text{hav } \bar{a} - \text{hav}(\varphi - \delta)}{\cos\varphi\cos\delta}. \tag{5.21}$$

[55] [Andrew 1805, 206–207]. Andrew also provides a solution that employs logarithms, which we omit here.

First proposed by H. B. Goodwin in 1899 and related to (5.20),[56] this identity also can be derived directly from the spherical Law of Cosines.[57] Goodwin suggested that it would be handy to apply with tables that display haversines and their logarithms beside each other. Percy Davis implemented this idea a few years later in his *Requisite Tables*, and it quickly became the standard approach.[58] In fact, in some contexts the cosine-haversine formula is named after Davis.

The sight reduction techniques that we have seen so far were known as **direct methods** since they rely on a single formula. Another major category was the so-called **short methods**, which relied on dividing the *PZX* triangle into two right triangles by dropping a perpendicular from one vertex to the opposite side. Although this adds a step to the calculation, it allowed the navigator to apply the formulas for right-angled spherical triangles, which are better suited to use with logarithms. This practice, begun in the late nineteenth century, became prominent in the early twentieth century. One example of these methods, composed by Sinkiti Ogura of the Japanese navy, became widely adopted under the title "A and K tables." Originally published in Tokyo in 1920, they were reprinted in an English edition four years later.[59]

Suppose we know the time of day P, the local latitude φ, and the sun's declination δ; we wish to calculate the sun's altitude a (figure 5.11), equal to $90° - ZX$. Drop a perpendicular from Z onto PX, splitting the astronomical triangle into two right-angled triangles, and let K be the distance from Y to the equator. Then from the right-angled triangle identities applied to *PZY* (the **time triangle**) we have $\cos P = \tan PY \cot \overline{\varphi}$, or (since $PY = \overline{K} = 90° - K$)

$$\cot K = \cos P \cot \varphi; \tag{5.22}$$

this gives us a value for K. Again from the right-angled triangle identities, $\sin ZY = \sin P \cos \varphi$, or

$$\sin ZY = \sin P \cos \varphi; \tag{5.23}$$

this gives us ZY. Finally, from the spherical Pythagorean Theorem applied to *XYZ* (the **altitude triangle**), $\cos ZX = \cos ZY \cos XY$, which transforms to

$$\csc a = \sec ZY \sec(K \sim \delta),^{60} \tag{5.24}$$

---

[56] [Goodwin 1899], especially pp. 529–531.

[57] See [Cotter 1974b, 540–541], [Robusto 1957], [Clough-Smith 1978, 85–86], and [Van Brummelen 2013, 161–162]. See also [Cotter 1982], a survey of direct methods of sight reduction.

[58] [Davis 1905].

[59] [Hydrographic Department, Tokyo, 1924]. What follows is the common derivation of Ogura's method, not found in Ogura's original work.

[60] The symbol ~ refers to the absolute value of the difference between two quantities. Cosecants and secants were often preferred to sines and cosines since their values are greater than one and therefore avoid problems with roundoff error.

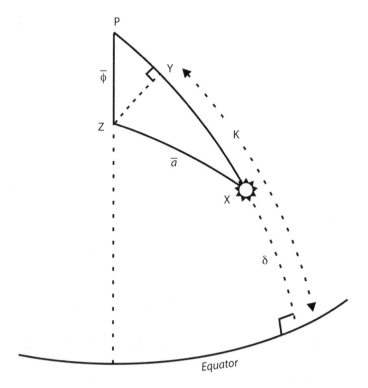

Figure 5.11
Ogura's method of sight reduction.

and we have the sought quantity a. Ogura's method was advantageous not so much because of these equations but because of his method of tabulation. Let $A = \cos ZY$, and tabulate the two functions $\log 1/A$ and K (both functions of P and φ, according to (5.22) and (5.23)), the "A and K tables." Take the logarithm of (5.24), and we have

$$\log \csc a = \log \frac{1}{A} + \log \sec(K \sim \delta).$$ (5.25)

For a navigator to find a, they needed only to subtract δ from K, use a log secant table, add the result to $\log \frac{1}{A}$, and use a log cosecant table. These tables were commonly available.[61]

One might wonder why Ogura would want to determine the sun's altitude from the time of day in the first place since the sun's position can be

---

[61] On Ogura's method see [Cotter 1974a, 396–398], [Cotter 1968, 331–333], and [Clough-Smith 1978, 91–93]. [Cotter 1974a] also surveys other short methods.

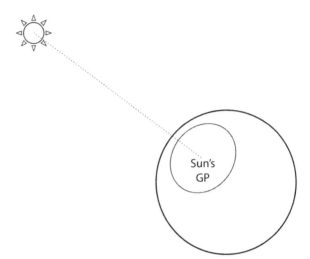

Figure 5.12
The sun's geographic
position (GP).

measured easily. The answer lies in a new approach called *line of position navigation* that arose in the nineteenth century from an initial discovery by Thomas Hubbard Sumner and a development of the idea by Marcq Saint Hilaire.[62] The idea is simple. Observe the altitude of a single star. If the star is not directly above you (in which case you are at the star's *geographic position* or GP), then you must be somewhere on a circle centered at the star's GP (figure 5.12). This circle is usually so large that the part of it near the ship is effectively a straight line, the *line of position*. If you observe a second star, you have two lines of position; their intersection is called the *fix*, the ship's position.

Briefly, Saint Hilaire's method works as follows. Start with a rough estimate of the ship's position, usually obtained by dead reckoning. From this, calculate what the altitude of the star should be at that position, and compare it with the star's actual observed altitude. The difference between these two altitudes turns out to be equal to the distance from the ship's estimated location to the line of position. So, navigators using line of position methods still needed to compute the sun's altitude, even though they were also observing it.

Although navigation has been taken over largely by computers and GPS devices, celestial navigation is still a thriving practice among enthusiasts. In addition, the U.S. Naval Academy trains its cadets in this forgotten art to deal with a situation where an enemy has disabled the GPS satellite network.

---

[62] On Sumner's and Saint-Hilaire's methods, see [Vanvaerenbergh/Ifland 2003], which contains Sumner's original 1843 publication as well as Saint-Hilaire's publications of 1873 and 1875. See also [Cotter 1968, 268–308], [Silverberg 2007], and [Van Brummelen 2013, 151–171].

▨ Tables

Time and again we have seen that one of the primary motives for theoretical advances in trigonometry has been the construction of tables of the sine and other functions for use in applications such as astronomy and navigation.[63] From the beginning with Hipparchus up to the fifteenth century, calculators had relied on trigonometric identities, that is, geometric theorems; and when the identities failed, various approximation techniques (either geometric or algebraic) had bridged the gaps. The first major revolution in table construction had replaced geometry with infinite series for the sine and cosine functions, discovered as early as the fourteenth to fifteenth centuries by the Madhava school in Kerala. These series were rediscovered in seventeenth-century Europe and quickly began to transform European table construction, as we saw beginning with the work of Abraham Sharp in chapter 3. In conjunction with finite difference interpolation methods, series became the preferred method of constructing tables from scratch.

But change does not come wholesale overnight, and the calculation of a trigonometric table from scratch was not an enterprise to be undertaken lightly. A small industry of table makers was taking shape in the late eighteenth century and flourished in the nineteenth, but as has been the case since mathematical tables began, many of them borrowed their entries from other sources. In the renowned 1873 *Report of the Committee on Mathematical Tables*, J. W. L. Glaisher complained:

> It is thus seen that, with the exception of the *Tables du Cadastre*, and the second half of Mr. Sang's table [which we shall discuss below], every one of the hundreds of tables that have appeared has been copied from Briggs or Vlacq. . . . In the preface to his tables (1849) Mr. Filipowski concludes by a sneering remark on the Chinese, saying that Mr. Babbage proved, "as had long been suspected, from what source those original inventors had derived their logarithms;" and we have noticed this tendency to ridicule the Chinese in this matter as detected plagiarists in others. In point of fact there is no more plagiarism than when Babbage or Callet publishes a table of logarithms without the name of Vlacq on the titlepage.[64]

Glaisher was referring here to the reprinting of Vlacq's tables (in particular, the *Arithmetica logarithmica* and *Trigonometria artificialis*) in the early

---

[63] Several historical and bibliographic surveys of trigonometric and logarithmic tables have been compiled in the last two centuries. See [Hutton 1811b, 1–124], [De Morgan 1868], [Glaisher et al. 1873], [Horsburgh 1914, 30–68], and [Archibald 1948].

[64] [Glaisher et al. 1873, 53, 54].

190   ❦❧

Natürliche oder hyperbolifche Logarithmen.							
1	0,000000	000000	000000	000000	000000	000000	000000	000000
2	0,693147	180559	945309	417232	121458	176568	075500	134360
3	1,098612	288668	109691	395245	236922	525704	647490	557823
4	1,386294	361119	890618	834464	242916	353136	151000	268721
5	1,609437	912434	100374	600759	333226	187639	525601	354269
6	1,791759	469228	055000	812477	358380	702272	722990	692183
7	1,945910	149055	313305	105352	743443	179729	637084	729582
8	2,079441	541679	835928	251696	364374	529704	226500	403081
9	2,197224	577336	219382	790490	473845	051409	294981	115645
10	2,302585	092994	045684	017991	454684	364207	601101	488629

Figure 5.13
The beginning of Wolfram's table of natural logarithms to 48 decimal places.

eighteenth century by various Chinese works.[65] Some of the best-known European tables of the late eighteenth century are known to derive from Vlacq. Georgio Vega's 1794 *Thesaurus logarithmorum completus*, itself the basis of several nineteenth-century tables, states directly in its subtitle that it is a version of Vlacq with a number of errors corrected.[66] The popular tables by François Callet (1795) were partly copied from Gardiner's *Tables* (1742) but also may owe a debt to Vlacq.[67]

The titles of many of these tables involve references to logarithms. In fact, logarithms and trigonometric functions continued the symbiotic relationship that they had shared since the invention of logarithms in 1614. Through the nineteenth century, tables of logarithms of trigonometric functions were at least as common as tables of natural trigonometric functions or even as common as straightforward logarithm tables, and different types of table often appeared together in the same volume.

Several late eighteenth-century collections of tables deserve special mention for various reasons. The first is a remarkable table of natural logarithms that appeared as part of the 1778 collection *Neue und Erweiterte Sammlung Logarithmischer, Trigonometrischer und anderer zum Gebrauch der Mathematik unentbehrlicher Tafeln,* edited by Johann Carl Schulze. This table, given to an astonishing 48 decimal places (see figure 5.13) for all whole numbers up to 2,200 and for all prime numbers (and some composite numbers)

[65] [Vlacq 1628] and [Vlacq 1633]; see [Roegel 2011a], [Roegel 2011b], and [Roegel 2011f]. Vlacq's reliance in turn on Henry Briggs has been discussed in chapter 2.
[66] [Vega 1794].
[67] [Callet 1795], [Gardiner 1742]. See [Lefort 1857] and [Glaisher et al. 1873, 54].

308

Rationale Trigonometrie.						
$t \frac{1}{2} \omega$	$t\frac{I}{2}\omega$ in Decim. Th.	ω.	Perp.	Hypoth	Bafis.	$90° - \omega$.
1:25	0,0400000	4 34 52	25	313	312	85 25 8
1:24	0,0416667	4 46 19	48	577	575	85 13 41
1:23	0,0434783	4 58 45	23	265	264	85 1 15
1:22	0,0454545	5 12 18	44	485	483	84 47 42
1:21	0,0476190	5 27 9	21	221	220	84 32 51
1:20	0,0500000	5 43 29	40	401	399	84 16 31
1:19	0,0526316	6 1 32	19	181	180	83 58 28
1:18	0,0555556	6 21 35	36	325	323	83 38 25
1:17	0,0588235	6 43 58	17	145	144	83 16 2
1:16	0,0625000	7 9 10	32	257	255	82 50 50

Figure 5.14
The beginning of the table of rational trigonometry in Schulze's tables.

up to 10,009, was assembled by Dutch artillery officer Isaac Wolfram.[68] Composing this extraordinary work took him six years; likely he used series for $\ln(1 + x)$ and $\ln(1/(1 - x))$. Due to its impressive accuracy, it was used as a basis for other tables during the nineteenth century.

The second of our curiosities is a small tradition of tables of "circular measure" that convert arcs measured in degrees to the lengths of those arcs in units of distance. Today we think of this as a conversion from degree to radian measure, so the tables contain nothing more than multiples of $\pi/180°$. One such table given to 27 decimal places is found in Schulze's *Sammlung*, the same treatise in which Wolfram's tables were first published.[69] Of course, Schulze's readers did not have electronic aids to computation, so these tables were good time savers for the same reason that logarithms had been invented: to save the reader from performing complicated multiplications.

Our third curiosity is the tables of "rational trigonometry." We saw in chapter 1 that François Viète had composed a 45-page list of Pythagorean triples and arranged them so that they could be used as a trigonometric table with perfect precision. The practice seems to have been reinvented in correspondence between Johann Heinrich Lambert and Simon Baum in the late eighteenth century. Another table of rational trigonometry appears in Schulze's increasingly

---

[68] [Schulze 1778, vol. 1, 189–258]. Of the more than 166,000 digits in this table, only 38 are known to be incorrect ([Archibald 1936a] and [Archibald 1950, 194]). Details of Isaac Wolfram's life are available in [Archibald 1936b], [Archibald 1950], and [Archibald 1955b]. A serious illness (possibly fatal) prevented him from completing the last few entries; they were completed by others in later editions.

[69] [Schulze 1778, vol. 2, 265–277].

remarkable set of tables, the same volume in which Wolfram's table of loga-rithms appeared.[70] Schulze's table (see figure 5.14) is nowhere near as ambitious as Viète's had been, but he does go beyond Viète in one way by including approximations for the angles in the triangles. Interest in rational trigonometry would be revived in 1864 by famed table maker Edward Sang (complete with a four-page table), on whom we shall have more to say soon.[71]

But the most ambitious table-building project ever completed before the nineteenth century never saw the light of day. Among the initiatives triggered by the French Revolution was the creation of a universal system of measurement depending entirely on reason and defined by natural phenomena. Relying on a base of 10, much of the metric system is standard around the world today. Certain innovations did not survive, such as the division of the day into ten "hours" and each hour into 100 "minutes," the month into three "weeks" of ten days, and a new calendar beginning in 1793. One of the metric units that would eventually pass into history was a new division of the circle. Rather than breaking a right angle into 90°, it was portioned into 100 "grades," later called "gradians."[72] Parts of gradians, of course, were represented using decimal fractions.

If successful, the conversion to gradians would render all existing trigo-nometric tables obsolete since the arguments would be entirely different. It was incumbent on the French government to replace them. Clearly the new tables would serve as one of the foundations for a new approach to knowledge, so they had to be unsurpassed. The government's instructions were

> not only to compose tables that left nothing to be desired as to ac-curacy, but to build the vastest monument to calculation and the most imposing that had ever been executed or even conceived.[73]

The person chosen to lead the vast undertaking was Gaspard Riche de Prony. An engineering educator, de Prony was appointed to the new École Polytechnique in 1794 and would work with the Bureau des Longitudes start-ing in 1801. During the 1790s de Prony established the Bureau de Cadastre, intended to record land holdings precisely for the purpose of fair taxation. It was in this role that de Prony was charged with overseeing the "vastest mon-ument to calculation" in history.[74]

---

[70] See [von Braunmühl 1900/1903, vol. 2, 147] and [Schulze 1778, vol. 2, 308–311]. For a modern reinvention of the spirit of rational trigonometry, see [Wildberger 2005].

[71] [Sang 1864]; the table is on pp. 757–760.

[72] Gradians were a feature of scientific calculators up to the late twentieth century and are still used occasionally in certain scientific subdisciplines.

[73] [de Prony 1804b, 49].

[74] On de Prony's life and work, see [M. Bradley 1994] and [M. Bradley 1998]. On the cultural context of the Cadastre project, see [Daston 1994].

De Prony's approach to the enterprise forever altered how tables were generated. Reading about the division of labor in the manufacture of pins, he decided to take a similar approach to his tables. No longer considering tables as a union of great minds with hard work, de Prony separated the intellectual process from the labor. Several mathematicians (including Legendre and Carnot) were tasked with designing the process, another group prepared the work to be done, and a workforce of about 70 assistants actually performed the calculations.[75] The tables, then, were constructed just as products are now manufactured in factories.

With 10,000 sine values, over 200,000 logarithms, and logarithms of various trigonometric functions to compute, the project needed to be planned with care. The laborers could be trusted to add and subtract numbers but not much beyond that, so the limited hours that the fewer skilled employees could work had to be managed prudently. The first sines to be computed were found for every ten gradians (10^g or $9°$) using infinite series. To generate the sines of every gradian, the formula

$$\sin(a+b) = 2\cos a \sin b + \sin(a-b) \tag{5.26}$$

(demonstrable from the sine sum and difference laws) was used. They were checked using Euler's identity

$$\sin x + \sin(40^g - x) + \sin(80^g + x) = \sin(40^g + x) + \sin(80^g - x). \tag{5.27}$$

At this point the rest of the entries (99 percent of them, for every 1/100 of a gradian) could be filled in using finite difference interpolation, which could be entrusted to the laborers. The success of the interpolation process could be judged by observing how the error accumulated over one hundred sine values, from one gradian to the next. Jean Baptiste Joseph Delambre reported that the error usually reached no more than one unit in the 22nd decimal place.[76] The work was done by two sets of computers working independently of each other, and the results were then compared for correctness.[77]

---

[75] See [Grattan-Guinness 1990b, 179] and [Grattan-Guinness 2003, 109]. Grattan-Guinness's report that many of these workers were hairdressers left unemployed by the downfall of the nobility is challenged in [Daston 1994, 190]. In any case, they were untrained laborers.

[76] In addition to his involvement in the Cadastre project, Delambre also completed a smaller set of tables using gradians in 1801 (to only six or seven decimal places), initiated by Jean Charles de Borda but interrupted by the latter's death two years earlier [Borda/Delambre 1801]. Two other sets of tables using gradians are noteworthy also for their accuracy: [Hobert/Ideler 1799] (sines and tangents to seven places) and [Callet 1795] (to 15 places), which was published in a number of editions during the nineteenth century.

[77] For more information on the methods of calculation by the mathematicians in charge of the project see [Delambre 1804], especially pp. 59–60, and an analysis of the accuracy of other tables in [de Prony 1804a]. See also [Grattan-Guinness 1990b, 182–183] and [Grattan-Guinness 2003, 115–116]. A debate arose between Edward Sang and F. Lefort concerning the accuracy

The Cadastre tables were completed within several years. In 1801 two complete manuscript sets, each consisting of 19 volumes, were ready, and the publisher Firmin Didot was assigned the task of printing what would have been a large 1,200-page volume. However, financial crises repeatedly delayed the publication. Eventually in 1819 English overtures were made to rescue the project, but the English wanted to move away from gradians, and this was not received positively. Almost a century after they had been composed, de Prony's logarithm tables and logarithmic trigonometric tables were finally published in 1891 in a diminished version, displaying its values to only eight places.[78]

In the end the centesimal division of the right angle, which would have rendered all other tables obsolete, rendered the Cadastre tables themselves obsolete as France returned to the use of degrees and minutes. In the past half century the use of minutes and seconds to represent fractions of degrees has gradually been replaced by decimal fractions, but the gradian itself remains mostly as an obscure relic.

Of course, trigonometric and logarithmic tables with the extreme precision of the Cadastre were not needed for most applications. Through the nineteenth century tables proliferated but with fewer decimal places for use in industry and in schools. Since by this time more precise values were a matter of public record, it is perhaps not surprising that entries in these new tables were not recomputed but borrowed or rounded from existing tables. One exception to this was the work of the extraordinary Scottish mathematician and engineer Edward Sang. Over a 25-year period with the aid of his daughters Flora and Jane, Sang compiled massive logarithmic and trigonometric tables filling 47 manuscript volumes. Logarithms from 1 to 20 000 were given to 28 decimal places and from 100,000 to 370,000 to 15 places. Sines were given to 33 places for every 1/20 of a gradian, and to 15 places for every 1/100 of a gradian (computed by interpolation). Sang's adoption of gradians was done with some passion, part of a larger desire to convert all units to decimal systems:

> There is no doing without decimals; when in making a proportion, we have to compare two quantities of one kind, we, as the arithmeticians say, bring them both to one denomination: 2 cwt. 3 qrs. 17 lbs. 11¼ oz. must be brought to quarter ounces, of which there are 20 845 in this quantity. That is to say, having found our old system

---

and methods of the tables; see [Lefort 1858], [Sang 1875a], [Lefort 1875], [Sang 1875b], [Craik 2002, 38], and [Craik 2003, 65–66].

[78] [Grattan-Guinness 2003, 111–113, 117–118]. De Prony's tables are [Service Géographique de l'Armée 1891].

to be unworkable, we have recourse to counting in tens; and, more-over, the troubling of converting our confused measures into decimals exceeds that of the real business in hand. Every such conversion is a protest in favour of uniformity.[79]

Whether Sang's tables match de Prony's for accuracy is a complicated question that has not been answered fully. At the very least they are competitive and are likely better.[80] Unfortunately, the parallels between de Prony and Sang extend beyond their use of the gradian and their great accuracy: Sang's tables also were fated never to be published, except in a reduced version.[81]

By this time, tables were becoming so accurate and filled with so many entries that further improvements were starting to seem daunting or impossible. But mechanical calculating devices were improving, and tables were thought by some scientists to be an ideal application. One of those scientists was Charles Babbage, who considered how to mechanize the production of tables of logarithms.[82] His difference engine, so named because of its ability to calculate finite differences, would have helped to put the next generation of human computers out of work. However, it never was completed, and competing devices that were finished did not have much of an immediate impact.[83]

Table making continued in the early twentieth century, with mechanical assistance casting a larger and larger shadow. Of the several large trigonometric table projects of this time, the three-volume set by Henri Andoyer (1915–1918) is especially notable. Entries of the sine and cosine are given to 20 decimal places for every 10″ of arc,[84] thereby matching the magnitude of Rheticus and Otho's great 1596 *Opus palatinum* but doubling the number of decimal places—astonishingly, without any mechanical aids. Andoyer's methods were standard: infinite series were used to generate a grid of entries, and interpolation filled in the gaps between them. In our age of computers, this feat of hand calculation will likely never be duplicated.[85]

---

[79] [Sang 1884, 534–535].

[80] See [Craik 2003, 68–73] for a discussion.

[81] [Sang 1871]. On Sang's life and work on the logarithmic tables, see [Craik 2002] and [Craik 2003].

[82] Babbage himself was the author of a popular table of logarithms ([Babbage 1827], not computed with his difference engine); see [Campbell-Kelly 1988].

[83] See [Williams 2003] for a description of Babbage's and other difference engines of the nineteenth century, including accounts of machines that could calculate logarithms as early as 1854.

[84] Tangents, cotangents, secants, and cosecants are given to 15 places.

[85] [Andoyer 1915–1918]; he describes his methods in vol. 1, pp. IX–XVIII. Andoyer published separately a table of logarithms [Andoyer 1911]. See also [Baillaud 1915], [Roegel 2012a], [Roegel 2012b], and [Roegel 2012c]. R. C. Archibald stated that "one had to wait for nearly three and a half centuries before the tables of Rheticus were finally superseded by those of Andoyer" [Archibald 1949b, 558].

In fact, as technology improved, it became clear that tables eventually could be rendered obsolete: a machine might be built that could compute the value of a trigonometric function for any given argument more or less instantly. As anyone with a pocket scientific calculator knows, this did occur. When it did, it provoked the most fundamental transformation in methods of calculating sines since Hipparchus: it replaced the entity doing the calculation—the human brain—with an electronic device, which has an entirely different collection of strengths. In the late 1950s, flight engineer Jack Volder was asked by his employer Corvair to replace the B-58 bomber's analog navigation system with a digital equivalent. The result was CORDIC (**CO**ordinate **R**otation **DI**gital **C**omputer), which quickly replaced series methods for computing sines. Variants of it are used today in calculators and computers.

Suppose a user wishes to find the sine of, say, 67.89°. The computer begins by breaking this value down into sums and differences of the following quantities:

$$\theta_0 = \tan^{-1} 1 = 45°, \; \theta_1 = \tan^{-1}\frac{1}{2} = 26.565°, \; \theta_2 = \tan^{-1}\frac{1}{4} = 14.036°,$$

and so on. (To achieve ten decimal places of accuracy, this process continues up to θ_{39}.) We then consider our angle 67.89° as a series of rotations θ_n of the unit vector (1,0): counterclockwise if θ_n is added or clockwise if it is subtracted. In the case of 67.89°, θ_0 and θ_1 turn out to be added while θ_2 is subtracted. Each rotation corresponds to the matrix

$$\begin{bmatrix} \cos\theta_n & -\sin\theta_n \\ \sin\theta_n & \cos\theta_n \end{bmatrix}$$

(with the negative sign moving from top right to bottom left if θ_n is subtracted), so the matrix (1,0) rotated counterclockwise by 67.89° is

$$\begin{bmatrix} \cos\theta_{39} & \sin\theta_{39} \\ -\sin\theta_{39} & \cos\theta_{39} \end{bmatrix} \Big[\cdots \Big] \begin{bmatrix} \cos(\theta_2) & \sin(\theta_2) \\ -\sin(\theta_2) & \cos(\theta_2) \end{bmatrix}$$

$$\begin{bmatrix} \cos\theta_1 & -\sin\theta_1 \\ \sin\theta_1 & \cos\theta_1 \end{bmatrix} \begin{bmatrix} \cos\theta_0 & -\sin\theta_0 \\ \sin\theta_0 & \cos\theta_0 \end{bmatrix} \begin{bmatrix} 1 \\ 0 \end{bmatrix}. \tag{5.28}$$

The x coordinate of this vector is cos 67.89°; the y coordinate is sin 67.89°.

If this were all there was to CORDIC, we would have reduced the problem of calculating one sine/cosine pair of values into the problem of calculating 40 sine/cosine pairs of values, which is not helpful. However, the rotation matrices in (5.28) can be simplified to easily render their evaluation by a binary computing device:

$$\begin{bmatrix} \cos\theta_n & -\sin\theta_n \\ \sin\theta_n & \cos\theta_n \end{bmatrix} = \cos\theta_n \begin{bmatrix} 1 & -\tan\theta_n \\ \tan\theta_n & 1 \end{bmatrix}$$

$$= \cos\theta_n \begin{bmatrix} 1 & -2^{-n} \\ 2^{-n} & 1 \end{bmatrix}. \tag{5.29}$$

Since $\cos\theta_n = 1/\sec\theta_n = 1/\sqrt{1+\tan^2\theta_n} = 1/\sqrt{1+2^{-2n}}$, the entire matrix can be written without any recourse to trigonometry:

$$\begin{bmatrix} \cos\theta_n & -\sin\theta_n \\ \sin\theta_n & \cos\theta_n \end{bmatrix} = \frac{1}{\sqrt{1+2^{-2n}}} \begin{bmatrix} 1 & -2^{-n} \\ 2^{-n} & 1 \end{bmatrix}. \tag{5.30}$$

Now, the quantities $\dfrac{1}{\sqrt{1+2^{-2n}}}$ are identical regardless of the angle whose sine we are evaluating, so they can be gathered together, multiplied, and hard-wired into the calculator's design. For a calculation that ends with θ_{39}, the product turns out to be 0.607252935. Therefore, the sine and cosine of 67.89° may be evaluated using the expression

$$0.607252935 \begin{bmatrix} 1 & 2^{-39} \\ -2^{-39} & 1 \end{bmatrix} \cdots \begin{bmatrix} 1 & 2^{-2} \\ -2^{-2} & 1 \end{bmatrix}$$

$$\begin{bmatrix} 1 & -2^{-1} \\ 2^{-1} & 1 \end{bmatrix} \begin{bmatrix} 1 & -2^0 \\ 2^0 & 1 \end{bmatrix} \begin{bmatrix} 1 \\ 0 \end{bmatrix}, \tag{5.31}$$

an expression tailor-made for implementation by a computer.

Originally designed for trigonometric functions, CORDIC's applications have spread to include a number of other mathematical operations. Clearly, the rise of the computer has changed the criteria whereby computational algorithms are judged, by replacing the human actor with a mechanical one.[86]

■ **Fourier Series**

In chapter 3, within the context of celestial mechanics, we saw Leonhard Euler convert an expression that he found difficult to handle into a more manageable sum of infinitely many cosine terms. This was just one example of inter-

---

[86] [Volder 1959] is the original report of CORDIC; [Volder 2000] is an account of its birth by its inventor. For mathematical accounts of the method, see also [Sultan 2009] and [Eklund 2001].

actions between physics and analysis that were leading in similar directions. Since Newton's time, and even earlier, scientists and mathematicians had been able to represent functions as infinite series of multiples of power functions, which had led to a considerable expansion of the power of calculus. Now they began to find that they could represent functions as series of multiples of trigonometric functions. These new expressions were much more helpful when dealing with certain types of physical problems, especially those involving partial differential equations.

The earliest of these problems was as follows. Take a string in a certain initial position curve $y = u(x)$ and set it in motion. The string's position is then also a function of time. Our goal is to determine the string's position over time, $u(x, t)$, given its initial position $u(x, 0)$. The **wave equation** governing this motion was first derived by Jean le Rond d'Alembert in the 1740s:

$$\frac{\partial^2 u}{\partial t^2} = a^2 \frac{\partial^2 u}{\partial x^2}, \tag{5.32}$$

where a is a constant depending on the string's density and tension. In the case that $a = 1$, he demonstrated that solutions of this equation are given by

$$u = \Psi(t+x) + \Gamma(t-x), \tag{5.33}$$

where Ψ and Γ are arbitrary twice differentiable functions with periods equal to twice the string's length L. In his first paper on the subject he dealt with the special case that the string begins in its equilibrium position ($u(x, 0) = 0$ for all x) and that the string is fixed in place at its endpoints ($u(0, t) = u(L, t) = 0$ for all t). In his second paper he dealt with the more general case that the string's initial position $u(x, 0)$ is some function $f(x)$ and its initial velocity $\frac{\partial}{\partial t} u(x, 0)$ is some other function $g(x)$. He found that f and g must be odd functions with period $2L$.[87]

What exactly are these functions f and g? This was not a question with an obvious answer at the time. For d'Alembert they had to be analytic expressions, twice differentiable to satisfy the wave equation. In a paper responding to d'Alembert, Euler allowed the string's initial shape $f(x)$ to be either "regular and contained in a certain equation, or irregular and mechanical"—much closer to

---

[87] D'Alembert's two papers are [d'Alembert 1749a] and [d'Alembert 1749b]. A number of modern authors have written on the vibrating string controversy, including [Truesdell 1960], [Ravetz 1961], [Grattan-Guinness 1970, 1–21], [Youschkevitch 1976–1977, 64–69], [Demidov 1982], [Bottazzini 1986, 21–33], [Wheeler/Crummett 1987], and [Kahane/Lemarié-Rieusset 1995, 23–25]. On recently discovered correspondence between d'Alembert and Euler on the topic, see [Hug/Steiner 2015] and [Jouve 2017]. For the historical background see [Cannon/Dostrovsky 1981], and for a biography of d'Alembert see [Hankins 1970].

the notion of a function in mathematics today as the function does not require it to be represented by an equation.[88] Euler's revision triggered a fierce debate over the meanings of the words "function" and "continuity" and was one of the causes that led eventually to the development of modern analysis.

One of those who entered into the fray was Daniel Bernoulli. Relying as it does on the equality of two second partial derivatives, the wave equation immediately suggests the possibility of trigonometric solutions, but d'Alembert and Euler did not mention them directly. Daniel Bernoulli was more of a physical scientist than the other two and had done research in acoustics. Based on his understanding of the superposition of musical tones, he asserted that the general solution of the wave equation might be represented as a sum of products of sines and cosines.[89] In particular, he proposed but did not prove that the solutions must look like

$$u(x,t) = \alpha \sin\frac{\pi x}{L}\cos\frac{\pi at}{L} + \beta \sin\frac{2\pi x}{L}\cos\frac{2\pi at}{L}$$
$$+ \gamma \sin\frac{3\pi x}{L}\cos\frac{3\pi at}{L} + \cdots. \tag{5.34}$$

Neither Euler nor d'Alembert were impressed with Bernoulli's lack of mathematical precision; in the absence of a demonstration, they were unwilling to accept this solution.

The debate continued. Italian-French analyst Joseph-Louis Lagrange was still in his early twenties when he joined the discussion later that decade—although, as we saw in chapter 3, he had already been in correspondence with Euler for several years. His solution to the vibrating string problem, developed from scratch and resulting from over 50 pages of analytic calculations, is

$$u = \frac{2}{L}\int_0^L \sum_{r=1}^{\infty} \sin\frac{r\pi X}{L}\sin\frac{r\pi x}{L}\cos\frac{r\pi at}{L} f(X)dX$$
$$+ \frac{2}{\pi a}\int_0^L \sum_{r=1}^{\infty}\frac{1}{r}\sin\frac{r\pi X}{L}\sin\frac{r\pi x}{L}\sin\frac{r\pi at}{L}g(X)dX. \tag{5.35}$$

Something rather curious happens if one inserts $t=0$ into this equation. Of course, u becomes the string's initial position $f(x)$ and the second term disappears. This leaves

$$f(x) = \frac{2}{L}\int_0^L \sum_{r=1}^{\infty}\sin\frac{r\pi X}{L}\sin\frac{r\pi x}{L}f(X)dX. \tag{5.36}$$

[88] [Euler 1749a].
[89] [Bernoulli 1755].

This expression is quite close (although not identical) to the Fourier sine series representation of the function $f(x)$, fifty years before Fourier. Lagrange, however, was focused on finding solutions to the wave equation, and if he noticed the possible mathematical implications of (5.36), he did not pursue them.[90]

The debate over the legitimacy of proposed solutions to the vibrating string problem continued throughout the rest of the eighteenth century. To pursue it here would take us away from trigonometry and toward analysis, so we return to the narrative several decades later to observe the most critical chapter in the story of trigonometric series. The debate arose again in the first decade of the nineteenth century, this time within the context of another physical problem. Jean Baptiste Joseph Fourier (1768–1830) was not a typical mathematician, even for his time. As a young man he trained for the priesthood, but he eventually became a mathematics teacher at a military school. Involved in the French Revolution and entangled in various factions, he was at one point imprisoned and later released when the political tide turned. He joined Napoleon in Egypt between 1798 and 1801, and his reputation today is due partly to his contributions to *Description de l'Égypte*. Some of his experiments in the last decade of his life have been associated with the discovery of the greenhouse effect.

Upon his return to France, still serving as an administrator, he began his research on the diffusion of heat through a medium. By 1807 he felt he had solved the problem by formulating a partial differential equation from Newton's laws and then solving that equation—an approach that, while standard practice today, still held novelty at the time. He completed his paper in 1807 and submitted it to the Institut de France, but Lagrange was critical. A revised version of the paper won an Institut prize competition on the subject in 1812, but it was not until 1822 that Fourier's work was finally published in his monumental *Théorie Analytique de la Chaleur*.[91] His faith in the power of analysis to solve physical problems could not have been stated more clearly:

> Problems relative to the uniform propagation or to the varied movement of heat in the interior of solids, are reduced . . . to problems of pure analysis, and the progress of this part of physics will depend in consequence upon the advance which may be made in the art of

---

[90] [Lagrange 1759]; for the derivation, see pp. 26–64. See also [Grattan-Guinness 1970, 13–17], [Delsedime 1971], and [Bottazzini 1986, 31–33].

[91] [Fourier 1822], reprinted in [Fourier 1988]; also the first volume of [Fourier (Darboux) 1888/1890]. An English translation is available in [Fourier (Freeman) 1878]. The original 1807 paper was published with extensive commentary in [Grattan-Guinness/Ravetz 1972].

analysis. The differential equations which we have proved contain the chief results of the theory . . . and they connect for ever with mathematical science one of the most important branches of natural philosophy.[92]

To see how Fourier applied analysis to physics, consider the following example. Suppose that we have a three-dimensional solid prism of infinite length. Knowing the heat on the prism's boundaries, how does the heat propagate through it? Let $v(x, y, z, t)$ be the temperature at any point (x, y, z) within the prism at time t, and imagine that the temperature has reached equilibrium everywhere. Pick an arbitrary, infinitesimally small cuboid within the prism, where x ranges from x to $x+dx$, y from y to $y+dy$, and z from z to $z+dz$. Consider the prism's face corresponding to the initial x value. The heat flowing into the cuboid through this face per unit volume is proportional to the rate of change of temperature with respect to x, $-K\dfrac{\partial v}{\partial x}$ (where K is a constant of conductivity); since the face's area is $dy\ dz$, the quantity of heat inflow is $-K\ dy\ dz\dfrac{\partial v}{\partial x}$. On the opposite face (corresponding to $x+dx$), the cuboid loses $-K\ dy\ dz\dfrac{\partial v}{\partial x} - K\ dy\ dz\ d\left(\dfrac{\partial v}{\partial x}\right)$ (where d is the differential with respect to x). The difference between these, $K\ dy\ dz\ d\left(\dfrac{\partial v}{\partial x}\right) = K\ dx\ dy\ dz\dfrac{\partial^2 v}{\partial x^2}$, is the heat gained through these two faces. Of course the same is true in the y and z directions, so the total heat gained is

$$K\ dx\ dy\ dz\left(\frac{\partial^2 v}{\partial x^2} + \frac{\partial^2 v}{\partial y^2} + \frac{\partial^2 v}{\partial z^2}\right).$$

Since the system is at equilibrium we know that this quantity is zero, and we have arrived at our differential equation,[93]

$$\frac{\partial^2 v}{\partial x^2} + \frac{\partial^2 v}{\partial y^2} + \frac{\partial^2 v}{\partial z^2} = 0. \tag{5.37}$$

A little later, Fourier considers a two-dimensional example. In figure 5.15 we have a rectangular prism extending upward to infinity whose width is π. The dashed vertical line down the middle is the x axis, and the horizontal line along the base is the y axis. The bottom boundary is maintained at a constant temperature of one while the two vertical boundaries have a constant

[92] [Fourier (Freeman) 1878, 131].
[93] [Fourier (Freeman) 1878, 98–99].

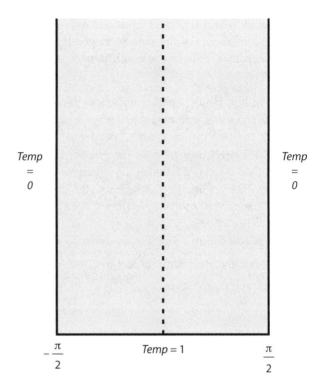

Figure 5.15
Fourier's example of
heat diffusion.

Temp
=
0

Temp
=
0

$-\dfrac{\pi}{2}$ Temp = 1 $\dfrac{\pi}{2}$

temperature of zero. Once the temperatures have reached an equilibrium, what is the temperature function $v(x, y)$?

Fourier's approach, known at the time and now called "separation of variables," was to hope that v is a product of a function of x and another function of y; that is, $v(x, y) = F(x) f(y)$. If we substitute this expression into (5.37) (first removing the z term, since we are in two dimensions) and then divide by $v(x, y)$, we get

$$\frac{F''(x)}{F(x)} + \frac{f''(y)}{f(y)} = 0. \tag{5.38}$$

Setting the two fractions in this expression to the constants m and $-m$, Fourier reduces the temperature problem to two simple ordinary differential equations. Example solutions are $F(x) = e^{-mx}$ and $f(y) = \cos my$, so one solution to (5.37) is $v(x, y) = e^{-mx} \cos my$. Now, m must be positive (otherwise the temperature would rise exponentially to infinity as x increases); and since $v\left(0, \pm\dfrac{\pi}{2}\right) = 0$, $\cos\left(\pm\dfrac{m\pi}{2}\right)$ must be zero, so m must be an odd whole number.

Now, by inspection, the solutions to (5.37) are closed under linear combinations, so Fourier pieces together the general solution

$$v = ae^{-x}\cos y + be^{-3x}\cos 3y + ce^{-5x}\cos 5y + \cdots. \qquad (5.39)$$

But $v(0, y)$ (the temperature on the bottom edge) must be 1, so

$$1 = a\cos y + b\cos 3y + c\cos 5y + \ldots. \qquad (5.40)$$

To find the coefficients $a, b, c, \ldots$, Fourier differentiates (5.40) infinitely many times, substituting $y = 0$ into each result. Eventually he concludes that $a = \dfrac{4}{\pi}$, $b = -\dfrac{4}{3\pi}$, $c = \dfrac{4}{5\pi}$, and so on, so that

$$\frac{\pi}{4} = \cos y - \frac{1}{3}\cos 3y + \frac{1}{5}\cos 5y - \cdots. \qquad (5.41)$$

Text 5.3

Jean Baptiste Joseph Fourier, A Trigonometric Series as a Function

The second member [the right hand side of (5.41)] is a function of y, which does not change in value when we give to the variable y a value included between $-\frac{1}{2}\pi$ and $+\frac{1}{2}\pi$. It would be easy to prove that this series is always convergent. . . . Without stopping for a proof, which the reader may supply, we remark that the fixed value which is continually approached is $\frac{1}{4}\pi$, if the value attributed to y is included between 0 and $\frac{1}{2}\pi$, but that it is $-\frac{1}{4}\pi$, if y is included between $\frac{1}{2}\pi$ and $\frac{3}{2}\pi$; for, in this second interval, each term of the series changes in sign. . . .

The equation

$$y = \cos x - \frac{1}{3}\cos 3x + \frac{1}{5}\cos 5x - \frac{1}{7}\cos 7x + \&c.$$

belongs to a line which having x for abscissa and y for ordinate, is composed of separated straight lines, each of which is parallel to the axis, and equal to the circumference. These parallels are situated alternately above and below the axis, and the distance $\frac{1}{4}\pi$, and joined by perpendiculars which themselves make part of the line.[94]

Explanation: Note that Fourier switches the independent variable from y to x part of the way through this passage. In (5.41), Fourier had demonstrated that the function

---

[94] [Fourier (Freeman) 1878, 143–144].

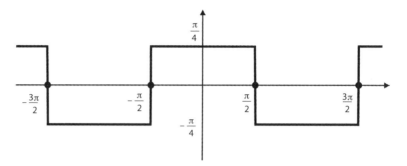

Figure 5.16
Fourier's step function. Although for x values that are odd multiples of π the function is discontinuous (it jumps from one step to another with a single function value on the x axis), Fourier describes the entire curve, including the vertical segments.

$$y = \cos x - \frac{1}{3}\cos 3x + \frac{1}{5}\cos 5x - \frac{1}{7}\cos 7x + \dots$$

is equal to the constant $\frac{\pi}{4}$ if $x \in \left(-\frac{\pi}{2}, \frac{\pi}{2}\right)$. But if $x \in \left(\frac{\pi}{2}, \frac{3\pi}{2}\right)$, the cosines all change sign, so $y = -\frac{\pi}{4}$. (Fourier does not make a note of the fact that $y(0) = 0$.) The rest of the passage attempts to describe the graph of the function (figure 5.16). The curve is a step function with function values on the x axis at the discontinuities, but Fourier describes it as a continuously drawn curve where the plateaus and the valleys are joined by vertical line segments. This is one of the earliest examples of a series of infinitely differentiable functions whose limit is not a continuous function in the modern sense. What exactly was meant by a continuous function was not clear at Fourier's time, and this example indicated that the question might be deeper than it seemed. Of course, Fourier also may have been concerned more with the physics than the mathematical implications of this strange curve.

Clearly Fourier was worried about the validity of these series expressions; they were likely what had led Lagrange to the criticisms that had delayed publication of his results by 15 years. Fourier's book revisits the matter several times, demonstrating that different mathematical arguments lead to the same solution. Extending the applications, he goes on to represent other functions $f(x)$ using series of sines and cosines. They are expressed most generally in the form

$$\frac{1}{2}a_0 + \sum_{n=1}^{\infty}a_n \cos(nx) + \sum_{n=1}^{\infty}b_n \sin(nx). \tag{5.42}$$

He quickly finds equivalents to the modern formulas for determining the coefficients,

$$a_0 = \frac{1}{\pi} \int_{-\pi}^{\pi} f(x)\, dx, \tag{5.43}$$

$$a_n = \frac{1}{\pi} \int_{-\pi}^{\pi} f(x)\cos(nx)\, dx, \tag{5.44}$$

and

$$b_n = \frac{1}{\pi} \int_{-\pi}^{\pi} f(x)\sin(nx)\, dx. \tag{5.45}$$

What other functions $f(x)$ can be represented using these series? Fourier was confident in his methods: "we can extend the same results to any functions, even to those which are discontinuous and entirely arbitrary."[95] Later he asserts that "the series arranged according to sines or cosines of multiple arcs are always convergent; that is to say, on giving to the variable any value whatever that is not imaginary, the sum of the terms converges more and more to a single fixed limit, which is the value of the developed function."[96] Although specific series of this sort had been examined as early as Euler, Fourier's two claims were to say the least rather bold. Evaluations followed quickly, especially in the work of Gustav Peter Lejeune Dirichlet.[97] Much of the fuel for expansion in nineteenth-century analysis came from attempts to understand to what extent Fourier was correct in these claims.[98] Meanwhile, Fourier series and the theory that developed from it soon became fundamental to partial differential equations and mathematical physics.

The controversy over Fourier's series came to play a crucial and surprising role later in the century. It was provoked by the following natural question: supposing that a function $f(x)$ can be represented with a Fourier series, is that series guaranteed to be unique? This problem had been studied already

---

[95] [Fourier (Freeman) 1878, 184].

[96] [Fourier (Freeman) 1878, 207].

[97] [Dirichlet 1829].

[98] A great deal has been written about Fourier, his series, and the developments in analysis that were triggered by them. A sampling of these works is [Grattan-Guinness 1969], [Grattan-Guinness 1970, 20–21, 96–98], [Grattan-Guinness/Ravetz 1972] (which we cited above for its edition of Fourier's original 1807 paper, but it also includes extensive commentaries), [Grattan-Guinness 1980, 104–109], [Bottazzini 1986, 59–81], [Grattan-Guinness 1990a, vol. 2, 583–632], [Yoshida/Takata 1991], [Kahane/Lemarié-Rieusset 1995] and other papers by Kahane, [Ferraro 2007a, 80–83], [Roy 2011, 412–417], [Herreman 2013] (identifying the *Analytic Theory of Heat* as an "inaugural text"), and [Gray 2015, 13–19].

by Eduard Heine, but he had been able to demonstrate uniqueness only with the assumption of certain properties. The young mathematician Georg Cantor wanted to remove as many of the restrictions on uniqueness as he could. In a series of three successive papers, his progress continued. Firstly, in 1870 he proved that the series is unique under the assumption that that the series converges for all values of x. The following year he showed that uniqueness holds even if the series fails to represent f or converge for a finite number of exceptional values of x. Finally, another year later, he was able to demonstrate uniqueness even if the number of exceptional values of x is infinite—as long as it is the right kind of infinity. But what exactly does this mean? It took Cantor some while to work this out, and his clarifications included a new definition of irrational numbers as well as the concept of "derived set" in point-set topology. This result, in fact, was Cantor's jumping-off point for his explorations of set theory and transfinite numbers.[99]

Concerns About Negativity

Clearly, by Fourier's time the trigonometric functions were accepted as simply another group of functions within the toolbox of analysis. However, to use them in this way, one must assume that $f(x) = \sin x$, for instance, taking on values for all real values of x, positive or negative. We have seen various authors, Euler among them, extend their notions of trigonometric functions in this way, usually implicitly. But what exactly is a negative magnitude if the object in question is a geometric entity (for instance, a line segment or an area)? The trigonometric functions had been geometric in nature since the beginning, and much of the theory of trigonometry had been developed on that understanding. If negative geometric magnitudes could not be considered sensible, then many developments in analysis would need to be rethought.

Such talk may seem strange today in a world where children work with negative numbers and coordinate systems from an early age. But this was a serious concern as late as the early nineteenth century, one that led to philosophical pondering. D'Alembert had attempted to deal with this issue in an article in his *Encyclopédie* by claiming that negative quantities are real, but Lazare Carnot (1753–1823) was not satisfied. A military leader in the French

---

[99] Cantor's three papers are [Cantor 1870], [Cantor 1871], and [Cantor 1872]. For more on Cantor's work on trigonometric series, see [Dauben 1971], [Dauben 1980] (slightly less technical), [Bottazzini 1986, 275–280], [Cooke 1993, 293–297], [Kahane/Lemarié-Rieusset 1995, 67–70], [Dauben 2005] (dealing more generally with Cantor's interactions with set theory), and [Gray 2015, 227–237] (a survey that also includes some of Cantor's successors). The standard biography of Cantor is [Dauben 1979].

revolutionary wars and one of Napoleon's generals, Carnot is one of very few mathematical scholars to serve as a head of state.[100] Apart from his political activities, his reputation lies mostly in his theoretical work in engineering.

Carnot's physical perspective was at the heart of his mathematical thinking. In geometry he worked out his objections to d'Alembert's stance and presented his own proposals in a pair of books: *De la Corrélation des Figures de Géométrie* (1801) and *Géométrie de Position* (1803), on projective geometry.[101] For Carnot, a geometric object had to live up to a higher standard than an algebraic one: it needed to be explicitly defined and to be realized in the physical universe. Therefore the quantity $x - y$ is genuine if $x > y$ but not if $y > x$. In the latter case, while one can perform algebraic manipulations on the expression $x - y$, that does not make it real. Raising the struggles that had befuddled mathematicians over the meaning of the logarithm of a negative number, Carnot made the following ontological claims:

> (1) Every isolated negative quantity is an object of reason, and those that we meet in calculation are only simple algebraic forms, unable to represent any real and effective quantity. (2) Each of these algebraic forms being taken, apart from its sign, is nothing other than the difference between two other absolute quantities, of which the larger one in the case on which the reasoning was established is the smallest in the case which we want to apply the results of the calculation.[102]

Trigonometry was an obvious test case for Carnot's claims:

> One of the most proper applications to get a sense of the play of correlative figures in geometry is the one that takes place among the linear-angular quantities, that is, the sine, cosine, tangent, etc. These quantities are neither lines nor angles, but abstract numbers that serve as intermediaries to establish the relation of these heterogeneous quantities; because the sine of an arc is not precisely the perpendicular dropped from one of the extremities of this arc onto the radius that passes through its other extremity, but the ratio of this perpendicular to the radius.[103]

---

[100] Carnot was the president of the French National Convention, a position that seems to have put many of its holders at risk of the guillotine, for the usual two-week term in 1794.

[101] [Carnot 1801] and [Carnot 1803]. The latter develops Carnot's views more fully, and we shall concentrate on it. See also [Carnot 1806].

[102] [Carnot 1803, xviii–xix]. Carnot was much more content with infinitesimal quantities and used them in his engineering work.

[103] [Carnot 1803, 126].

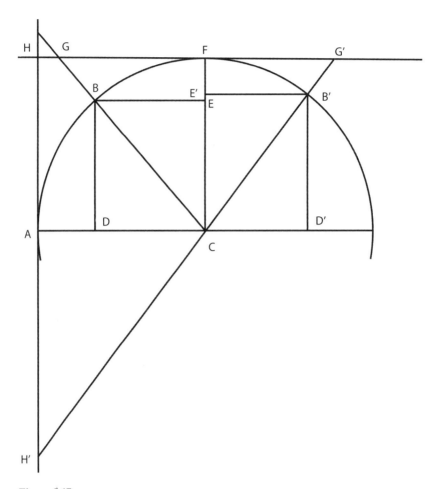

Figure 5.17
Carnot's primitive and correlative systems for the trigonometric functions.

So a sine is not actually a geometric entity in itself but an abstraction representing a ratio between two entities. But how can we work with them if we do not allow negative quantities to exist? Carnot begins to answer this question with figure 5.17, which in the top left quadrant of the circle contains the centuries-old definitions of the standard trigonometric functions. B is chosen arbitrarily on AF, so that $BD/BC = \sin \widehat{AB}$ and $BE/BC = \cos \widehat{AB}$. Tangents to the circle are drawn at A and F, and CB is extended to G and onward to H so that $AH/AC = \tan \widehat{AB}$, $FG/FC = \cot \widehat{AB}$, and so on. In this quadrant all the trigonometric quantities are positive; $ABCDEFGH$ therefore poses no problems and is called the ***primitive system***. Now suppose our arbitrary point is B' in the second quadrant. We perform the same constructions as before, leading to the ***correlative system*** *AB'CD'E'FG'H'*. The various arcs and lengths in this new system correlate with objects in the primitive system according to the obvious pairings. For each pair of correlated objects Carnot establishes a "correlation of the signs." For instance, the correlation between *BD* and *B'D'* (corresponding to the sine) is ***direct*** and is assigned the symbol + because as $\widehat{AB}$ increases and *B* moves beyond *F* (becoming *B'*), the value of *BD* does not "pass through 0, nor through ∞." However, *BE* (corresponding to the cosine) passes through zero when *B* passes through *F*, so the correlation of *BE* and *B'E'* is ***inverse*** and is assigned the symbol −. These correlations may be used to establish that $\sin(180° − θ) = \sin θ$ and $\cos(180° − θ) = −\cos θ$. The signs + and − are nothing more than indicators that the geometric argument is right or wrong concerning the correlation of the figures in question; they do not posit the existence of positive and negative quantities. If the argument is wrong (i.e., there is an inverse correlation), some adjustment to the conclusion is required, depending on the situation. Carnot extends the construction through the other two quadrants, generating two more correlative systems; later in the book he extends to arcs beyond 2π radians.

Having relegated + and − to mere signifiers of correlations, Carnot is able to establish the trigonometric identities without having to admit that negative quantities exist while at the same time extending the arguments of trigonometric functions as far as he likes. In the end nothing changes: the theorems all look as they did before. But in the beginning everything is different: geometry and trigonometry are not required to rely on an ontological claim that some found dubious.

This system, and the revisiting of many dozens of simple identities that it provoked, may seem to modern readers a tempest in a teapot. Historian of geometry Julian Lowell Coolidge wrote that "the whole idea seems to me vague and not worthy of the esteem in which it was held by contemporaries."[104] However, the fact that Carnot's colleagues (including Gauss, Chasles, and later Felix Klein[105]) treated the *Géometrie de Position* seriously is an example of the dictum in the history of mathematics that, if we read something that seems silly, we should read it again with more care. Von Braunmühl's judgment at the beginning of the twentieth century was that Carnot "for the first time proved the general validity of the [sine and cosine] addition formulas, which . . . implies all trigonometric formulas."[106]

[104] [Coolidge 1940, 92].
[105] [See Gillispie/Pisano 2014, 108].
[106] [von Braunmühl 1900/1903, vol. 2, 171]. On Carnot's theory of correlative figures see [von Braunmühl 1900/1903, vol. 2, 169–172], [Grattan-Guinness 1990a, vol. 1, 254–256], [Schubring 2005, 353–365], [Nabonnand 2011], [Nabonnand 2016], and [Gillispie/Pisano 2014, 109–117]. (The first part of this book consists of revised excerpts from Gillispie's earlier

Hyperbolic Trigonometry

In chapter 3 we saw that the practitioners of calculus in the late seventeenth and early eighteenth centuries, such as James Gregory and Roger Cotes, were noticing various parallels between certain mathematical constructions on the circle and similar constructions on the hyperbola and between trigonometric and logarithmic functions. These correspondences had been part of the inspiration for the birth of the calculus in the complex plane, including the unification of trigonometry and exponential functions within the calculus of Leonhard Euler. In fact, Euler's analytic prescience went so far at one point as to determine infinite products for the quantities $\dfrac{e^x - e^{-x}}{2}$ and $\dfrac{e^x + e^{-x}}{2}$, quantities which we now call the hyperbolic sine and cosine. However, Euler used these products only as an intermediate step toward finding infinite products for the sine and cosine. The hyperbolic cosine is also the solution to the problem of determining the shape of a hanging chain—the *catenary*. This had been demonstrated by Christian Huygens, Gottfried Wilhelm Leibniz, and Johann Bernoulli in 1691; however, the name "hyperbolic cosine" did not yet exist, and the function did not attract further attention until much later. Nevertheless, this ground was clearly fertile, and it would not be long after Euler's work that several independent scholars would bring forward related ideas that would lead to hyperbolic trigonometry.

In fact, the ground broke more than once. The first time was in 1757 within a collection of physical and mathematical essays by Vincenzo Riccati (1707–1775). Vincenzo was the son of Jacopo Riccati, the physicist and mathematician for whom the Riccati differential equation is named. Vincenzo had similar interests and concentrated on mechanical problems. His work on the hyperbolic functions appears in two of thirteen essays in the first volume of his *Opusculorum ad res physicas, et mathematicas pertinentium*.

Text 5.4
Vincenzo Riccati, The Invention of the Hyperbolic Functions

I will set out the construction, as is necessary, in order that I might speak first about the sines and cosines with the circles, then a few words about the hyperbolae. In a circle with radius *CA* let a certain arc *AF* be taken, and from point *F* let perpendicular *FD* be dropped onto *CA*. One notes that the radius or the semiaxis *CA* is called the total sine, the perpendicular *FD* the sine of arc *AF*,

biography of Lazare Carnot, [Gillispie 1971].) [Dhombres/Dhombres 1997] is a detailed biography of Carnot; on his geometry of position see especially pp. 514–521.

and the intercept *CD* the cosine. But by this, sector *ACF* is equal to half of the rectangle from the total sine and from arc *AF*; to say the same thing, double the sector *ACF* divided by the total sine is equal to arc *AF*. Thus *FD* can be called the sine, and *CD* the cosine, of the double of the sector divided by the total sine.

By analogy, the sine and the cosine can be carried over from the circle to the equilateral hyperbola. Let the equilateral hyperbola be described, whose semiaxis *CA* is equal to the radius of the circle, and from the center *C* to an arbitrary point *F* on the curve join line *CF*, and from the same point *F* let *FD* be dropped on the extension of axis *CA*. If *CA* is called the total hyperbolic sine, *FD* could be called the hyperbolic sine and *CD* the hyperbolic cosine of the double the sector *ACF* divided by the total sine. Let these be the definitions of the names.[107]

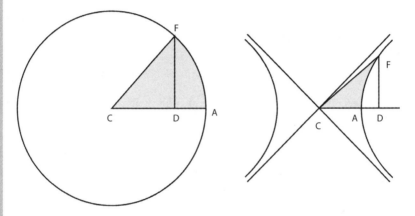

Figure 5.18
Riccati's definitions of the circular and hyperbolic functions.

Explanation: (See figure 5.18.) Riccati sets up two parallel constructions. At first (on the left diagram) he defines the circular functions in the conventional way; *FD* is called the sine of $\widehat{AF}$ while *CD* is the cosine. The circle can have any radius r, so $r = CA$ is the "total sine"; our modern sine function is equal to FD/CA. The argument of the sine and cosine is $\widehat{AF}$. But in preparation for the analogy he is about to make with the hyperbolic functions, Riccati notes that the area of sector *ACF* is $\frac{1}{2}r \cdot \widehat{AF}$,[108] so the argument $\widehat{AF}$ of the sine and cosine may be thought of as $2 \cdot \mathrm{Area}(ACF)/r$.

In the second paragraph Riccati mimics this construction on the right diagram of figure 5.18, replacing the circle $x^2 + y^2 = r^2$ with the equilateral hyper-

---

[107] [Riccati 1757/1762, vol. 1, 68].

[108] $\mathrm{Area}(ACF) = \frac{1}{2}r^2 \cdot \measuredangle C = \frac{1}{2}r^2 \cdot (\widehat{AF}/r) = \frac{1}{2}r \cdot \widehat{AF}.$

bola $x^2 - y^2 = r^2$. Everything follows as before. Riccati's definition of the argument of the circular sine and cosine as $2 \cdot \text{Area}(ACF)/r$ carries across to the argument of the hyperbolic trigonometric functions. Other than the scaling for the radius, this is the same argument that we use today.

Riccati did not make the connection between the hyperbolic functions and non-Euclidean geometry; that was still far in the future.

Riccati's purpose here is to use the new functions to help in the solutions of various cubic equations. Along the way he identifies some of the fundamental identities of the hyperbolic functions, beginning with

$$\cosh^2 x - \sinh^2 x = 1, \tag{5.46}$$

$$\cosh(x + y) = \cosh x \cosh y + \sinh x \sinh y, \tag{5.47}$$

and[109]

$$\sinh(x + y) = \cosh x \sinh y + \cosh y \sinh x. \tag{5.48}$$

In another article in the same volume, he reports power series expansions for the hyperbolic sine and cosine, which had been derived by Josepho Suzzio.[110]

The appearance of Riccati's work and its similarity to modern presentations might suggest that it was from him that we became aware of the hyperbolic functions, but this appears not to be the case. The path that led to their dissemination begins a couple of years later by Daviet de Foncenex (1734–1799), a student of Lagrange who wrote a paper discussing paradoxical results on the use of imaginary quantities in analysis, especially the controversy over the correct definition of the logarithm of a negative number. Within the paper he observes that the y values of $x^2 - y^2 = r^2$ and $x^2 + y^2 = r^2$, for the same x, are in the ratio of 1 to $\sqrt{-1}$.[111] This curiosity was noticed by Johann Lambert, and there the story really began.

One of Lambert's most famous papers is his 1761 "Mémoire due quelques propriétés remarquables des quantités transcendentes circulaires et logarithmiques,"[112] in which he proves for the first time that π is irrational.

---

[109] [Riccati 1757/1762, vol. 1, 71]. His abbreviations for the hyperbolic cosine and sine are *Ch* and *Sh* (as opposed to *Cc* and *Sc* for the circular cosine and sine). Our modern expressions of these functions employ a semiaxis of unit length; Riccati's expressions are in terms of an arbitrary semiaxis *r*.

[110] [Riccati 1757/1762, vol. 1, 115]. On Riccati's work with hyperbolic functions see [Naux 1966/1971, vol. 2, 126–136] and [Barnett 2004, 26–28].

[111] [de Foncenex 1759, 128–129].

[112] [Lambert 1768]. The article was published in the *Mémoires de l'Académie Royale des Sciences et Belles-Lettres* for the year 1761 but appeared in 1768.

Elsewhere in the same paper, he notes that the power series for $\dfrac{e^v - e^{-v}}{2}$ and $\dfrac{e^v + e^{-v}}{2}$ are identical to the power series for the sine and cosine once one makes the signs of all the terms positive. This can be seen easily enough by substituting $v = iu$, essentially the same substitution that Euler had made when he had briefly encountered these functions earlier. Working with these power series and with continued fractions similar to those he had used to prove that π is irrational, he suggests from his results that e is not just irrational but transcendental. Recognizing the "affinity" between his "logarithmic quantities" (hyperbolic functions) and the circular quantities (sine and cosine), Lambert pursued the similarities between them, especially involving their differentials. Hunting very different mathematical quarry than Riccati had been seeking several years earlier, Lambert's apparent ignorance of Riccati's footsteps on this terrain is perhaps not surprising.

It is one thing to notice an affinity between the circular and hyperbolic domains but quite another to classify the hyperbolic quantities as trigonometric themselves. Of course, Riccati had already taken this step, but Lambert did not do so until seven years later in his 1768 paper innocuously titled "Observations trigonométriques." He starts, incongruously, with a problem from spherical astronomy: suppose we have a star with a certain declination δ above the equator and we know the number of hours since it passed the meridian. Can we calculate the star's altitude above the horizon?

In figure 5.19, we are outside the celestial sphere, observing it from a position directly east of it. Over the day, the star travels on circle SM, which is the angular distance δ above the equator. As we can see, this star spends most (but not all) of its time above the horizon. Draw perpendiculars SB and MA onto the vertical line AC between the zenith and nadir. Lambert represents the star's daily journey with circle $AGBF$; the star is above the horizon when it is within arc FAG and below the horizon when it is within arc GBF. Arc AQ represents the number of hours since the star passed the meridian. Drop a perpendicular from Q onto FG; this segment QK is the sine of the star's altitude. It is a straightforward spherical trigonometric problem to solve for it.

But what happens if δ is so large that S is above the horizon or so negative that M is below it? Then circle AB will be so high or so low that it will not cross the horizon at all, and Lambert tells us that the intersections points F and G with the horizon will become "imaginary." Nevertheless the star still has an altitude. Lambert shows that there is another way to compute it, but the alternate formula is complicated. Can we proceed, instead, with the original geometry and use an imaginary diurnal arc?

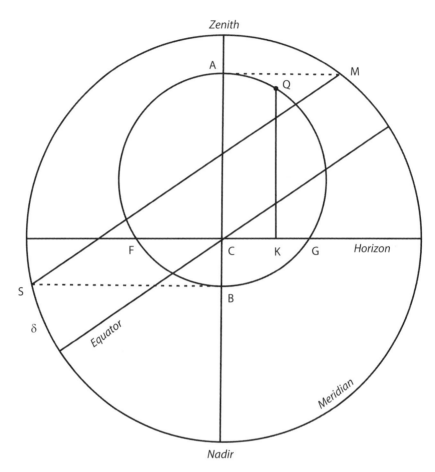

Figure 5.19
Lambert on finding the altitude of a star.

We know that imaginary sines become real quantities when one transfers them to the hyperbola. It is thus necessary to examine in what way the hyperbolic functions analogous to the circular functions can be put into use in trigonometry, and what are the advantages that result from it.[113]

On the next page Lambert introduces the terms "hyperbolic sine" and "hyperbolic cosine" and their definitions. Along the way to solving this astronomical problem by using his novel creations, he lists 25 identities of the new

---

[113] [Lambert 1770, 329].

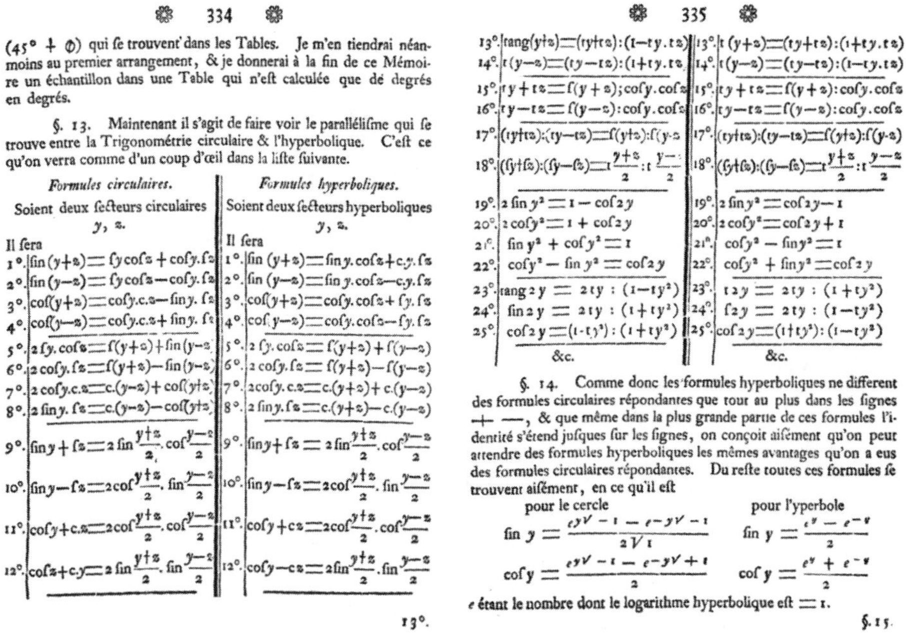

🌼 334 🌼

(45° + φ) qui fe trouvent dans les Tables. Je m'en tiendrai néanmoins au premier arrangement, & je donnerai à la fin de ce Mémoire un échantillon dans une Table qui n'eft calculée que dé degrés en degrés.

§. 13. Maintenant il s'agit de faire voir le parallélifme qui fe trouve entre la Trigonométrie circulaire & l'hyperbolique. C'eft ce qu'on verra comme d'un coup d'œil dans la lifte fuivante.

Formules circulaires.

Soient deux fecteurs circulaires
y, z.

Il fera

1°. $\sin(y+z) = \cos z + \cos y \cdot \sin z$
2°. $\sin(y-z) = \cos z - \cos y \cdot \sin z$
3°. $\cos(y+z) = \cos y \cdot \cos z - \sin y \cdot \sin z$
4°. $\cos(y-z) = \cos y \cdot \cos z + \sin y \cdot \sin z$
5°. $2 \sin y \cdot \cos z = \sin(y+z) + \sin(y-z)$
6°. $2 \cos y \cdot \sin z = \sin(y+z) - \sin(y-z)$
7°. $2 \cos y \cdot \cos z = \cos(y-z) + \cos(y+z)$
8°. $2 \sin y \cdot \sin z = \cos(y-z) - \cos(y+z)$
9°. $\sin y + \sin z = 2 \sin \frac{y+z}{2} \cdot \cos \frac{y-z}{2}$
10°. $\sin y - \sin z = 2 \cos \frac{y+z}{2} \cdot \sin \frac{y-z}{2}$
11°. $\cos y + \cos z = 2 \cos \frac{y+z}{2} \cdot \cos \frac{y-z}{2}$
12°. $\cos z + \cos y = 2 \sin \frac{y+z}{2} \cdot \sin \frac{y-z}{2}$

Formules hyperboliques.

Soient deux fecteurs hyperboliques
y, z.

Il fera

1°. $\sin(y+z) = \sin y \cdot \cos z + \cos y \cdot \sin z$
2°. $\sin(y-z) = \sin y \cdot \cos z - \cos y \cdot \sin z$
3°. $\cos(y+z) = \cos y \cdot \cos z + \sin y \cdot \sin z$
4°. $\cos(y-z) = \cos y \cdot \cos z - \sin y \cdot \sin z$
5°. $2 \cos y \cdot \cos z = \cos(y+z) + \cos(y-z)$
6°. $2 \cos y \cdot \sin z = \cos(y+z) - \cos(y-z)$
7°. $2 \cos y \cdot \cos z = \cos(y+z) + \cos(y-z)$
8°. $2 \sin y \cdot \sin z = \cos(y+z) - \cos(y-z)$
9°. $\sin y + \sin z = 2 \sin \frac{y+z}{2} \cdot \cos \frac{y-z}{2}$
10°. $\sin y - \sin z = 2 \cos \frac{y+z}{2} \cdot \sin \frac{y-z}{2}$
11°. $\cos y + \cos z = 2 \cos \frac{y+z}{2} \cdot \cos \frac{y-z}{2}$
12°. $\cos y - \cos z = 2 \sin \frac{y+z}{2} \cdot \sin \frac{y-z}{2}$

13°.

🌼 335 🌼

13°. $\tan(y+z) = (\tan y + \tan z):(1 - \tan y \cdot \tan z)$
14°. $\tan(y-z) = (\tan y - \tan z):(1 + \tan y \cdot \tan z)$
15°. $\tan y + \tan z = \sin(y+z):\cos y \cdot \cos z$
16°. $\tan y - \tan z = \sin(y-z):\cos y \cdot \cos z$
17°. $(\tan y + \tan z):(\tan y - \tan z) = \sin(y+z):\sin(y-z)$
18°. $(\sin y + \sin z):(\sin y - \sin z) = \tan \frac{y+z}{2} : \tan \frac{y-z}{2}$
19°. $2 \sin y^2 = 1 - \cos 2y$
20°. $2 \cos y^2 = 1 + \cos 2y$
21°. $\sin y^2 + \cos y^2 = 1$
22°. $\cos y^2 - \sin y^2 = \cos 2y$
23°. $\tan 2y = 2\tan y : (1 - \tan y^2)$
24°. $\sin 2y = 2\tan y : (1 + \tan y^2)$
25°. $\cos 2y = (1 - \tan y^2):(1 + \tan y^2)$
&c.

13°. $\tan(y+z) = (\tan y + \tan z):(1 + \tan y \cdot \tan z)$
14°. $\tan(y-z) = (\tan y - \tan z):(1 - \tan y \cdot \tan z)$
15°. $\tan y + \tan z = \sin(y+z):\cos y \cdot \cos z$
16°. $\tan y - \tan z = \sin(y-z):\cos y \cdot \cos z$
17°. $(\tan y + \tan z):(\tan y - \tan z) = \sin(y+z):\sin(y-z)$
18°. $(\sin y + \sin z):(\sin y - \sin z) = \tan \frac{y+z}{2} : \tan \frac{y-z}{2}$
19°. $2 \sin y^2 = \cos 2y - 1$
20°. $2 \cos y^2 = \cos 2y + 1$
21°. $\cos y^2 - \sin y^2 = 1$
22°. $\cos y^2 + \sin y^2 = \cos 2y$
23°. $\tan 2y = 2\tan y : (1 + \tan y^2)$
24°. $\sin 2y = 2\tan y : (1 - \tan y^2)$
25°. $\cos 2y = (1 + \tan y^2):(1 - \tan y^2)$
&c.

§. 14. Comme donc les formules hyperboliques ne different des formules circulaires répondantes que tout au plus dans les fignes + —, & que même dans la plus grande partie de ces formules l'identité s'étend jufques fur les fignes, on conçoit aifément qu'on peut attendre des formules hyperboliques les mêmes avantages qu'on a eus des formules circulaires répondantes. Du refte toutes ces formules fe trouvent aifément, en ce qu'il eft

pour le cercle

$$\sin y = \frac{e^{y\sqrt{-1}} - e^{-y\sqrt{-1}}}{2\sqrt{1}}$$

$$\cos y = \frac{e^{y\sqrt{-1}} + e^{-y\sqrt{-1}}}{2}$$

pour l'hyperbole

$$\sin y = \frac{e^y - e^{-y}}{2}$$

$$\cos y = \frac{e^y + e^{-y}}{2}$$

e étant le nombre dont le logarithme hyperbolique eft $= 1$.

§. 15.

Figure 5.20
Lambert's identities for the hyperbolic functions.

functions in a table alongside their circular counterparts (figure 5.20).[114] The paper concludes with the first table of hyperbolic sines and cosines, their logarithms, and four other related functions.[115]

Not many years later, Lambert dealt with the problem of parallel lines and Euclid's fifth postulate in a book entitled *Theory of Parallel Lines* that would not be published until over a century after his death.[116] Whether inspired by his considerations of imaginary arcs, in this work Lambert approached non-Euclidean geometry without quite reaching it. He attempted to prove and came close to proving that Euclidean geometry was the only consistent geometry but was unable to demonstrate that a geometry where all triangles have an angle sum less than 180° is inconsistent. He did show that such triangles have an area proportional to the *defect* (the difference between

[114] [Lambert 1770, 334–335].
[115] On Lambert's work on the hyperbolic functions, see [von Braunmühl 1900/1903, vol. 2, 133–135], [Gray/Tilling 1978, 32–33], and [Barnett 2004, 20–26].
[116] The work appears in [Engel/Stäckel 1895, 152–207].

180° and the sum of the angles), a result that parallels Girard's Theorem in spherical geometry. Lambert remarked that this geometry "almost" might be realized on a sphere with an imaginary radius.[117]

In fact, he might have removed the word "almost." The idea of an imaginary sphere reappeared about 50 years later in the work of Franz Adolph Taurinus (1794–1874) of Cologne. His uncle F. K. Schweikart, a professor of law, had developed an "astral geometry" that was a predecessor of hyperbolic geometry and had exchanged correspondence with Gauss on the topic. In his first book, *Die Theorie der Parallellinien* (1825), Taurinus was more than a little skeptical of the possibility of non-Euclidean geometry. However, only a year later he was writing about "logarithmic-spherical" geometry in *Geometriae prima elementa*.[118] His approach differed from others who were working on new geometries. Rather than beginning with an alternate axiom to replace Euclid's parallel postulate, Taurinus began with trigonometry. If we rewrite the spherical Law of Cosines to apply to any radius R, we have

$$\cos\frac{c}{R} = \cos\frac{a}{R}\cos\frac{b}{R} + \sin\frac{a}{R}\sin\frac{b}{R}\cos C. \tag{5.49}$$

Then, if we replace radius R with the imaginary radius iR and perform a little algebra using the exponential definitions of the hyperbolic functions, we arrive at

$$\cosh\frac{c}{R} = \cosh\frac{a}{R}\cosh\frac{b}{R} - \sinh\frac{a}{R}\sinh\frac{b}{R}\cos C. \tag{5.50}$$

All the circular functions of side lengths have been replaced by their hyperbolic twins, and setting $R = 1$, we have the Law of Cosines in hyperbolic trigonometry. Now, what exactly is meant by a sphere with imaginary radius was never made clear by Taurinus, other than that it is not realized on a plane. Taurinus's trigonometric approach, then, showed how hyperbolic trigonometry might work if the geometry were to exist, but he did not explain how the geometry itself might be realized.[119]

---

[117] On Lambert's work on what eventually became hyperbolic geometry, see [Bonola 1912, 44–51], [Gray/Tilling 1978, 31–34], [Gray 1979a, 63–68], [Gray 2007, 84–89], and [Papadopoulos/ Théret 2014].

[118] [Taurinus 1825] and [Taurinus 1826] are both very difficult to find. Only a few copies of the *Geometriae prima elementa* are known to exist, since Taurinus, disappointed with the book's reception and perhaps with a lack of response from Gauss, burned his remaining copies. Excerpts of both books appear in [Engel/Stäckel 1895, 255–283].

[119] On Taurinus's work especially in *Geometriae prima elementa*, see [Engel/Stäckel 1895, 246– 252], [Bonola 1912, 77–83], [Gray 1979a, 87–95], [Gray 1979b, 242–243], [Houzel 1992, 7–9], and [Gray 2007, 93–97].

Clearly non-Euclidean geometry was in the air, so it is not surprising that three different mathematicians independently working within a decade of Taurinus's conclusions took what most consider to be the decisive step in its discovery. One of these three, Carl Friedrich Gauss, had worked on the topic privately and never published his results.[120] He had been in correspondence with Taurinus's uncle on related matters as well as with the father Farkas of one of the other two discoverers, Hungarian mathematician János Bolyai (1802–1860). Farkas famously had begged János to avoid the theory of parallel lines, but that did not stop János from publishing his findings in an appendix to his father's 1832 textbook *Tentamen juventutem studiosam in elementa matheseos purae*.[121] It was not until about 15 years later that he realized that someone else had published something similar in 1829.

Nicolai Lobachevsky (1792–1856), a professor at Kazan University, had given a presentation on his version of the new geometry in 1826, the year before he became rector of the university. He also had a connection with Gauss as his teacher Martin Bartels had been Gauss's teacher years before. Lobachevsky published his work in a series of five articles in the *Kazan Messenger* in 1829. Unsurprisingly the European mathematical community was not in the practice of reading that particular publication, so his discoveries remained unknown. He published on the topic several more times—one of his papers appeared in *Crelle's Journal* in 1837—but his work that received the most attention was his 1840 pamphlet *Geometrische Untersuchungen zur Theorie der Parallellinien*, which in many cases parallels Bolyai's work; we shall follow a sample from it.[122]

Given point A and line BC in figure 5.21, draw perpendicular AD onto BC, and call its length p. The perpendicular AE to AD does not cross BC, but of course AD itself does. If one imagines drawing another line through A, say AG, AH, or AF, it may or may not cross BC. Assume that AG does not cross it; then any line through A between AG and AE will not cross it either. Likewise, if AF does cross it, then any line through A between AD and AF will also cross it. This implies that there is a boundary line AH, for which any line

[120] On the claim that Gauss was one of the discoverers of non-Euclidean geometry, see [Gray 2006].

[121] [Bolyai 1832/1833]; an English translation of János's appendix by George Bruce Halsted appears in [Bonola 1912, 1–71 (separately paginated)], and another in [Kárteszi 1987]. On János Bolyai and his geometry, see [Bonola 1912, 96–113], [Gray 1979a, 96–116], [Rosenfeld 1988, 212–214], [Gray 2004], [Prékopa 2006], and [Gray 2007, 101–114].

[122] [Lobachevsky 1837], [Lobachevsky 1840]. The latter appears in an English translation by George Bruce Halsted in [Bonola 1912, 1–45 (separately paginated)] and in [Lobachevsky 1914]. A new English translation with extensive commentary may be found in [Braver 2011]. Lobachevsky's 1855 *Pangeometry* is also now available in English translation; see [Lobachevsky (Papadopoulos) 2010].

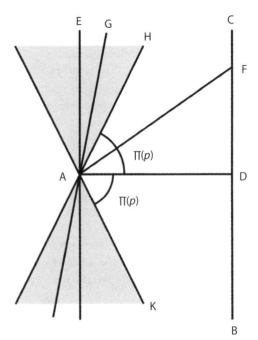

Figure 5.21
Lobachevsky's angle of
parallelism.

through A on one side of AH will cross BC and any line on the other side will
not cross BC. Lobachevsky defines this line as the ***parallel*** to BC, and he calls
the angle $\Pi(p)$ that it makes with AD the ***angle of parallelism***.[123] If we draw
line AK at the same angle to AD but on the other side, we get a shaded region
within which any line through A does not cross BC but any other line through A
does cross it. Eventually Lobachevsky derives a formula for the angle of parallelism, a fundamental theorem in his "imaginary" (non-Euclidean) geometry:

$$\tan \frac{1}{2}\Pi(p) = e^{-p}. \tag{5.51}$$

This formula had been anticipated by Taurinus and had been found by Bolyai
as early as 1823.[124] If $\Pi(p) < \pi/2$, then the angle sum of any triangle is $< \pi$
(or 180°), and we are in what is now called hyperbolic geometry.

Using $\Pi(p)$, Lobachevsky demonstrates a number of trigonometric identities for an arbitrary triangle that are true regardless of which geometry is in
play, Euclidean or not. Two of these[125] are

$$\sin A \tan \Pi(a) = \sin B \tan \Pi(b) \tag{5.52}$$

[123] [Lobachevsky 1840, 8–9].
[124] [Lobachevsky 1840, 53]; [Engel/Stäckel 1895, 275]; [Gray 1979a, 107].
[125] [Lobachevsky 1840, 58].

and

$$\cos A \cos \Pi(b) \cos \Pi(c) + \frac{\sin \Pi(b) \sin \Pi(c)}{\sin \Pi(a)} = 1. \tag{5.53}$$

These are the Law of Sines and Law of Cosines respectively in hyperbolic geometry.[126] If we allow the side lengths in these formulas to converge to zero and approximate these quantities by second-order Taylor polynomials, we get the planar Laws of Sines and Cosines. Finally, as Lobachevsky himself points out at the end of his book, if we replace the sides a, b, and c with ai, bi, and ci, we get the spherical Laws of Sines and Cosines.[127]

The models that we use today to visualize hyperbolic geometry, such as Eugenio Beltrami's pseudosphere and the Poincaré disk, did not yet exist: at this time, Lobachevsky's geometry was truly imaginary. Perhaps because of this, and because there were no immediate applications, the new geometry was not immediately received with enthusiasm. However, the new models started to appear in the late nineteenth century, and work by Bernhard Riemann and others that worked toward systematizing geometry helped to bring these ideas into the mainstream. In the twentieth and twenty-first centuries, non-Euclidean geometries have found roles in physics and cosmology; for instance, they have become fundamental in the study of the possible shapes of the universe.

▨ Education

The question of who should learn trigonometry received various different answers in Europe and the Americas from the eighteenth century to the dawn of the twentieth. Political and social conditions varied considerably from one nation to the next, and these inevitably created different situations and attitudes. Nevertheless a couple of general trends are apparent. We have seen already that the applications of trigonometry had expanded well beyond the classical discipline of astronomy toward the practitioners' domains of surveying and altimetry from the end of the sixteenth century. The rise of logarithms in the seventeenth century had extended the reach of trigonometry in these new areas even further, and it was not long before navigation and architecture were making substantial use of the subject as well. By the end of

---

[126] This may be seen by substituting the equalities $\sinh p = 1 / \tan \Pi(p)$, $\cosh p = 1 / \sin \Pi(p)$, and $\tanh p = \cos \Pi(p)$. Note that Lobachevsky never uses the hyperbolic functions directly.

[127] On Lobachevsky's imaginary geometry, begin with [Bonola 1912, 84–93], [Gray 1979a, 96–116], [Rosenfeld 1988, 206–212, 221–231], and [Gray 2007, 115–127]. There is a substantial literature on Lobachevsky, much of it in Russian.

the seventeenth century there was more demand for trigonometry among the practitioners' trades than there was from astronomers, and those preparing for employment in these trades needed to be taught. Schools in the classical tradition were not teaching trigonometry at the turn of the eighteenth century, although it did appear to a limited extent in university curricula. Instead, schools focused on the professions that arose as alternatives to classical institutions, emphasizing preparation for naval, military, surveying, cartography, or other technical occupations depending on the nation's demands. Although not always private, these schools kindled a new marketplace for textbooks that did not rely on a scholarly style of presentation. In stark contrast to Euclid's *Elements*, the new books emphasized quantitative measurement in geometry from the start, a natural context for trigonometry to be taught. From around the middle of the nineteenth century, public secondary schools also began to include practical, scientific, and engineering topics in their curricula. As part of this movement, trigonometry (plane, and to a lesser extent spherical) gradually entered the school curriculum. By the end of the century, most schools in Europe and the Americas were teaching at least some trigonometry. Although the boundary between classical and practical education never vanished entirely, one had to look harder to find it at the end of the nineteenth century than at the beginning. Originally a classical subject in service of astronomy, trigonometry ironically found its place in the general curriculum through its applications in the trades, sciences, and engineering.[128]

The nature of trigonometric training that these practitioners received changed markedly between 1750 and 1900 as trigonometry itself was being reinvented from its foundations. We have seen already that the analytic revolution, the signature event of eighteenth-century mathematics, reshaped the basis of trigonometry through the work of Euler, Lambert, and others. No longer a quantification of Euclid's geometry, trigonometry was becoming the study of particular kinds of functions. Geometry was now merely the starting point, and algebraic manipulations of infinite series and other representations of sines, cosines, and the like were taking over the research environment. Indeed, geometry itself was being freed from Euclid. Alexis Clairaut's influential *Élémens de Géométrie* (1741) opens with neither a definition or an axiom but rather with a discussion of measurements of lengths—as he calls it, a "natural geometry."[129] Over the next century, due partly to the rise of professional schools, Euclid's hold on the curriculum gradually decreased. It

---

[128] The complexities of the developments of mathematics education in the various political environments in Europe and the Americas are beyond our scope. A good place to start to explore the substantial literature on these topics is [Karp/Schubring 2014]. For a book-length treatment of the history of English mathematics education, see [Howson 1982].

[129] [Clairaut 1741, 1–2].

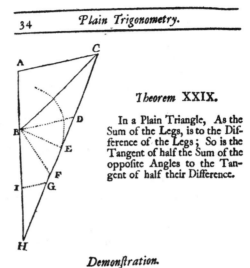

Figure 5.22
The Law of Tangents (first half of proof) from Benjamin Martin's *Young Trigonometer's Compleat Guide* (1736).

was some of the alternative geometry textbooks that Lewis Carroll, preferring the logical structure of the *Elements*, conjured Euclid himself to speak against in his dramatic polemic *Euclid and His Modern Rivals*.[130]

We may witness the effects of these sea changes on the mathematics classroom in a before-and-after comparison of two treatments of the Law of Tangents. The first (figure 5.22) is from the well-known mathematical instrument maker Benjamin Martin's *Young Trigonometer's Compleat Guide* (1736); the second (figure 5.23) is from Swiss American surveyor Ferdinand Rudolph Hassler's *Elements of Analytic Trigonometry* (1826). Several differences are conspicuous. Firstly, Martin works primarily from a given geometric figure while Hassler derives his result entirely algebraically from the Law

---

[130] [Dodgson 1879].

the two sides and the included angle, to find the two remaining angles.

The sum of the three angles of a plane triangle is always equal to two right angles; (elementary geometry;) in this case then we have given not only the two sides, and the angle included, but also the sum of the two angles sought; all then that is necessary to determine each angle separately, is to find their difference; for the largest is equal to half the sum increased by half the difference, and the least to half the sum diminished by half the difference.

In the same triangle that has been used in the first problem, having given A, b, and c, we have, as has been demonstrated,

$$b : c = \sin B : \sin C$$

And by composition of this proportion,

$$b + c : b - c = \sin B + \sin C : \sin B - \sin C$$

Substituting from series N, No. 3,

$$b + c : b - c = \tan \tfrac{1}{2} (B + C) : \tan \tfrac{1}{2} (B - C)$$

whence :

$$\tan \tfrac{1}{2} (B - C) = \tan \tfrac{1}{2} (B + C) \frac{b - c}{b + c}$$

And since the three angles, $A + B + C = 180°$,

or $\tfrac{1}{2} A + \tfrac{1}{2} (B + C) = 90°$,

we have : $90° - \tfrac{1}{2} A = \tfrac{1}{2} (B + C)$

and $\tan \tfrac{1}{2} (B + C) = \tan (90° - \tfrac{1}{2} A) = \cot \tfrac{1}{2} A$

The formula becomes :

$$\tan \tfrac{1}{2} (B - C) = \cot \tfrac{1}{2} A \frac{b - c}{b + c}$$

2

And calling $\dfrac{d}{2} = \dfrac{B - C}{2}$; we have the two angles, avoid-

Figure 5.23
The Law of Tangents in Ferdinand Rudolph Hassler's *Elements of Analytic Trigonometry* (1826).

of Sines and another identity given earlier in his book. This example is no exception. Martin's book is littered with diagrams on almost every page; Hassler's entire book (which includes spherical trigonometry) contains only 17 altogether.

Secondly, Martin's style of presentation is taken almost entirely from Euclid. It begins with the statement of a theorem stated in general terms and proceeds with a geometric demonstration. He does not follow this pattern consistently, but all the important results are first established in pure geometry, and practical content emerges from it afterward. Hassler's treatment relies on geometry only to the extent that the theorems he quotes may be traced backward to rest on geometric principles. This recalls Lagrange's work, which we saw earlier in this chapter, in which he reduced all of spherical trigonometry to analysis other than one geometric theorem, the Law of Cosines.

Some of the differences between the two books are not quite as obvious from these examples. Martin states his theorem as an equality of ratios of line segments: as this is to that, so this is to that. This mode of expression, typical of its time, reflects the ancient view that line segments are geometric magnitudes rather than numbers. One can place them in ratio to each other and set two ratios equal to each other, but to solve for one of the magnitudes in that equality is a qualitatively different exercise. Hassler's expression of the Law of Tangents in figure 5.23, isolating from the equality of ratios the expression $\tan\frac{1}{2}(B - C)$, is natural to us but would not have been to Martin.

The difference in styles led to a difference in content. Paging through Hassler's book, one finds dozens of identities listed in sequence, each derived algebraically from those that came before. These include many formulas involving sums, differences, and products of trigonometric and inverse trigonometric functions. On the other hand, Martin describes only the identities he needs for other purposes. In the sections on calculating trigonometric tables, Martin begins with the standard geometric theorems used for this purpose ever since Ptolemy, while Hassler turns immediately to Taylor series.

Hassler, one of the earliest of textbook writers to incorporate analysis, opens his book with a strongly worded justification of the new approach:

> Mathematical science must, from its very nature, have taken its rise in the simple inspection of geometric figures. The abstractions, upon which the calculus is founded, and whose great extension and generalization has produced the analytic method, must have arisen at a later period, as the product of a higher cultivation of the powers of the mind. . . . Analysis, so bold in its steps and so universal in its methods, which has carried mathematical science to results the

most general . . . , has naturally changed the mode of proceeding in trigonometry, as well as in other departments of mathematics. It is therefore necessary now, in order to study trigonometry in a truly scientific way to treat of it in the most general manner; and proceeding from principles the most general, yet at the same time the most simple and elementary, to found upon them a complete system; whose results may be fitted from universal application.[131]

Clearly Hassler's book required an entirely different collection of skills and ways of thinking than those that had gone before. Although he was one of the first to introduce analytic geometry to education, he was not alone. In fact, in 1849 Augustus de Morgan went so far as to describe trigonometry in passing as "a branch of algebra."[132] While mathematical researchers had made the transition a number of decades earlier, practitioners of navigation, surveying, and the like—and the authors of their textbooks—were not immediately convinced that the conversion was right for them. One of the most important English textbook authors, Thomas Keith, was not shy about his opinion of analysis:

The improvements made by the French in the various branches of mathematics, though highly extolled, will be found, in many cases, to be more specious than real; the modern analysis so *universally* adopted (and which in some instances is certainly illogical, if not unscientifical) has contributed greatly to vitiate the taste and explode that solid and accurate method of reasoning which is so conspicuous in the writings of the ancients. A flimsy mode of demonstration, grounded on a dexterous management of algebraical characters, has frequently been substituted for perspicuity and logical exactness.[133]

Others simply ignored the new approach entirely and carried on with their geometry. However, they could not stand forever. By the end of the nineteenth century the geometric and analytic approaches had merged into a unified subject that modern students would find somewhat familiar. In recent decades, proofs of any sort (geometric or analytic) have fallen even further into the background and in many cases have disappeared altogether.

Echoes of new approaches were felt more and more in textbooks over the course of the nineteenth century, starting with the definitions of the trigonometric functions themselves. Since Hipparchus, the fundamental quantities (the chord, sine, versed sine, cosine, tangent, and so on) had been lengths of line

[131] [Hassler 1826, 6].
[132] [De Morgan 1849, iii].
[133] [Keith 1810, xii].

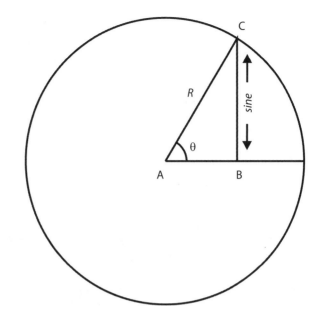

Figure 5.24
Defining the sine.

segments within or outside of a base circle, whose radius might be 60, or 3,438, or 60,000, or 10,000,000, or 1, or any other value. Especially in the works of Euler and Lambert, the sine was considered to be the ratio of BC to the radius R of the circle rather than the length of BC (figure 5.24). This new conception makes $R = 1$ obviously superior to other values: in this case the sine's values are the same whether one takes it to be a length or a ratio.

However, it took some while for the ratio definition of the trigonometric quantities to be proposed explicitly in textbooks. It did appear early, in an isolated case, in Georg Simon Klügel's 1770 *Analytische Trigonometrie* (figure 5.25).[134] Klügel's contributions to mathematics were varied; his encyclopedic knowledge led to his *Mathematisches Wörterbuch*, which influenced the definitions of terms in real analysis,[135] and his dissertation critiquing various attempts to prove the parallel postulate was also well read. In the *Analytische Trigonometrie*, we find the sine, cosine, tangent, and cotangent defined explicitly as ratios. This holds a couple of advantages over the then-conventional definitions: the tangent and the cotangent may be defined without needing to draw additional line segments in figure 5.24, and the base circle itself becomes superfluous. The title of the opening chapter, "Von den trigonometrische

---

[134] [Klügel 1770, 5].
[135] [Klügel 1803–1808]. [Schubring 2005] contains several discussions of Klügel's role in analysis.

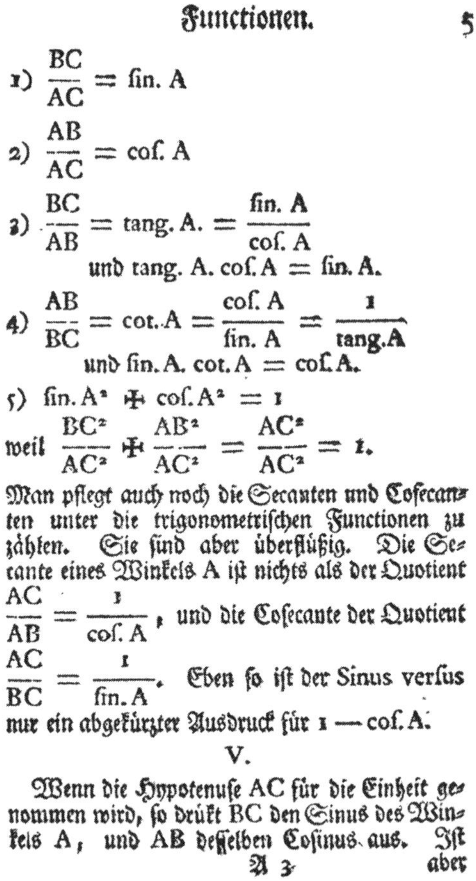

Figure 5.25
The first definition of the trigonometric quantities as ratios in G. S. Klügel's *Analytische Trigonometrie* (1770).

1) $\dfrac{BC}{AC}$ = fin. A

2) $\dfrac{AB}{AC}$ = cof. A

3) $\dfrac{BC}{AB}$ = tang. A. = $\dfrac{\text{fin. A}}{\text{cof. A}}$

  und tang. A. cof. A = fin. A.

4) $\dfrac{AB}{BC}$ = cot. A = $\dfrac{\text{cof. A}}{\text{fin. A}}$ = $\dfrac{1}{\text{tang. A}}$

  und fin. A. cot. A = cof. A.

5) fin. A² ⊞ cof. A² = 1

weil $\dfrac{BC^2}{AC^2}$ ⊞ $\dfrac{AB^2}{AC^2}$ = $\dfrac{AC^2}{AC^2}$ = 1.

Man pflegt auch noch die Secanten und Cosecanten unter die trigonometrischen Functionen zu zählen. Sie sind aber überflüßig. Die Secante eines Winkels A ist nichts als der Quotient $\dfrac{AC}{AB}$ = $\dfrac{1}{\text{cof. A}}$, und die Cosecante der Quotient $\dfrac{AC}{BC}$ = $\dfrac{1}{\text{fin. A}}$. Eben so ist der Sinus verſus nur ein abgekürzter Ausdruck für 1 — cof. A.

V.

Wenn die Hypotenuſe AC für die Einheit genommen wird, ſo drükt BC den Sinus des Winkels A, und AB deſſelben Coſinus aus. Iſt

A 3   aber

functionen," explicitly states that trigonometric quantities may be considered functions like any other.

Benefits notwithstanding, Klügel's innovations seem to have fallen on deaf ears, and it took several decades before the ratio definition was to reappear. Analytic trigonometry itself received a boost in later editions of Charles Hutton's hugely popular *A Course of Mathematics*, printed in many editions from 1798 onward. By 1811 it contains a chapter entitled "Plane Trigonometry Considered Analytically," which opens as follows:

> There are two methods which are adopted by mathematicians in investigating the theory of Trigonometry: the one *Geometrical*, the other *Algebraical*. . . . In the latter, the nature and properties of the linear-angular quantities (sines, tangents, &c,) being first defined, some general relation of these quantities, or of them in connection

with a triangle, is expressed by one or more algebraical equations; and then every other theorem or precept, of use in this branch of science, is developed by the simple reduction and transformation of the primitive equation. Each of these general methods has its peculiar advantages. The geometrical method carries conviction at every step; and by keeping the objects of enquiry constantly before the eye of the student, serves admirably to guard him against the admission of error: the algebraic method, on the contrary, requiring little aid from first principles, but merely at the commencement of its career, is more properly mechanical than mental, and requires frequent checks to prevent any deviation from truth.[136]

Hutton still defines the trigonometric quantities as line segments, but he quickly derives Klügel's definitions as consequences of the geometry and proceeds from there.

Analytical topics crept into textbooks over the next couple of decades, often for the purpose of generating identities. The first reappearance of the ratio definitions of trigonometric functions after Klügel was Hassler's 1826 *Elements of Analytic Trigonometry*, perhaps the first major innovation in trigonometry to come from North America. Soon after this, some textbooks adopted the ratio method while others resisted. Uptake varied in different jurisdictions, but by the late nineteenth century the skeptics had either given in or died.[137]

The textbooks' incorporation of the language of functions, coordinate graphs, and Taylor series implied that students needed to be exposed to the possibility that both angles and trigonometric functions could take on negative values. Early textbooks devoted much space trying carefully to explain what a negative ratio could mean, usually by appealing to a change of direction in a line segment. For instance, from Hutton's textbook,

It follows that an obtuse angle (measured by an arc greater than a quadrant) has *the same sine and cosine as its supplement*; the cosine however, being reckoned subtractive or negative, because it is situated contrariwise with regard to the centre C.[138]

---

[136] [Hutton 1811a, vol. 3, 53–54].

[137] See [Cajori 1890, 135] and [Van Sickle 2011a, 14–15]. [Van Sickle 2011a] is a study of trigonometry education in America before 1900; see especially pp. 103–145, as well as [Van Sickle 2011b], on the ratio method. On analysis and synthesis in American geometry instruction, see [Ackerberg-Hastings 2000]. [Allen 1977] describes the history of trigonometry education in America after 1890. For a survey of trigonometry education in England in the first three decades of the twentieth century, see [Price 1981, 309–315].

[138] [Hutton 1811a, vol. 3, 59].

17. The magnitudes of the cotangents, secants, and cose-cants, may be traced in like manner; and the results of the 13th, 14th, and 15th articles, recapitulated and tabulated as below.

	0°	90°	180°	270°	360°	
Sin.	O	R	O	−R	O	
Tan.	O	∞	O	−∞	O	
Sec.	R	∞	−R	−∞	R	(VI.)
Cos.	R	O	−R	O	R	
Cot.	∞	O	−∞	O	∞	
Cosec.	∞	R	−∞	−R	∞	

The changes of sines are these :

					sin.	cos.	tan.	cot.	sec.	cosec.	
1st.	5th.	9th.	13th.	quadrants	+	+	+	+	+	+	
2d.	6th.	10th.	14th.		+	−	−	−	−	+	(VII.)
3d.	7th.	11th.	15th.		−	−	+	+	−	−	
4th.	8th.	12th.	16th.		−	+	−	−	+	−	

We have been thus particular in tracing the mutations, both with regard to value and algebraic signs, of the princi-pal trigonometrical quantities, because a knowledge of them is absolutely necessary in the application of trigonometry to the solution of equations, and to various astronomical and physical problems.

Figure 5.26
In Charles Hutton's *A Course of Mathematics* (1811), a summary of the signs of the six trigonometric functions in different quadrants. Notice the distinction between the values ∞ and −∞.

Hutton continues along these lines for several pages, explaining in detail how the six major trigonometric functions change their sign as the argument takes on values in different quadrants. Although he does not discuss negative angles, he does allow the angle to continue well beyond a full circle; as we see in figure 5.26, he displays the signs of the trigonometric functions for arguments up to the sixteenth quadrant.

Another effect of analysis on the nineteenth-century classroom was the choice of units to measure angles. The rise of derivatives and integrals had rendered measurement in degrees (or in the French centesimal degrees) im-practical and ungainly, which helped lead to the invention of radian measure with Roger Cotes in the early eighteenth century. However, the replacement of degrees by radians in textbooks was much slower in coming than either the ratio definitions of trigonometric functions or the acceptance of negative

16. *De quelques arcs dont la tangente se calcule faci-lement.* — On voit aisément que le côté du polygone régu-lier de n côtés , circonscrit au cercle de rayon 1, a pour valeur 2 tang$\frac{\pi}{n}$. On peut, par la géométrie, trouver le côté du carré, de l'hexagone et du triangle équilatéral circon-scrits, et par conséquent les valeurs de tang $\frac{\pi}{4}$, tang $\frac{\pi}{6}$ et tang $\frac{\pi}{3}$; on trouve ainsi :

$$\text{tang } \frac{\pi}{4} = 1,$$
$$\text{tang } \frac{\pi}{3} = \sqrt{3},$$
$$\text{tang } \frac{\pi}{6} = \frac{1}{\sqrt{3}}.$$

Figure 5.27
Serret's *Traité de Trigonométrie* (1850), using radians. Note that a struggling student has written degree measures (incorrectly) above the radian measures.

values. We might not be surprised that Klügel's 1770 textbook, as far ahead of its time as it was, defines radian measure quickly, stating that it should be used "when the angle itself comes into the calculation. . . . [Degree measure] is very convenient for the construction of instruments; however, it is quite ar-bitrary, and in calculation, when angles come in, do not use it."[139] In Sylves-tre Lacroix's popular 1798 textbook *Traité Élémentaire de Trigonométrie Rec-tiligne et Sphérique*, he presents an alternative unit equal to twice a radian (before proposing and using yet another unit, the right angle).[140] However, most early nineteenth-century textbook authors, including Hutton and Has-sler, introduced angular measurement without referring to radians.

In the middle of the nineteenth century, the uptake of radians increased markedly in Europe; in fact, some textbooks mentioned degree measure only as an afterthought (figure 5.27).[141] These authors were therefore able to take advan-tage of analytic techniques immediately. English-language textbooks that intro-duced radians tended to describe them as an alternative to degrees, sometimes

[139] [Klügel 1770, 15–16].
[140] [Lacroix 1798, 14].
[141] See for instance [Serret 1850], [Weiss 1851], and [Le Cointe 1858]. Especially in the first half of the nineteenth century, the French system of centesimal degrees (now called gradians) was also introduced, with the French name "grade" (see, for instance, [Lacroix 1798, 13]).

calling them "circular measures."[142] The word "radian" itself was coined somewhat later, probably simultaneously and not in print, by mathematician Thomas Muir in 1869 and by physicist James T. Thomson, brother of Lord Kelvin, around 1871. Muir and philologist Alexander Ellis agreed on the term "radian" as an abbreviation for "radial angle" around 1874, a year after meetings between Thomson and Muir.[143] The earliest appearances found to date of the word in print are in two sources both dated 1874; in both cases the word is attributed to Thomson by his brother William (later Lord Kelvin).[144] The name was quickly adopted, and it continued to spread near the end of the century. That did not imply universal approval; as late as 1905 there were still detractors:

> The use of "number of radians" as a definition seems unnatural and unsatisfactory. The *radian* is an uninteresting angle—Lord Kelvin introduced the word merely as a convenience in lecturing, to avoid the long phrase "angle whose circular measure is."[145]

The teaching of spherical trigonometry went through fewer transformations. The efforts of Lagrange and others to minimize the subject's reliance on geometry made an impact on some textbooks. However, the rise of practitioners' schools, especially for navigation, had the effect of shifting many textbooks away from analytical elegance and toward practical applications at sea and in the heavens. As a result, formulas derived for easier use with logarithms gained a stronger foothold, but the analytical retreat from the visual did not carry as much weight. Spherical trigonometry continued to be important in the education of those practitioners who needed it, but its prominence in general mathematics education gradually faded. However, it was not until the mid-1950s that spherical trigonometry disappeared from textbooks forever.

▨ Concluding Remarks

Alas, as witnessed by its absence in the Mathematics Subject Classification,[146] trigonometry is no longer an active area of research. This does not mean that trigonometry does not inspire current research. Often, however, it happens

---

[142] See for instance [Thomson 1825] (father of Lord Kelvin and of James T. Thomson) and [Hind 1828]. By the fifth edition Hind was using the term "circular measure." Other names included the "natural" or "intrinsic" unit, or "π-measure" or "arcual measure" [Jones 1953, 421].

[143] See [Cooper 1992] for a summary of what is known about these interactions.

[144] Until now the earliest known reference was from an 1877 textbook, quoted in [Cooper 1992, 100]. The two notices from 1874 are [Thomson 1874b, 366] and [Thomson 1874a, 222].

[145] [Roseveare 1905, 133]. Roseveare seemed unaware that Lord Kelvin simply promoted the name introduced by his brother.

[146] There is still a category for plane and spherical trigonometry, but it is within mathematics education. A MathSciNet search for this category produces six results in total: four proofs without words and two historical articles.

through discoveries of algebraic structures related to those found in plane trigonometry but in a context that goes beyond geometry in some way. There is a long tradition of such findings; the most well-known example is within the study of elliptic integrals where Abel, Jacobi, and Legendre were among those who identified functions that have similar periodicity, identities, and even derivatives as the sine and the cosine.[147] Much more recently, more exotic creations have been explored: for instance, gyrotrigonometry is part of an approach to hyperbolic geometry that employs concepts from vector spaces, and fractional trigonometry is used to help solve problems in fractional calculus and differential equations. Closer to home, rational trigonometry has found a rebirth in a slightly different form in the work of Norman Wildberger, who advocates replacing measurements of distance and angle with the related notions of quadrance and spread. The advantage of this is that transcendental functions like the sine are not needed.[148]

Although trigonometry is no longer a research focus and now inspires only an occasional development in mathematical theory, it has become fundamental to daily practice in both science and the trades. The sixteenth-century profession of the mathematical practitioner is alive and active today in the work of surveyors, carpenters, navigators, architects, chemists, astronomers, and other skilled specialists. This ensures that trigonometry will continue to play an active role in education and in culture, likely long after the next survey of the history of trigonometry is written.

---

[147] See [Rice 2008] for a historical survey.
[148] [Wildberger 2005] develops this theory.

Bibliography

For sources in Chinese languages, see the extensive bibliographies in [Chen 2010] and [Chen 2015].

Académie Royale des Sciences. *Divers Ouvrages de Mathématique et de Physique.* Paris: Imprimerie Royale, 1693.

Ackerberg-Hastings, Amy. *Mathematics in a Gentleman's Art: Analysis and Synthesis in American College Geometry Teaching, 1790–1840.* PhD diss., Iowa State University, 2000.

Alexander, Amir. *Harriot and Dee on Exploration and Mathematics: Did Scientific Imagery Make for New Scientific Practice?* Cambridge, MA: MIT Press, 2005.

Allen, Harold Don. *The Teaching of Trigonometry in the United States and Canada: A Consideration of Elementary Course Content and Approach and of Factors Influencing Change, 1890–1970.* PhD diss., Rutgers University, 1977.

Andoyer, Henri. *Nouvelles Tables Trigonométriques Fondamentales Contenant les Logarithmes des Lignes Trigonométriques.* Paris: Hermann et fils, 1911.

Andoyer, Henri. *Nouvelles Tables Trigonométriques Fondamentales.* 3 vols. Paris: A. Hermann, 1915–1918.

Andrew, James. *Astronomical and Nautical Tables.* London: Plummer, 1805.

Ang Tian Se; and Swetz, Frank. A Chinese mathematical classic of the third century: *The Sea Island Mathematical Manual* of Liu Hui, *Historia Mathematica* **13** (1986), 99–117.

Anonymous. An account of the book, intituled, *Harmonia mensuram . . .* , *Philosophical Transactions* **32** (1722–23), 139–150.

Apian, Peter. *Cosmographicus.* Birckman, 1533a.

Apian, Peter. *Introductio geographica.* Ingolstadt, 1533b.

Apian, Peter. *Instrumentum sinuum seu primi mobilis.* Nuremberg: Johannes Petreius, 1541.

Archibald, Raymond Clare. Errors found by Duarte in Wolfram's table of natural logarithms, *Scripta Mathematica* **4** (1936a), 293.

Archibald, Raymond Clare. Wolfram's table, *Scripta Mathematica* **4** (1936b), 99–100.

Archibald, Raymond Clare. Tables of trigonometric functions in non-sexagesimal arguments, *Mathematical Tables and Other Aids to Computation* **1** (1943), 33–44.

Archibald, Raymond Clare. *Mathematical Table Makers.* New York: Scripta Mathematica, 1948.

Archibald, Raymond Clare. Bartholomäus Pitiscus (1561–1613), *Mathematical Tables and Other Aids to Computation* **3** (1949a), 390–397.

Archibald, Raymond Clare. Rheticus, with special reference to his *Opus palatinum*, *Mathematical Tables and Other Aids to Computation* **3** (1949b), 552–561.

Archibald, Raymond Clare. New information concerning Isaac Wolfram's life and calculations, *Mathematical Tables and Other Aids to Computation* **4** (1950), 185–200.

Archibald, Raymond Clare. The *Canon Doctrinae Triangulorvm* (1551) of Rheticus (1514–1576), *Mathematical Tables and Other Aids to Computation* **7** (1953), 131.

Archibald, Raymond Clare. The first published table of logarithms to the base ten, *Mathematical Tables and Other Aids to Computation* **9** (1955a), 62–63.

Archibald, Raymond Clare. Wolfram, Vega, and Thiele, *Mathematical Tables and Other Aids to Computation* **9** (1955b), 21.

Atzema, Eisso. "A most elegant property": On the early history of Lexell's Theorem. In *Research in History and Philosophy of Mathematics: The CSHPM 2016 Annual Meeting in Calgary, Alberta*, ed. Maria Zack and Dirk Schlimm, pp. 117–132. Cham: Birkhäuser, 2017.

Aubrey, John. *Brief Lives*, ed. R. Barber. Woodbridge: Boydell Press, 1982.

Auger, Léon. *Un Savant Méconnu: Gilles Personne de Roberval*. Paris: Blanchard, 1962.

Ayoub, Raymond. What is a Napierian logarithm? *American Mathematical Monthly* **100** (1993), 351–364.

Babbage, Charles. *Table of Logarithms of the Natural Numbers, from 1 to 108000*. London: J. Mawman, 1827.

Bachmakova, I. G.; and Slavutin, E. I. "*Genesis triangulorum*" de François Viète et ses recherches dans l'analyse indéterminée, *Archive for History of Exact Sciences* **16** (1977), 289–306.

Badolati, Ennio. On the history of Kepler's equation, *Vistas in Astronomy* **28** (1985), 343–345.

Bai Limin. Mathematical study and intellectual transition in the early and mid-Qing, *Late Imperial China* **16** (1995), 23–61.

Baillaud, Benjamin. Review of Henri Andoyer, *Nouvelles Tables Trigonométriques Fondamentales*, vol. 1, *Bulletin Astronomique* **32** (1915), 225–229.

Baker, Roger, ed. *Euler Reconsidered: Tercentenary Essays*. Heber City, UT: Kendrick Press, 2007.

Baldwin, Robert. John Dee's interest in the application of nautical science, mathematics and law to English naval affairs. In *John Dee: Interdisciplinary Studies in Renaissance Thought*, ed. Stephen Clucas, pp. 97–130. New York: Springer, 2006.

Barbin, Évelyne; and Commission inter-IREM histoire et épistémologie des mathématiques, eds. *Histoires de Logarithmes*. Paris: Ellipses, 2006.

Barnett, Janet Heine. Enter stage center: The early drama of the hyperbolic functions, *Mathematics Magazine* **77** (2004), 15–30.

Baron, Margaret E. *The Origins of the Infinitesimal Calculus*. Oxford: Pergamon Press, 1969.

Barrow, Isaac. *Lectiones geometricae*. London: Godbid, 1670.

Barrow, Isaac. *Geometrical Lectures*, transl. Edmund Stone. London: Austen, 1735.

Barrow, Isaac. *The Geometrical Lectures of Isaac Barrow*, transl. J. M. Child. Chicago/London: Open Court, 1916.

Beckmann, Petr. *A History of π.* 2nd ed. New York: St. Martin's Press, 1971.

Beer, A.; Ho Ping-yu; Lu Gwei-djen; Needham, J.; Pulleyblank, E. G.; and Thompson, G. An 8th-century meridian line: I-HSING's chain of gnomons and the prehistory of the metric system, *Vistas in Astronomy* **4** (1961), 3–28.

Beery, Janet. Navigating between triangular numbers and trigonometric tables: How Thomas Harriot developed his interpolation formulas, *Proceedings of the Canadian Society for History and Philosophy of Mathematics 32nd Annual Meeting* (Montreal) **20** (2007), 37–47.

Beery, Janet. "Ad calculum sinuum": Thomas Harriot's sine table interpolation formulas, *Proceedings of the Canadian Society for History and Philosophy of Mathematics 33rd Annual Meeting* (Vancouver) **21** (2008), 24–35.

Beery, Janet. Formulating figurate numbers, *BSHM Bulletin* **24** (2009), 78–91.

Beery, Janet; and Stedall, Jacqueline. *Thomas Harriot's Doctrine of Triangular Numbers: The 'Magisteria Magna.'* Zürich: European Mathematical Society, 2009.

Bekken, Otto. Viète's generation of triangles. In *Around Caspar Wessel and the Geometric Representation of Complex Numbers*, ed. Jesper Lützen, pp. 121–124. Copenhagen: Reitzels, 2001.

Bellhouse, David. *Abraham de Moivre: Setting the Stage for Classical Probability and Its Applications.* Boca Raton, FL: CRC Press, 2011.

Bellhouse, David; and Genest, Christian. Maty's biography of Abraham de Moivre, translated, annotated and augmented, *Statistical Science* **22** (2007), 109–136.

Belyj, Ju. A.; and Trifunovic, Dragan. Zur geschichte der logarithmentafeln Keplers, *NTM-Schriftenreihe für Geschichte der Naturwissenschaften, Technik, und Medizin* **9** (1972), 5–20.

Bennett, Jim. *The Mathematical Science of Christopher Wren.* Cambridge, UK: Cambridge University Press, 1982.

Bennett, Jim. *The Divided Circle: A History of Instruments for Astronomy, Navigation and Surveying.* Oxford: Phaidon Christie's, 1987.

Bennett, Jim. The challenge of practical mathematics. In *Science, Belief, and Popular Culture in Renaissance Europe*, ed. S. Pumfrey, P. Rossi, and M. Slawinski, pp. 176–190. Manchester: Manchester University Press, 1991a.

Bennett, Jim. Geometry and surveying in early-seventeenth-century England, *Annals of Science* **48** (1991b), 345–354.

Bennett, Jim. Projection and the ubiquitous virtue of geometry in the Renaissance. In *Making Space for Science: Territorial Themes in the Shaping of Knowledge*, ed. C. Smith and J. Agar, pp. 27–38. Basingstoke: Macmillan, 1998.

Bennett, Jim. Mathematics, instruments and navigation. In *Mathematics and the Historian's Craft*, ed. G. Van Brummelen and M. Kinyon. New York: Springer, 2005.

Bennett, Jim. Early modern mathematical instruments, *Isis* **102** (2011), 697–705.

Bennett, Jim. *Navigation: A Very Short Introduction.* Oxford: Oxford University Press, 2017.

Bennett, Jim. Mathematicians on board: Introducing lunar distances to life at sea, *British Journal for History of Science* **52** (2019), 65–83.

Bernoulli, Daniel. Réflexions et éclaircissements sur les nouvelles vibrations des cordes exposées dans les *Mémoires de l'Academie* de 1747 & 1748, *Histoire de l'Academie Royales des Sciences et Belles Lettres à Berlin* (1755), 147–172.

Bissell, Christopher. Cartesian geometry: The Dutch contribution, *Mathematical Intelligencer* **9** (4) (1987), 38–44.

Blagrave, John. *The Mathematical Jewel*. London: Venge, 1585.

Blundeville, Thomas. *His Exercises*. London: Windet, 1594.

Blundeville, Thomas. *His Exercises Containing Six Treatises*. Amsterdam: Theatrum Orbis Terrarum, 1971.

Bockstaele, Paul. Adrianus Romanus and the trigonometric tables of Georg Joachim Rheticus. In *Amphora: Festchrift für Hans Wussing zu seinem 65. Geburtstag*, ed. S. S. Demidov, M. Folkerts, D. E. Rowe, and C. J. Scriba, pp. 55–66. Basel: Birkhäuser, 1992.

Bogolyubov, N. N.; Mikhaïlov, G. K.; and Youschkevitch, A. P., eds. *Euler and Modern Science*. Washington, DC: Mathematical Association of America, 2007.

Bolyai, János. Appendix scientiam spatii absolute veram exhibens. In *Tentamen juventutem studiosam in elementa matheosis purae elementaris ac sublimioris methodo intuitiva evidentiaque huic propria, introducendi*, 2 vols., ed. W. Bolyai, vol. 1, pp. 1–26 (separately paginated). Maros Vásárhelyini: Typis Collegii Reformatorum per Josephum et Simeonem Kali, 1832/1833.

Bonola, Roberto. *Non-Euclidean Geometry*, transl. H. S. Carslaw. Chicago: Open Court, 1912. Reprint, New York: Dover, 1955.

Borda, J. C.; and Delambre, J. B. J. *Tables Trigonométriques Décimales*. Paris: Imprimerie République, 1801 (An IX).

Bosmans, H. La notion des "indivisibles" chez Blaise Pascal, *Archeion* **4** (1923), 369–379.

Bosmans, H. La trigonométrie d'Albert Girard, *Mathesis* **40** (1926), 337–348, 385–392, 433–439.

Bottazzini, Umberto. *The Higher Calculus: A History of Real and Complex Analysis from Euler to Weierstrass*. New York: Springer, 1986.

Boyer, Carl. History of the derivative and integral of the sine, *Mathematics Teacher* **40** (1947), 267–275.

Boyer, Carl. *The Concepts of the Calculus*. New York: Hafner, 1949. Reprint, *The History of the Calculus and Its Conceptual Development*, New York: Dover, 1959.

Boyer, Carl. The foremost textbook of modern times, *American Mathematical Monthly* **58** (1951), 223–226.

Bradley, Harry C. The graphical solution of spherical triangles, *American Mathematical Monthly* **27** (1920), 452–460.

Bradley, Margaret. Prony the bridge-builder: The life and times of Gaspard de Prony, educator and scientist, *Centaurus* **37** (1994), 230–268.

Bradley, Margaret. *Career Biography of Gaspard Clair François Marie Riche de Prony, Bridge-Builder, Educator and Scientist*. Lewiston, NY: Edwin Mellen, 1998.

Bradley, Robert E. Euler, d'Alembert and the logarithm function. In *Leonhard Euler: Life, Work and Legacy*, ed. R. E. Bradley and C. E. Sandifer, pp. 255–276. Amsterdam: Elsevier, 2007.

Bradley, Robert E.; D'Antonio, Lawrence A.; and Sandifer, C. Edward. *Euler at 300: An Appreciation*. Washington, DC: Mathematical Association of America, 2007.

Bradley, Robert E.; and Sandifer, C. Edward. *Leonhard Euler: Life, Work and Legacy*. Amsterdam: Elsevier, 2007.

Brahe, Tycho. *Tychonis Brahe Dani Opera Omnia*, ed. J. L. E. Dreyer. 15 vols. Copenhagen: Hauniae, 1913–1929.

Braver, Seth. *Lobachevski Illuminated*. Washington, DC: Mathematical Association of America, 2011.

Bréard, Andrea. War und ist Mathematik in China anders? Ein Ausblick auf historische Narration und neue Trends, *Mitteilungen der Mathematische Gesellschaft in Hamburg* **33** (2013), 15–35.

Bressieu, Maurice. *Metrices astronomicae*. Paris, 1581.

Briggs, Henry. *Logarithmorum chilias prima*. London, 1617.

Briggs, Henry. *Arithmetica logarithmica*. London: Jones, 1624.

Briggs, Henry. *Arithmetica logarithmica*, transl. Ian Bruce. 2006. http://www.17centurymaths.com/contents/albriggs.html.

Briggs, Henry; and Gellibrand, Henry. *Trigonometria Britannica*. Gouda: Rammaseijn, 1633.

Brioist, Jean-Jacques. Entre astrologie et cartographie, la genèse des logarithmes, *Journal de la Renaissance* **2** (2004), 63–74.

Brockey, Liam Matthew. *Journey to the East: The Jesuit Mission to China, 1579–1724*. Cambridge, MA: Harvard University Press, 2007.

Brown, B. M. Review of J. D. H. Donnay, *Spherical Trigonometry after the Cesàro Method*, *Mathematical Gazette* **30** (288) (1946), 52.

Brown, P. Hume. John Napier of Merchiston. In *Napier Tercentenary Volume*, ed. C. G. Knott, pp. 33–51. London: Longmans, Green, 1915.

Brown, Richard C. *The Tangled Origins of the Leibnizian Calculus*. Hackensack, NJ: World Scientific, 2012.

Bruce, Ian. Napier's logarithms, *American Journal of Physics* **68** (2000), 148–155.

Bruce, Ian. The agony and the ecstasy—the development of logarithms by Henry Briggs, *Mathematical Gazette* **86** (2002), 216–227.

Bruce, Ian. Henry Briggs: The *Trigonometria Britannica*, *Mathematical Gazette* **88** (2004), 457–474.

Bruins, Evert. On the history of logarithms: Bürgi, Napier, Briggs, de Decker, Vlacq, Huygens, *Janus* **67** (1980), 241–260.

Burckhardt, Johann Karl. Aus mehreren Briefen des Dr. Burckhardt, *Allgemeine Geographische Ephemeriden* **3** (1799), 179–193.

Bürgi, Joost. *Arithmetische und Geometrische Progress Tabulen*. Prague: Sessen, 1620.

Burmeister, Karl Heinz. *Georg Joachim Rheticus, 1514–1574. Eine Bio-Bibliographie*. 3 vols. Wiesbaden: Pressler-Verlage, 1967–1968.

Busard, H. L. L. *De arte mensurandi: A Geometrical Handbook of the Fourteenth Century*. Stuttgart: Steiner, 1998.

Buzengeiger, Karl Heribert Ignatz. Vergleichung zweier sehr kleinen Dreiecke von Gleichen Seiten, wovon das eine sphärisch, das andere eben ist, *Zeitschrift für Astronomie und verwandte Wissenschaften* **6** (1818), 264–270.

Cagnoli, Antonio. *Trigonometria piana e sferica*, Paris: Francesco Ambrogio Didot, 1786.

Cairns, W. D. Napier's logarithms as he developed them, *American Mathematical Monthly* **35** (1928), 64–67.

Cajori, Florian. *The Teaching and History of Mathematics in the United States*. Washington, DC: Government Printing Office, 1890.

Cajori, Florian. History of the exponential and logarithmic concepts. I. From Napier to Leibniz and Jean Bernoulli I, *American Mathematical Monthly* **20** (1913a), 5–14.

Cajori, Florian. History of the exponential and logarithmic concepts. III. The creation of a theory of logarithms of complex numbers by Euler, *American Mathematical Monthly* **20** (1913b), 75–84.

Cajori, Florian. Past struggles between symbolists and rhetoricians in mathematical publications. In *Proceedings of the International Mathematical Congress, 1924*, ed. J. C. Fields, pp. 937–941. Vol. 2. Toronto: University of Toronto Press, 1928.

Cajori, Florian. *A History of Mathematical Notations*. 2 vols. Chicago: Open Court Press, 1928–1929.

Cajori, Florian. History of determinations of the heights of mountains, *Isis* **12** (1929), 482–514.

Calinger, Ronald S. Leonhard Euler: The Swiss years, *Methodology and Science* **16** (1983), 69–89.

Calinger, Ronald S. Leonhard Euler: The first St. Petersburg years (1727–1741), *Historia Mathematica* **23** (1996), 121–166.

Calinger, Ronald S. *Leonhard Euler: Mathematical Genius in the Enlightenment*. Princeton, NJ: Princeton University Press, 2016.

Callet, François. *Tables Portatives de Logarithmes*. An 3. Paris: Firmin Didot, 1795.

Campbell-Kelly, Martin. Charles Babbage's table of logarithms (1827), *Annals of the History of Computing* **10** (1988), 159–169.

Cannon, John T.; and Dostrovsky, Sigalia. *The Evolution of Dynamics: Vibration Theory from 1687 to 1742*. New York: Springer, 1981.

Cantor, Georg. Beweis, daß eine für jeden reellen Wert von x durch eine trigonometrische Reihe gegebene Funktion $f(x)$ sich nur auf eine einzige Weise in dieser Form darstellen läßt, *Journal für die Reine und Angewandte Mathematik* **72** (1870), 139–142.

Cantor, Georg. Notiz zu dem Aufsatz: Beweis, daß eine für jeden reellen Wert von x durch eine trigonometrische Reihe gegebene Funktion $f(x)$ sich nur auf eine einzige Weise in dieser Form darstellen läßt, *Journal für die Reine und Angewandte Mathematik* **73** (1871), 294–296.

Cantor, Georg. Über die Ausdehnung eines Satzes aus der Theorie der trigonometrischen Reihen, *Mathematische Annalen* **5** (1872), 123–132.

Cantor, Moritz. *Vorlesungen über Geschichte der mathematik*, 2nd ed. 4 vols. Leipzig: Teubner, 1913.

Capobianco, Giovanni; Enea, Maria Rosaria; and Ferraro, Giovanni. Geometry and analysis in Euler's integral calculus, *Archive for History of Exact Sciences* **71** (2017), 1–38.

Cardano, Girolamo. *De subtilitate*. Nuremberg: Petreius, 1550.

Cardano, Girolamo. *The* De subtilitate *of Girolamo Cardano*, ed. J. M. Forrester. 2 vols. Tempe, AZ: Arizona Center for Medieval and Renaissance Studies, 2013.

Carnot, Lazare. *De la Corrélation des Figures de Géométrie*. Paris: Crapelet, 1801.

Carnot, Lazare. *Géométrie de Position*. Paris: Crapelet, 1803.

Carnot, Lazare. Digression sur la nature des quantités dites negatives. In Lazare Carnot, *Mémoire sur la Relations qui Existe entre les Distances Respectives de Cinq quelconques Pris dans l'Espace*, pp. 96–111. Paris: Courcier, 1806.

Carslaw, Horatio. Napier's logarithms: The development of his theory, *London, Edinburgh, and Dublin Philosophical Magazine and Journal of Science* (6) **32** (1916), 476–486.

Casey, John. *A Treatise on Spherical Trigonometry*. Dublin: Hodges, Figgis; London: Longmans, Green, 1889.

Cavalieri, Bonaventura. *Directorium generale uranometricum*. Bologna: Tebaldini, 1632.

Cesàro, Giuseppe. Nouvelle méthode pour l'établissement des formules de la trigonométrie sphérique, *Académie Royale de Belgique Bulletins de la Classes des Sciences* (1905a), 434–454.

Cesàro, Giuseppe. Les formules de la trigonométrie sphérique déduites de la projection stéréographique du triangle—Emploi de cette projection dans les recherches sur la sphère, *Académie Royale de Belgique Bulletins de la Classes des Sciences* (1905b), 560–583.

Chemla, Karine. The background to Gergonne's treatment of duality: Spherical trigonometry in the late 18th century. In *The History of Modern Mathematics*. Vol. 1: *Ideas and Their Reception*, ed. D. E. Rowe and J. McCleary, pp. 331–359. Boston: Academic Press, 1989.

Chemla, Karine. Le rôle joué par la sphère dans la maturation de l'idée de dualité au début du XIX° siècle. Les articles de Gergonne entre 1811 et 1827. In *Actes de la Quatrième Université d'Été d'Histoire des Mathématiques* (Lille, 1990), pp. 57–72. Lille: IREM, 1994.

Chemla, Karine. Lazare Carnot et la généralité en géometrie. Variations sur le théorème dit de Menelaus, *Revue d'Histoire des Mathématiques* **4** (1998), 163–190.

Chemla, Karine. Euler's work in spherical trigonometry: Contributions and applications. In *Leonhardi Euleri Opera Omnia*, 3rd series. Vol. 10: *Commentationes Physicae ad Theoriam Caloris Electricitatis et Magnetismi Pertinentes*, ed.

Patricia Radelet-de Grave and David Speiser, pp. 125–187. Basel: Birkhäuser, 2004.

Chemla, Karine; and Guo Shuchun. *Les Neufs Chapîtres: Le Classique Mathématique de la Chine Ancienne et ses Commentaires*, Paris: Dunod, 2005.

Chemla, Karine; and Pahaut, Serge. Préhistoires de la dualité: Explorations algébriques en trigonométrie sphérique (1753–1825). In *Sciences à l'Époque de la Révolution Française*, ed. Roshdi Rashed, pp. 151–200. Paris: Blanchard, 1988.

Chen, Ann-ping; and Freeman, Mansfield. *Tai Chen on Mencius: Explorations in Words and Meanings*. New Haven, CT: Yale University Press, 1990.

Chen, Jiang-Ping Jeff. The evolution of transformation media in spherical trigonometry in 17th- and 18th-century China, and its relation to "Western learning," *Historia Mathematica* **37** (2010), 62–109.

Chen, Jiang-Ping Jeff. Re-examining Dai Zhen's mathematics in terms of construction and mathematical principles, *Ziran kexueshi yanjiu* (*Studies in the History of Natural Sciences*) **30** (1) (2011), 28–44.

Chen, Jiang-Ping Jeff. Trigonometric tables: Explicating their construction principles in China, *Archive for History of Exact Sciences* **69** (2015), 491–536.

Chen Meidong, A study of some astronomical data in Muslim Calendar. In *History of Oriental Astronomy*, ed. G. Swaup, A. K. Bag, and K. S. Shukla, pp. 169–174. Cambridge, UK: Cambridge University Press, 1987.

Child, J. M. Isaac Barrow: The drawer of tangents, *Open Court* **30** (1916), 65–69.

Chisholm, Grace. *Algebraisch-gruppentheoretische Untersuchungen zur Sphärischen Trigonometrie*. Göttingen: Dieterich'schen Univ.-Buchdruckerei, 1895.

Christianson, Gale E. *In the Presence of the Creator: Isaac Newton and His Times*. New York: The Free Press, 1984.

Chu Pingyi. Remembering our grand tradition: The historical memory of the scientific exchanges between China and Europe, 1600–1800, *History of Science* **12** (2003), 193–215.

Clairaut, Alexis. *Élémens de Géométrie*. Paris: Lambert & Durand, 1741.

Clark, Kathleen. *Jost Bürgi's Aritmetische und Geometrische Progreß Tabulen: Edition and Commentary*. New York: Springer, 2015.

Clark, Kathleen; and Montelle, Clemency. Priority, parallel discovery, and pre-eminence: Napier, Bürgi, and the early history of the logarithm relation, *Revue d'Histoire des Mathématiques* **18** (2012), 223–270.

Clarke, Alexander Ross. *Account of the Observations and Calculations of the Principal Triangulation*. London: Ordnance Trigonometrical Survey of Great Britain and Ireland, 1858.

Clavius, Christoph. *Theodosii Tripolitae sphaericorum*, Rome: D. Basae, 1586.

Clavius, Christoph. *Tabulae sinuum, tangentium et secantium*, Mainz: Albini, 1587.

Clavius, Christoph. *Astrolabium*. Rome: Gabiani, 1593.

Clavius, Christoph. *Geometria practica*. Rome: Zannetti, 1604.

Clough-Smith, Nigel. *An Introduction to Spherical Trigonometry, with Practical Examples, for Students of Navigation, Hydrographic Surveying and Nautical Astronomy*. 2nd ed. Glasgow: Brown, Son & Ferguson, 1978.

Cohen, I. Bernard; and Smith, George, eds. *The Cambridge Companion to Newton*. Cambridge, UK: Cambridge University Press, 2002.

Colwell, Peter. *Solving Kepler's Equation over the Centuries*. Richmond, VA: Willmann-Bell, 1993.

Connor, Elizabeth. Abraham Sharp, 1653–1942, *Publications of the Astronomical Society of the Pacific* **54** (1942), 237–243.

Cooke, Roger. Uniqueness of Fourier series and descriptive set theory: 1870–1985, *Archive for History of Exact Sciences* **51** (1993), 251–334.

Coolidge, Julian Lowell. *A History of Geometrical Methods*. Oxford: Clarendon Press, 1940.

Coolidge, Julian Lowell. The story of tangents, *American Mathematical Monthly* **58** (1951), 449–462.

Cooper, Michael. Who named the radian? *Mathematical Gazette* **76** (1992), 100–101.

Copernicus, Nicolaus. *De lateribus et angulis triangulorum*. Wittenberg: Lufft, 1542.

Cormack, Leslie. The commerce of utility: Teaching mathematical geography in early modern Europe, *Science & Education* **15** (2006), 305–322.

Costabel, Pierre. Notes relatives à l'influence de Pascal sur Leibniz, *Revue d'Histoire des Sciences et de leurs Applications* **15** (1962), 369–374.

Costabel, Pierre. Leibniz et les séries numériques, *Studia Leibnitiana* Suppl., **17** (1978), 81–101.

Cotes, Roger. Logometria, *Philosophical Transactions* **29** (1714), 5–45.

Cotes, Roger. *Harmonia mensuram*. Cambridge, UK, 1722.

Cotter, Charles H. *A History of Nautical Astronomy*. New York: American Elsevier, 1968.

Cotter, Charles H. A study of a species of short-method table, *Journal of Navigation* **27** (1974a), 395–401.

Cotter, Charles H. Sines, versines and haversines in nautical astronomy, *Journal of Navigation* **27** (1974b), 536–541.

Cotter, Charles H. Edmund Gunter (1581–1626), *Journal of Navigation* **34** (1981), 363–367.

Cotter, Charles H. Direct methods of sight reduction: An historical review, *Journal of Navigation* **35** (1982), 260–273.

Craig, H. V. Review of J. D. H. Donnay, *Spherical Trigonometry after the Cesàro Method*, *American Mathematical Monthly* **53** (1946), 32–33.

Craik, Alex D. D. Edward Sang (1805–1890): Calculator extraordinary, *British Society for History of Mathematics Newsletter* **17** (45) (2002), 32–43.

Craik, Alex D. D. The logarithmic tables of Edward Sang and his daughters, *Historia Mathematica* **30** (2003), 47–84.

Crane, Nicholas. *Mercator: The Man Who Mapped the Planet*. London: Weidenfeld and Nicolson, 2002.

Crippa, Davide. *The Impossibility of Squaring the Circle in the 17th Century: A Debate Among Gregory, Huygens and Leibniz*. Cham: Birkhäuser, 2019.

Croarken, Mary. Tabulating the heavens: Computing the *Nautical Almanac* in 18th-century England, *IEEE Annals of the History of Computing* **25** (2003), 48–61.

Croarken, Mary; and Campbell-Kelly, Martin. Beautiful numbers: The rise and decline of the British Association Mathematical Tables Committee, *IEEE Annals of the History of Computing* **22** (2000), 44–61.

Cudworth, William. *Life and Correspondence of Abraham Sharp.* London: Sampson Low, Marston, Searle & Rivington, 1889.

Cullen, Christopher. An eighth century Chinese table of tangents, *Chinese Science* **5** (1982), 1–33.

Cullen, Christopher. *Astronomy and Mathematics in Ancient China: The Zhou bi suan jing*, Cambridge, UK: Cambridge University Press, 1996.

Cullen, Christopher. *The Suan shu shu, "Writings on Reckoning": A Translation of a Chinese Mathematical Collection of the Second Century BC, with Explanatory Commentary.* Cambridge, UK: Needham Research Institute, 2004.

D'Alembert, Jean le Rond. Recherches sur la courbe que forme une corde tendüe mise en vibration, *Histoire de l'Academie Royale des Sciences et Belles Lettres* (1749a), 214–219.

D'Alembert, Jean le Rond. Suite des recherches sur la courbe que forme und corde tendüe, mise en vibration, *Histoire de l'Academie Royale des Sciences et Belles Lettres* (1749b), 220–249.

D'Alembert, Jean le Rond. Sur les logarithmes des quantités négatives. In Jean le Rond d'Alembert, *Opuscules Mathématiques*, Vol. I, pp. 180–230. Paris: David, 1761.

Danielson, Dennis. *The First Copernican: Georg Joachim Rheticus and the Rise of the Copernican Revolution.* New York: Walker Books, 2006.

Daston, Lorraine. Enlightenment calculations, *Critical Inquiry* **21** (1994), 182–202.

Dauben, Joseph W. The trigonometric background to Georg Cantor's theory of sets, *Archive for History of Exact Sciences* **7** (1971), 181–216.

Dauben, Joseph W. *Georg Cantor: His Mathematics and Philosophy of the Infinite.* Cambridge, MA: Harvard University Press, 1979. Reprint, Princeton, NJ: Princeton University Press, 1990.

Dauben Joseph W. The development of Cantorian set theory. In *From the Calculus to Set Theory*, ed. Ivor Grattan-Guinness, pp. 181–219. London: Duckworth, 1980.

Dauben, Joseph W. The battle for Cantorian set theory. In *Mathematics and the Historian's Craft: The Kenneth O. May Lectures*, ed. G. Van Brummelen and M. Kinyon, pp. 221–241. New York: Springer, 2005.

Dauben, Joseph W. Chinese mathematics. In *The Mathematics of Egypt, Mesopotamia, China, India, and Islam: A Sourcebook*, ed. Victor J. Katz, pp. 187–384. Princeton, NJ: Princeton University Press, 2007.

Dauben, Joseph W. *Suan shu shu*, a book on numbers and computations: English translation with commentary, *Archive for History of Exact Sciences* **62** (2008), 91–178.

Davis, Percy L. H. *Requisite Tables.* London: J. D. Potter, 1905.

Debnath, Lokenath. *The Legacy of Leonhard Euler: A Tricentennial Tribute.* London: Imperial College Press, 2010.

de Foncenex, Daviet. Réflexions sur les quantités imaginaires, *Miscellanea Philosophico-Mathematica Societatis Privatae Taurinensis* **1** (1759), 113–146.

Dehn, Max; and Hellinger, E. D. On James Gregory's *Vera quadratura*. In *James Gregory Tercentenary Memorial Volume*, ed. Herbert Westren Turnbull, pp. 468–478. London: Bell & Sons, 1939.

Dehn, Max; and Hellinger, E. D. Certain mathematical achievements of James Gregory, *American Mathematical Monthly* **50** (1943), 149–163.

Delambre, J. B. J. Rapport sur les grandes tables trigonométriques décimales du cadastre, *Mémoires de l'Institut National des Sciences et Arts* **5** (An 12) (1804), 56–66.

Delambre, J. B. J. Remarques sur les formules précédentes, appendix to M. Henri, Formules pour calculer les parallaxes de longitude et de latitude et le demi-diamètre apparent d'une planète quelconque, *Connaissance de Temps* 1809 (publ. April 1807), 443–448.

Delambre, J. B. J. *Base du Système Métrique Décimal.* Vol. 3. Paris: Baudouin, 1810.

Delambre, J. B. J. *Histoire de l'Astronomie du Moyen Âge.* Paris: Courcier, 1819.

Delambre, J. B. J. *Histoire de l'Astronomie Moderne.* 2 vols. Paris: Courcier, 1821.

Delambre, J. B. J. *Histoire de l'Astronomie au Dix-Huitième Siècle.* Paris: Bachelier, 1827.

Delevsky, Jacques. L'invention de la projection de Mercator et les enseignements de son histoire, *Isis* **34** (1942), 110–117.

de Lorenzo, Javier. Pascal y los indivisibles, *Theoria* **1** (1985), 87–120.

Delsedime, P. La disputa delle corde vibranti ed una lettera inedita di Lagrange a Daniel Bernoulli, *Physis* **13** (1971), 117–146.

de Merez, Salomon. *Vie de Maurice Bressieu.* Valencia: Chenevier & Pessieux, 1880.

Demidov, S. S. Mathématiques et réalité physique dans la discussion sur la vibration d'une corde en XVIIIe siècle. In *Theoria cum praxi: On the Relationship of Theory and Praxis in the Seventeenth and Eighteenth Centuries*, Vol. 4, pp. 138–142. Wiesbaden: Steiner, 1982.

Demidov, S. S. Leonhard Euler, Treatise on the differential calculus (1755). In *Landmark Writings in Western Mathematics, 1640–1940*, ed. Ivor Grattan-Guinness, pp. 191–198. Amsterdam: Elsevier, 2005.

de Moivre, Abraham. A method of extracting the root of an infinite equation, *Philosophical Transactions* **20** (1698), 190–193.

de Moivre, Abraham. Aequationum quarundam potestatis tertiae, quintae, septimae, nonae, & superiorum, ad infinitum usque pergendo, in terminis finitis, ad instar regularum pro cubicis quae vocantur Cardani, resolutio analytica, *Philosophical Transactions* **25** (1707), 2368–2371.

de Moivre, Abraham. De sectione anguli, *Philosophical Transactions* **32** (1722), 228–230.

de Moivre, Abraham. *Miscellanea analytica de seriebus et quadraturis.* London: Tonson & Watts, 1730.

De Morgan, Augustus. On the invention of the circular parts, *Philosophical Magazine* (3) **22** (1843), 350–353.

De Morgan, Augustus. On the almost total disappearance of the earliest trigonometrical canon, *Philosophical Magazine* (3) **26** (1845), 517–526.

De Morgan, Augustus. On the first introduction of the words *tangent* and *secant*, *Philosophical Magazine* (3) **28** (Jan-June 1846), 382–387.

De Morgan, Augustus. *Trigonometry and Double Algebra*. London: Taylor, Walton, and Maberly, 1849.

De Morgan, Augustus. Table. In *English Cyclopaedia, Division of Arts and Sciences*, Vol. 7, ed. Charles Knight, columns 976–1016. London: Bradbury, Evans, 1868.

de Prony, G. F. C. M. Riche. Éclaircissemens sur un point d'histoire des tables trigonométriques, *Mémoires de l'Institut National des Sciences et Arts* **5** (An XII, 1804a), 67–93.

de Prony, G. F. C. M. Riche. Notice sur les grandes tables logarithmiques et trigonométriques, calculées au bureau du cadaster sous la direction du citoyen Prony, *Mémoires de l'Institut National des Sciences et Arts* **5** (An 12) (1804b), 49–55.

Descartes, René. *Discours de la Méthode* (appendices *La Dioptrique, Les Meteores, La Geometrie*). Leiden: Maire, 1637.

Descartes, René. *Geometria*, transl. Frans van Schooten. Leiden: Maire, 1649.

Descartes, René. *Geometria*, 2nd ed., transl. Frans van Schooten. 2 vols. Amsterdam: Ludovicum & Elzevier, 1659/1661.

Descartes, René. *The Geometry of Rene Descartes*, transl. David Eugene Smith and Marcia L. Latham. Chicago/London: Open Court Press, 1925. Reprint, New York: Dover, 1954.

de Siebenthal, Jean. *Les Mathématiques dans l'Occident Médiéval*. Lausanne: Éditions Terre Haute, 1993.

D'Hollander, Raymond. *Loxodromie et Projection de Mercator*. Paris: Institut Océanographique, 2005.

Dhombres, Jean; and Dhombres, Nicole. *Lazare Carnot*. Paris: Fayard, 1997.

Dietrich, Urs; and Girstmair, Kurt. John Napiers trigonometrie—ein Blick zurück, *Mathematische Semesterberichte* **56** (2009), 215–232.

Dijksterhuis, E. J. James Gregory and Christiaan Huygens. In *James Gregory Tercentenary Memorial Volume*, ed. Herbert Westren Turnbull, pp. 478–486. London: Bell & Sons, 1939.

Dijksterhuis, F. J. *Lenses and Waves: Christiaan Huygens and the Mathematical Science of Optics in the Seventeenth Century*. Dordrecht: Kluwer, 2004.

Dirichlet, Gustav Peter Lejeune. Sur la convergence des series trigonométriques que servent à representer une fonction arbitraire entre deux limites données, *Journal für die Reine und Angewandte Mathematik* **4** (1829), 157–169.

Dodgson, Charles L. *Euclid and His Modern Rivals*. London: Macmillan, 1879.

Donnay, J. D. H. *Spherical Trigonometry after the Cesàro Method*. New York: Interscience, 1945.

Drake, Stillman. Tartaglia's squadra and Galileo's compasso, *Annali dell'Istituto e Museo di Storia della Scienza di Firenze* **2** (1977), 35–54.

Duan Yao-Yong; and Li Wen-Lin. The influence of Indian trigonometry on Chinese calendar-calculations in the Tang dynasty. In *Ancient Indian Leaps into Mathematics*, ed. B. S. Yadav and Man Mohan, pp. 45–54. New York: Birkhäuser, 2011.

Duhamel, J.-M.-C. Note sur la méthode des tangentes de Roberval, *Mémoires Présentés par Divers Savants à l'Académie Royale des Sciences de l'Institut de France* **5** (1838), 257–266.

Dunham, William. *Euler: The Master of Us All*. Washington, DC: Mathematical Association of America, 1999.

Dunham, William, ed. *The Genius of Euler: Reflections on his Life and Work*. Washington, DC: Mathematical Association of America, 2007.

Eberhard-Bréard, Andrea; Dauben, Joseph W.; and Xu Yibao. The history of Chinese mathematics: The past 25 years, *LLULL* **26** (2003), 429–474.

Edwards, C. H., Jr. *The Historical Development of the Calculus*. New York: Springer, 1979.

Eklund, Neil. CORDIC: Elementary function computation using recursive sequences, *College Mathematics Journal* **32** (2001), 330–333.

Elman, Benjamin. *From Philosophy to Philology: Intellectual and Social Aspects of Change in Late Imperial China*. 2nd rev. ed. Los Angeles: UCLA, 2001.

Elman, Benjamin. New perspectives on the Jesuits and science in China: 1600–1800. In *Points of Contact Crossing Cultural Boundaries*, ed. Amy Golahny, pp. 37–49. Lewisburg: Bucknell University Press, 2004.

Elman, Benjamin. *On Their Own Terms: Science in China 1550–1900*. Cambridge, MA: Harvard University Press, 2005.

Encyclopaedia Britannica, or, Dictionary of arts, sciences, and general literature, 7th ed. 24 vols. Edinburgh: Adam and Charles Black, 1842.

Eneström, Gustav. Sur un théorème de Kepler équivalent à l'intégration d'une fonction trigonométrique, *Bibliotheca Mathematica* (new series) **3** (1889), 65–66.

Eneström, Gustav. Sur la découverte de l'intégrale complète des équations différentielles linéaires à coefficients constants, *Bibliotheca Mathematica* (new series) **11** (1897), 43–50.

Eneström, Gustav. Der Briefwechsel zwischen Leonhard Euler und Johann I Bernoulli. III. 1739–1746, *Bibliotheca Mathematica* (3) **6** (1905), 16–87.

Eneström, Gustav. Über die angebliche Integration einer trigonometrischen Function bei Kepler, *Bibliotheca Mathematica* (3) **13** (1912–13), 229–241.

Engel, Friedrich; and Stäckel, Paul. *Die Theorie der Parallellinien con Euklid bis auf Gauss*. Leipzig: Teubner, 1895.

Engel, Wolfgang, ed. *Festakt und Wissenschaftliche Konferenz aus Anlass des 200. Todestages von Leonhard Euler, 15./16. September 1983 in Berlin*. Berlin: Akademie-Verlag, 1983.

Engelfriet, Peter. *Euclid in China: The Genesis of the First Translation of Euclid's Elements in 1607 and Its Reception up to 1723*. Leiden: Brill, 1998.

Euler, Leonhard. De minimis oscillationibus corporum tam rigidorum quam flexibilium. Methodus nova et facilis (E40), *Commentarii academiae scientiarum Petropolitanae* **7** (1740a), 99–122.

Euler, Leonhard. De summis serierum reciprocarum (E41), *Commentarii academiae scientiarum Petropolitanae* **7** (1740b), 123–134.

Euler, Leonhard. De integratione aequationum differentialium altiorum graduum (E62), *Miscellanea Berolinensia* **7** (1743a), 193–242.

Euler, Leonhard. Theoremata circa reductionem formularum integralium ad quadraturam circuli (E59), *Miscellanea Berolinensia* **7** (1743b), 91–129.

Euler, Leonhard. *Methodus inveniendi lineas curvas maximi minimive proprietate gaudentes, sive solutio problematis isoperimetrici latissimo sensu accepti* (E65). Lausanne & Geneva: Marcum-Michaelem Bousquet & Socios., 1744.

Euler, Leonhard. *Introductio ad analysin infinitorum* (E101 and E102), 2 vols. Lausanne: Marcum-Michaelem Bousquet & Socios., 1748.

Euler, Leonhard. De vibratione chordarum exercitatio (E119), *Nova Acta Eruditorum* (1749a), 512–527. Reprint, French transl., Sur la vibration des cordes (E140), *Histoire de l'Academie Royale des Sciences et Belles Lettres* (1750), 69–85.

Euler, Leonhard. Recherches sur la question des inégalités du mouvement de Saturne et de Jupiter (E120). In *Pièce qui a Remporté le Prix de l'Académie Royale des Sciences en 1748 sur les Inégalités du Mouvement de Saturne et de Jupiter*, pp. 1–123. Paris: Martin: Coignard & Gurein, 1749b.

Euler, Leonhard. Methodus facilis computandi angulorum sinus ac tangentes tam naturales quam artificiales (E128), *Commentarii academiae scientiarum Petropolitanae* **11** (1750a), 194–230.

Euler, Leonhard. De novo genere oscillationum (E126), *Commentarii academiae scientiarum Petropolitanae* **11** (1750b), 128–149.

Euler, Leonhard. De seriebus quibusdam considerationes (E130), *Commentarii academiae scientiarum Petropolitanae* **12** (1750c), 53–96.

Euler, Leonhard. De la controverse entre Mrs. Leibniz et Bernoulli sur les logarithmes des nombres negatifs et imaginaires (E168), *Memoires de l'Académie des Sciences de Berlin* **5** (1751), 139–179.

Euler, Leonhard. Elemens de la trigonométrie sphéroïdique tirés de la méthode des plus grands et plus petits (E215), *Mémoires de l'Academie des Sciences en Berlin* **9** (1755a), 258–293.

Euler, Leonhard. *Institutiones calculi differentialis cum eius usu in analysi finitorum ac doctrina serierum* (E212). St. Petersburg: Academiae Imperialis Scientiarum Petropolitanae, 1755b.

Euler, Leonhard. Principes de la trigonométrie sphérique tirés de la méthode des plus grands et plus petits (E214), *Mémoires de l'Academie des Sciences en Berlin* **9** (1755c), 223–257.

Euler, Leonhard. Subsidium calculi sinuum (E246), *Novi commentarii academiae scientiarum Petropolitanae* **5** (1760), 164–204.

Euler, Leonhard. *Institutionum calculi integralis.* Vol. 1 (E342). St. Petersburg: Academiae Imperialis Scientiarum, 1768.

Euler, Leonhard. *Institutionum calculi integralis.* Vol. 2 (E366). St. Petersburg: Academiae Imperialis Scientiarum, 1769a.

Euler, Leonhard. Investigatio perturbationum quibus planetarum motus ob actionem eorum mutuam afficiuntur (E414). In *Recueil des Pièces qui ont Remporté les Prix de l'Académie des Sciences*, Vol. 8. Paris: Gabriel Martin, 1769b.

Euler, Leonhard. *Institutionum calculi integralis*. Vol. 3 (E385). St. Petersburg: Academiae Imperialis Scientiarum, 1770.

Euler, Leonhard. De mensura angulorum solidorum (E514), *Acta Academiae Scientiarum Imperialis Petropolitanae* **2** (1781), 31–54.

Euler, Leonhard. Trigonometria sphaerica universa, ex primis principiis breviter et dilucide derivata (E524), *Acta Academiae Scientiarum Imperialis Petropolitanae* **3** (1782), 72–86.

Euler, Leonhard. Variae speculationes super area triangulorum sphaericorum (E698), *Nova Acta Academiae Scientiarum Imperialis Petropolitanae* **10** (1797), 47–62.

Euler, Leonhard. *Zwei Abhandlungen über Sphärische Trigonometrie*, transl. Erwin Hammer. Leipzig: Wilhelm Engelmann, 1896.

Euler, Leonhard. Sur les logarithmes des nombres négatifs et imaginaires (E807). In Leonhard Euler, *Opera omnia*, Series I, Vol. 19, pp. 417–438. Basel: Birkhäuser, 1932.

Euler, Leonhard. *Introduction to Analysis of the Infinite, Book I*, transl. John D. Blanton. New York: Springer, 1988.

Euler, Leonhard. *Introduction to Analysis of the Infinite, Book II*, transl. John D. Blanton. New York: Springer, 1990.

Euler, Leonhard. *Foundations of Differential Calculus*, transl. John D. Blanton. New York: Springer, 2000.

Euler, Leonhard. *Euler's Institutionum calculi integralis*, transl. Ian Bruce. 2010. http://www.17centurymaths.com/contents/integralcalculus.html.

Euler, Leonhard. *Euler's Institutionum calculi differentialis*, transl. Ian Bruce. 2011. http://www.17centurymaths.com/contents/differentialcalculus.htm.

Euler, Leonhard. *Euler's Introductio in analysin infinitorum*, transl. Ian Bruce. 2013a. http://www.17centurymaths.com/contents/introductiontoanalysisvol1.htm.

Euler, Leonhard. Spherical trigonometry all derived briefly and clearly from first principles, transl. Ian Bruce. 2013b. http://www.17centurymaths.com/contents /euler/e524tr.pdf.

Euler, Leonhard. Principles of spherical trigonometry drawn from the method of the maxima and minima, transl. George Heine. 2014. http://eulerarchive.maa.org /docs/translations/E214en.pdf.

Euler, Leonhard. Elements of spheroidal trigonometry drawn from the method of the maxima and minima, transl. George Heine. 2015. http://eulerarchive.maa.org /docs/translations/E215en.pdf.

Everest, George. *An Account of the Measurement of Two Sections of the Meridional Arc of India*. London: East-India Company, 1847.

Fabri, Honoré. *Opusculum geometricum de linea sinuum et cycloide*. Rome: Corbelletti, 1659.

Fabri, Honoré. *Synopsis geometrica*. Leiden: Molin, 1669.

Faulhaber, Martin. *Zehntausend logarithmi*, Augsburg, 1631.

Feingold, Mordechai. *The Mathematicians' Apprenticeship: Science, Universities and Society in England 1560–1640.* Cambridge, UK: Cambridge University Press, 1984.

Feingold, Mordechai. *Before Newton: The Life and Times of Isaac Barrow.* Cambridge, UK: Cambridge University Press, 1990.

Feingold, Mordechai. Newton, Leibniz and Barrow too: An attempt at a reinterpretation, *Isis* **84** (1993), 310–338.

Fellmann, Emil A. Die mathematischen werke von Honoratus Fabry, *Physis* **1–2** (1959), 6–25, 69–102.

Fellmann, Emil A. *Leonhard Euler.* Basel: Birkhäuser, 2007.

Fellmann, Emil A., ed. *Leonhard Euler 1707–1783: Beiträge zu Leben und Werk.* Basel: Birkhäuser, 1983.

Fernández Garcia, Francisco; Jiménez Alcón, Francisco; and Muñoz Prieto, Luis Carlos. Mercator: the intuition of a genius [in Spanish], *Epsilon* **17** (2001), 327–336.

Ferraro, Giovanni. Functions, functional relations, and the laws of continuity in Euler, *Historia Mathematica* **27** (2000), 107–132.

Ferraro, Giovanni. Analytical symbols and geometrical figures in eighteenth-century calculus, *Studies in History and Philosophy of Science* **32** (2001), 535–555.

Ferraro, Giovanni. Differentials and differential coefficients in the Eulerian foundations of the calculus, *Historia Mathematica* **31** (2004), 34–61.

Ferraro, Giovanni. Convergence and formal manipulation in the theory of series from 1730 to 1815, *Historia Mathematica* **34** (2007a), 62–88.

Ferraro, Giovanni. Euler's treatises on infinitesimal analysis: *Introductio in analysin infinitorum, Institutiones calculi differentialis, Institutionum calculi integralis.* In *Euler Reconsidered: Tercentenary Essays,* ed. Roger Baker, pp. 40–102. Heber City, UT: Kendrick Press, 2007b.

Ferraro, Giovanni. *The Rise and Development of the Theory of Series up to the Early 1820s.* New York: Springer, 2008.

Ferzola, Anthony. Euler and differentials, *College Mathematics Journal* **25** (1994), 102–111.

Fierz, Markus. *Isaac Newton als Mathematiker.* Zürich: Kommissionsverlag Leemann AG, 1972.

Fincke, Thomas. *Geometriae rotundi.* Basel: Henric Petri, 1583.

Fine, Oronce. *De sinibus libri II.* Paris: Simon de Colines, 1542.

Fishman, R. S. Perish, then publish: Thomas Harriot and the sine law of refraction, *Archives of Ophthamology* **118** (2000), 405–409.

Fletcher, Alan; Comrie, Leslie; Miller, Jeffery; and Rosenhead, Louis. *An Index of Mathematical Tables.* 2nd ed. Reading, MA: Addison Wesley, 1962.

Folkerts, Menso. Regiomontanus als Mathematiker, *Centaurus* **21** (1977), 214–245.

Folkerts, Menso. Dei trigonometrie bei Apian. In *Peter Apian: Astronomie, Kosmographie und Mathematik am Beginn der Neuzeit,* ed. Karl Rötter, pp. 223–228. Buxheim/Eichstätt: Polygon-Verlag, 1997.

Folkerts, Menso; Launert, Dieter; and Thom, Andreas. Jost Bürgi's method for calculating sines, *Historia Mathematica* **43** (2016), 133–147.

Forbes, Eric G. Tobias Mayer's contributions to the development of lunar theory, *Journal for History of Astronomy* **1** (1970), 144–154.

Fourier, Jean Baptiste Joseph. *Théorie Analytique de la Chaleur.* Paris: Firmin Didot, 1822.

Fourier, Jean Baptiste Joseph. *The Analytical Theory of Heat*, transl. Alexander Freeman. Cambridge, UK: Cambridge University Press, 1878. Reprint, New York: Dover, 2003; reprint, Cambridge, UK: Cambridge University Press, 2009.

Fourier, Jean Baptiste Joseph. *Théorie Analytique de la Chaleur.* Paris: Jacques Gabay, 1988.

Fourier, Jean Baptiste Joseph. *Oeuvres de Fourier*, ed. Jean Gaston Darboux. 2 vols. Paris: Gauthier-Villars et Fils, 1888/1890. Reprint, Cambridge, UK: Cambridge University Press, 2013.

Fox, Robert, ed. *Thomas Harriot: An Elizabethan Man of Science.* Brookfield, VT: Ashgate, 2000.

Fox, Robert, ed. *Thomas Harriot and His World: Mathematics, Exploration, and Natural Philosophy in Early Modern England.* Brookfield, VT: Ashgate, 2012.

Fraser, D. C. Newton's interpolation formulas, *Journal of the Institute of Actuaries* **51** (1918), 77–106.

Fraser, D. C. Newton's interpolation formulas - further notes, *Journal of the Institute of Actuaries* **51** (1919), 211–232.

Fraser, D. C. Newton's interpolation formulas - further notes, *Journal of the Institute of Actuaries* **52** (1920), 117–135.

Fraser, D. C. Newton's interpolation formulas - an unpublished manuscript of Sir Isaac Newton, *Journal of the Institute of Actuaries* **58** (1927), 53–95.

Friedelmeyer, Jean-Pierre. Contexte et raisons d'une "mirifique" invention. In *Histoires de Logarithmes,* ed. Évelyne Barbin et al., pp. 39–72. Paris: Ellipses, 2006.

Frischauf, Johannes. Legendre's Theorem, *Österreichische Zeitschrift für Vermessungswesen* **14** (1916), 65–71, 86–90.

Galilei, Galileo. *Operations of the Geometric and Military Compass*, transl. Stillman Drake. Washington, DC: Smithsonian Institution Press, 1978.

Gardiner, William. *Tables of Logarithms for All Numbers from 1 to 102100.* London: G. Smith, 1742.

Gauchet, L. Note sur la trigonométrie sphérique de Kouo Cheou-King, *T'oung Pao* (2) **18** (1917), 151–174.

Gauss, Carl Friedrich. Über eine Aufgabe der sphärischen astronomie, *Monatlichen Correspondenz zur Beförderung der Erd- und Himmels-kunde* **18** (1808), 277–293.

Gauss, Carl Friedrich. *Theoria motus corporum coelestium.* Hamburg: Frid. Perthes and I. H. Besser, 1809.

Gauss, Carl Friedrich. *Disquisitiones generales circa superficies curvas.* Göttingen: Dieterich, 1828.

George, F. Hariot's meridional parts, *Journal of Navigation* **9** (1956), 66–69.

George, F. Hariot's meridional parts, *Journal of Navigation* **21** (1968), 82–83.

Gergonne, Joseph Diaz. Recherche de quelques-unes des lois générales qui régissent les polyèdres, *Annales de Mathématiques Pures et Appliquées* **15** (1824), 157–164.

Gergonne, Joseph Diaz. Considérations philosophiques sur les élémens de la science de l'étendue, *Annales de Mathématiques Pures et Appliquées* **16** (1826), 209–231.

Gerl, Armin. *Trigonometrisch-Astronomisches Rechnen kurz von Copernicus: Der Briefwechsel Regiomontanus-Bianchini.* Stuttgart: Franz Steiner, 1989.

Gibson, George A. Napier and the invention of logarithms. In *Handbook of the Exhibition of Napier Relics and of Books, Instruments, and Devices for Facilitating Calculation*, ed. E. M. Horsburgh, pp. 1–16. Edinburgh: Royal Society of Edinburgh, 1914. Reprint, Los Angeles: Tomash, 1982.

Gibson, George A. Napier's logarithms and the change to Briggs's logarithms. In *Napier Tercentenary Volume*, ed. C. G. Knott, pp. 111–137. London: Longmans, Green, 1915.

Gillispie, Charles Coulston. *Lazare Carnot Savant.* Princeton, NJ: Princeton University Press, 1971.

Gillispie, Charles Coulston; and Pisano, Raffaele. *Lazare and Sadi Carnot: A Scientific and Filial Relationship.* 2nd ed. Dordrecht: Springer, 2014.

Girard, Albert. *Tables des Sinus, Tangentes & Secantes, selon le Raid de 100000 Parties.* The Hague: Elzevier, 1626.

Girard, Albert. *Invention Nouvelle d'Algèbre.* Amsterdam: Blaeuw, 1629.

Girard, Albert. *Invention Nouvelle d'Algèbre*, ed. D. Bierens de Haan. Leiden: Muré Frères, 1884.

Gjertsen, Derek. *The Newton Handbook.* London: Routledge, 1986.

Gladstone-Millar, Lynne. *John Napier: Logarithm John.* Edinburgh: National Museums of Scotland, 2006.

Glaisher, J. W. L. The earliest use of the radix method for calculating logarithms, with historical notices relating to the contributions of Oughtred and others to mathematical notation, *Quarterly Journal of Pure and Applied Mathematics* **46** (1915), 125–197.

Glaisher, J. W. L. et al. *Report of the Committee on Mathematical Tables.* London: Taylor and Francis, 1873.

Glowatzki, Ernst; and Göttsche, Helmut. *Die Tafeln des Regiomontanus: Ein Jahrhundertwerk*, Munich: Institut für Geschichte der Naturwissenschaften, 1990.

Glushkov, Stanislav. An interpretation of Viète's "calculus of triangles" as a precursor to the algebra of complex numbers, *Historia Mathematica* **4** (1977), 127–136.

Glushkov, Stanislav. Algebraic thought in geometry as a source of de Moivre's formula [in Russian]. In *History and Methodology of the Natural Sciences 29*, ed. D. I. Gordeev, pp. 51–56. Moscow: Moscow State University, 1982.

Goldstein, Joel A. A matter of great magnitude: The conflict over arithmetization in 16th-, 17th-, and 18th-century editions of Euclid's *Elements* Books I Through VI (1561–1795), *Historia Mathematica* **27** (2000), 36–53.

Goldstine, Herman H. *A History of Numerical Analysis from the 16th Through the 19th Century*. New York: Springer, 1977.

Goldstine, Herman H. *A History of the Calculus of Variations from the 17th Century Through the 19th Century*. New York: Springer, 1980.

Golland, Louise Ahrndt; and Golland, Ronald William. Euler's troublesome series: An early example of the use of trigonometric series, *Historia Mathematica* **20** (1993), 54–67.

Golvers, Noël. *The Astronomia Europaea of Ferdinand Verbiest, S. J. (Dillingen, 1687)*. Nettetal: Steyler, 1993.

Goodwin, H. B. The simplification of formulae in nautical astronomy, *The Nautical Magazine* **68** (1899), 526–540.

Goodwin, H. B. The haversine in nautical astronomy, *United States Naval Institute Proceedings* **36** (1910), 735–746.

Goulding, Robert. Chymicorum in morem: Refraction, matter theory, and secrecy in the Harriot-Kepler correspondence. In *Thomas Harriot and His World: Mathematics, Exploration, and Natural Philosophy in Early Modern England*, ed. Robert Fox, pp. 27–51. Burlington, VT: Ashgate, 2012.

Govi, G. Dei metodi proposti nel 1639 da Bonaventura Cavalieri per ottenere direttamente il logaritmo della somma della differenza di due numeri dei quali sono dati i logaritmi, e per risolvere mediante le funzioni circolari equazioni di 2° grado, *Atti della Reale Academia dei Lincei* (2) **3** (part 2) (1875–76), 173–178.

Gowing, Ronald. *Roger Cotes—Natural Philosopher*. Cambridge, UK: Cambridge University Press, 1983.

Gowing, Ronald. A study of spirals: Cotes and Varignon. In *The Investigation of Difficult Things: Essays on Newton and the History of the Exact Sciences in Honour of D. T. Whiteside*, ed. P. M. Harman and Alan E. Shapiro, pp. 371–381. Cambridge, UK: Cambridge University Press, 1992.

Gowing, Ronald. Halley, Cotes, and the nautical meridian, *Historia Mathematica* **22** (1995), 19–32.

Grattan-Guinness, Ivor. Joseph Fourier and the revolution in mathematical physics, *Journal of the Institute of Mathematics and its Applications* **5** (1969), 230–253.

Grattan-Guinness, Ivor. *The Development of the Foundations of Mathematical Analysis from Euler to Riemann*. Cambridge, MA: MIT Press, 1970.

Grattan-Guinness, Ivor, ed. *From the Calculus to Set Theory, 1630–1910: An Introductory History*. London: Duckworth, 1980.

Grattan-Guinness, Ivor. *Convolutions in French Mathematics, 1800–1840*. 3 vols. Basel: Birkhäuser, 1990a.

Grattan-Guinness, Ivor. Work for the hairdressers: The production of de Prony's logarithmic and trigonometric tables, *Annals of the History of Computing* **12** (1990b), 177–185.

Grattan-Guinness, Ivor. The computation factory: De Prony's project for making tables in the 1790s. In *The History of Mathematical Tables: From Sumer to Spreadsheets*, ed. Martin Campbell-Kelly, Mary Croarken, Raymond Flood, and Eleanor Robson, pp. 105–121. Oxford: Oxford University Press, 2003.

Grattan-Guinness, Ivor; and Ravetz, J. R. *Joseph Fourier 1768–1830.* Cambridge, MA: MIT Press, 1972.

Gravelaar, N. L. W. A. Pitiscus' *Trigonometria, Nieuw Archief voor Wiskunde* **3** (1898), 253–278.

Gray, Jeremy. *Ideas of Space: Euclidean, Non-Euclidean, and Relativistic.* Oxford: Clarendon Press, 1979a.

Gray, Jeremy. Non-Euclidean geometry—A re-interpretation, *Historia Mathematica* **6** (1979b), 236–258.

Gray, Jeremy. Möbius's geometrical mechanics. In *Möbius and His Band: Mathematics and Astronomy in Nineteenth-Century Germany*, ed. John Fauvel, Raymond Flood, and Robin Wilson, pp. 78–103. Oxford: Oxford University Press, 1993.

Gray, Jeremy. *János Bolyai: Non-Euclidean Geometry and the Nature of Space.* Cambridge, MA: MIT Press, 2004.

Gray, Jeremy. Gauss and non-Euclidean geometry. In *Non-Euclidean Geometries: János Bolyai Memorial Volume*, ed. András Prékopa and Emil Molnár, pp. 60–80. New York: Springer, 2006.

Gray, Jeremy. *Worlds Out of Nothing: A Course in the History of Geometry in the 19th Century.* London: Springer, 2007.

Gray, Jeremy. *The Real and the Complex: A History of Analysis in the 19th Century.* Cham: Springer, 2015.

Gray, Jeremy; and Tilling, Laura. Johann Heinrich Lambert, mathematician and scientist, 1728–1777, *Historia Mathematica* **5** (1978), 13–41.

Gregory, James. *Vera circuli et hyperbolae quadratura.* Padua: Frambotti, 1667.

Gregory, James. *Exercitationes geometriae.* London: Godbid/Pitt, 1668a.

Gregory, James. *Geometriae pars universalis.* Padua: Frambotti, 1668b.

Gridgeman, N. T. John Napier and the history of logarithms, *Scripta Mathematica* **29** (1973), 49–65.

Gronau, Detlef. *Johannes Kepler und die Logarithmen. Berichte der Mathematisch-Statistischen Sektion in der Forschungsgesellschaft Joanneum* **284**. Graz: Forschungszentrum Graz, 1987.

Gronau, Detlef. Die logarithmen, von der rechenhilfe über funktionalgleichungen zur funktion. In *Mathematik im Wandel*, Vol. 2, ed. Michael Toepell, pp. 127–145. Hildesheim: Franzbecker, 2001.

Grosholz, Emily. Some uses of proportion in Newton's *Principia*, Book I: A case study in applied mathematics, *Studies in History and Philosophy of Science* **18** (1987), 209–220.

Grunert, Johann August. *Sphäroidische Trigonometrie.* Berlin: Reimer, 1833.

Grunert, Johann August. Ueber sphärische Dreiecke, deren Seiten im Verhältniss zu dem Halbmesser der Kugel, auf welcher sie liegen, sehr klein sind, *Archiv der Mathematik und Physik* **9** (1847), 8–45.

Guicciardini, Niccolò. *Isaac Newton on Mathematical Certainty and Method.* Cambridge, MA: MIT Press, 2009.

Gunter, Edmund. *Canon triangulorum.* London: Jones, 1620.

Gunter, Edmund. *De Sector et Radio*. London: Jones, 1624.

Gunter, Edmund. *The Description and Use of the Sector, Cross-staffe, and Other Instruments*. 2nd ed. London: W. Jones, 1636.

Günther, Siegmund. Über eine merkwüurdige Beziehung zwischen Pappus und Kepler, *Bibliotheca Mathematica* (new series) **2** (1888), 81–87.

Guo, Shuchun, and Tian, Miao. Mathematics. In *A History of Chinese Science and Technology*, Vol. 1, ed. Lu Yongxiang, pp. 203–287. Heidelberg: Springer, 2015.

Gupta, R. C. India's contributions to Chinese mathematics through the eighth century C.E. In *Ancient Indian Leaps into Mathematics*, ed. B. S. Yadav and M. Mohan, pp. 33–44. Boston: Birkhäuser, 2011.

Haasbroeck, N. D. *Gemma Frisius, Tycho Brahe and Snellius and Their Triangulations*. Delft: Rijkscommissie voor Geodesie, 1968.

Hackmann, Willem. Mathematical instruments. In *Oxford Figures: 800 Years of the Mathematical Sciences*, ed. John Fauvel, Raymond Flood, and Robin Wilson, pp. 63–75. Oxford: Oxford University Press, 2000.

Hairault, Jean-Pierre. Calcul des logarithmes décimaux par Henry Briggs. In *Histoires de Logarithmes,* ed. Évelyne Barbin et al., pp. 113–129. Paris: Ellipses, 2006.

Hall, A. Rupert. *Isaac Newton: Adventurer in Thought*. Oxford: Blackwell, 1992.

Haller, S. Beitrag zur Geschichte der konstruktiven Auflösung sphärischer Dreiecke durch stereographische Projektion, *Bibliotheca mathematica* **13** (1899), 71–80.

Halley, Edmund. An easie demonstration of the analogy of the logarithmick tangents to the meridian line or sum of the secants: With various methods for computing the same to the utmost exactness, *Philosophical Transactions* **19** (1695), 202–214.

Hallyn, F. Kepler, Snell, and the law of refraction [in Dutch], *Mededelingen van de Koninklijke Academie voor Wetenschappen (België)* **56** (1994), 119–134.

Han Qi. Patronage scientifique et carrière politique: Li Guangdi entre Kangxi et Mei Wending, *Etudes Chinoises* **16** (1997), 7–37.

Han Qi. Knowledge and power: A social history of the transmission of mathematics between China and Europe during the Kangxi reign (1662–1722). In *Proceedings of the International Congress of Mathematicians*, Vol. 4, ed. S. Y. Yang, Y. R. Kim, D.-W. Lee, and I. Yie, pp. 1217–1229. Seoul: Kyung Moon, 2014.

Han Qi. Chinese literati's attitudes toward Western science: Transition from the late Kangxi period to the mid-Qiaolong period, *Historia Scientiarum* **24** (2015), 76–87.

Hankins, Thomas L. *Jean d'Alembert: Science and the Enlightenment*. Oxford: Clarendon Press, 1970. Reprint, New York: Gordon and Breach, 1990.

Hara, Kokiti. Pascal et Wallis au sujet de la cycloïde, *Annals of the Japan Association for Philosophy of Science* **3** (1969), 166–187.

Hara, Kokiti. Pascal et Wallis au sujet de la cycloïde (II), *Gallia* **11** (1971a), 231–249.

Hara, Kokiti. Pascal et Wallis au sujet de la cycloïde (III), *Japanese Studies in the History of Science* **10** (1971b), 95–112.

Harris, John. *Elements of Plain and Spherical Trigonometry*. London: Midwinter, 1706.

Hart, Roger. *Imagined Civilizations: China, the West, and Their First Encounter.* Baltimore, MD: Johns Hopkins University Press, 2013.

Hashimoto, Keizo; and Jami, Catherine. From the *Elements* to calendar reform: Xu Guangqi's shaping of mathematics and astronomy. In *Statecraft and Intellectual Renewal in Late Ming China: The Cross-Cultural Synthesis of Xu Guangqi (1562–1633)*, ed. Catherine Jami, Peter Engelfriet, and Gregory Blue, pp. 263–278. Leiden: Brill, 2001.

Hassler, Ferdinand Rudolph. *Elements of Analytic Trigonometry, Plane and Spherical.* New York: James Bloomfield, 1826.

Hattendorf, John J., ed. *The Oxford Encyclopedia of Maritime History.* 4 vols. Oxford: Oxford University Press, 2007.

Hauer, F. Zur Geschichte des Satzes von Legendre, *Zeitschrift für Vermessungswesen* **67** (1938), 577–595, 641–653.

Havil, Julian. *John Napier: His Life, Logarithms and Legacy.* Princeton, NJ: Princeton University Press, 2014.

Hawkins, William F. The mathematical work of John Napier (1550–1617), *Bulletin of the Australian Mathematical Society* **26** (1982a), 455–468.

Hawkins, William F. *The Mathematical Work of John Napier (1550–1617).* 3 vols. PhD diss., University of Auckland, 1982b.

Hawney, William. *The Doctrine of Plain and Spherical Trigonometry.* London: Darby et al., 1725.

Heine, George. Euler and the flattening of the Earth, *Math Horizons* **21** (1) (2013), 25–29.

Heinrich, Georg. James Gregorys "Vera circuli et hyperbolae quadratura," *Bibliotheca Mathematica* (3) **2** (1901), 77–85.

Hellmann, Martin. Pitiscus und seine kleine, *Trigonometrie, Mannheimer Geschichtsblätter* **4** (1997), 107–129.

Helmert, F. *Die Mathematischen und Physikalischen Theorieen der höheren Geodäsie.* 2 vols. Leipzig: Teubner, 1880/1884.

Henderson, J. *Bibliotheca Tabularum Mathematicarum. Part I. Logarithmic Tables.* Cambridge, UK: Cambridge University Press, 1926.

Henderson, Janice. Erasmus Reinhold's determination of the distance of the Sun from the Earth. In *The Copernican Achievement*, ed. Robert S. Westman, pp. 108–129. Berkeley: University of California Press, 1975.

Henri, M. Formules pour calculer les parallaxes de longitude et de latitude et le demi-diamètre apparent d'une planète quelconque, *Connaissance de Temps* 1809 (publ. April 1807), 441–448.

Henrion, Dénis. *Traicté des logarithmes.* Paris, 1626.

Herreman, Alain. L'inauguration des séries trigonométriques dans la *Théorie Analytique de la Chaleur* de Fourier et dans la controverse des cordes vibrantes, *Revue d'Histoire des Mathématiques* **19** (2013), 151–243.

Higton, Hester. Does using an instrument make you mathematical? Mathematical practitioners of the 17th century, *Endeavour* **25** (1) (2001), 18–22.

Higton, Hester. Instruments and illustration: The use of images in Edmund Gunter's *De sectore et radio*, *Early Science and Medicine* **18** (2013), 180–200.

Hill, Katherine. Neither ancient nor modern: Wallis and Barrow on the composition of continua, Part I: Mathematical styles and the composition of continua, *Notes and Records of the Royal Society of London* **50** (1996), 165–178.

Hill, Katherine. Neither ancient nor modern: Wallis and Barrow on the composition of continua, Part II: The seventeenth century context: The struggle between ancient and modern, *Notes and Records of the Royal Society of London* **51** (1997), 13–22.

Hilliker, David Lee. A study in the history of analysis up to the time of Leibniz and Newton in regard to Newton's discovery of the binomial theorem. IV. Contributions of Newton, *Mathematics Student* **42** (1974), 397–404.

Hind, John. *The Elements of Plane and Spherical Trigonometry*. 2nd ed. Cambridge, UK: J. Smith, 1828.

Hobert, Jean Phillipe; and Ideler, Louis. *Nouvelles Tables Trigonométriques Calculées pour la Division Décimale du Quart de Cercle*. Berlin: L'École Réelle, 1799.

Hobson, E. W. *John Napier and the Invention of Logarithms, 1614*. Cambridge, UK: Cambridge University Press, 1914.

Hofmann, Joseph E. Der junge Newton als Mathematiker (1665–1675), *Mathematische-Physikalische Semesterberichte* **2** (1951), 45–70.

Hofmann, Joseph E. *Leibniz in Paris 1672–1676: His Growth to Mathematical Maturity*. Cambridge, UK: Cambridge University Press, 1974.

Hollingdale, S. H. Some mathematicians of the Renaissance: The "Triparty" of 1484, *Institute of Mathematics and Its Applications* **20** (1984), 130–136.

Hood, Thomas. *The Making and Use of the Geometrical Instrument, Called a Sector*. London: Windet, 1598.

Houzel, Christian. The birth of non-Euclidean geometry. In *1830–1930: A Century of Geometry*, ed. L. Boi, D. Flament, and J.-M. Salanskis, pp. 3–21. Berlin: Springer, 1992.

Horng Wann-Sheng. *Li Shanlan: The Impact of Western Mathematics in China During the Late 19th Century*. PhD diss., City University of New York, 1991.

Horsburgh, E. M., ed. *Modern Instruments and Methods of Calculation: A Handbook of the Napier Tercentenary Exhibition*. London: G. Bell and Sons/Royal Society of Edinburgh, 1914.

Horváth, Miklós. On the Leibnizian quadrature of the circle, *Annales Universitatis Scientiarum Budapestinensis. Sectio Computatorica* **4** (1983), 75–83.

Howarth, Richard J. History of the stereographic projection and its early use in geology, *Terra Nova* **8** (1996), 499–513.

Howson, Geoffrey. *A History of Mathematics Education in England*. Cambridge, UK: Cambridge University Press, 1982.

Huang, Xiang. The trading zone communication of scientific knowledge: An examination of Jesuit science in China (1582–1773), *Science in Context* **18** (2005), 393–427.

Hug, Vanja; and Steiner, Thomas. Une lettre d'Euler à d'Alembert retrouvée, *Historia Mathematica* **42** (2015), 84–94.

Hughes, Barnabas. The companion curves of Gilles Personne de Roberval, *Cubo* **4** (2002), 43–57.

Hummel, Arthur W., ed. *Eminent Chinese of the Ch'ing Period (1644–1912).* 2 vols. Washington, DC: United States Printing Office, 1943–1944.

Hunrath, Karl. Des Rheticus *Canon doctrinae triangulorum* und Vieta's *Canon mathematicus, Zeitschrift für Mathematik und Physik* **44** (suppl.) (1899), 211–240.

Hutton, Charles. *A Course of Mathematics.* 3 vols. London: Rivington et al., 1811a.

Hutton, Charles. *Mathematical Tables.* 5th ed. London: Rivington et al., 1811b.

Hydrographic Department, Tokyo. *New Altitude and Azimuth Tables Between Latitudes 65° N. and 65° S. for the Determination of the Position Line at Sea.* Glasgow: Brown, Son & Ferguson, 1924.

Hymers, John. *A Treatise on Plane and Spherical Trigonometry.* 4th ed. London: Longman, Brown, Green, Longmans, and Roberts, 1858.

Iannaccone, Isaia. The *Geyuan baxian biao* and some remarks about the scientific collaboration between Schall von Bell, Rho, and Schreck. In *Western Learning and Christianity in China: The Contribution and Impact of Johann Adam Schall von Bell, 1592–1666*, ed. Roman Malek, pp. 701–716. Nettetal: Steyler Verlag, 1998.

Inman, James. *Navigation and Nautical Astronomy, for the Use of British Seamen.* 3rd ed. London: Rivington, 1835.

Itard, Jean. Matériaux pour l'histoire des nombres complexes, *Bibliothèque d'Information sur l'Enseignement Mathématique* (1968).

Jagger, Graham. The making of logarithm tables. In *The History of Numerical Tables: From Sumer to Spreadsheets*, ed. M. Campbell-Kelly, M. Croarken, R. Flood, and E. Robson, pp. 49–78. Oxford: Oxford University Press, 2003.

Jagger, Graham. From Napier's to Napierian logarithms—The contribution of John and Euclid Speidell (chapter). PhD diss., n.d.

Jami, Catherine. Une histoire chinoise du "nombre Π," *Archive for History of Exact Sciences* **38** (1988a), 39–50.

Jami, Catherine. Western influence and Chinese tradition in an eighteenth century Chinese mathematical work, *Historia Mathematica* **15** (1988b), 311–331.

Jami, Catherine. *Les Méthodes Rapides pour la Trigonométrie et le Rapport Précis du Cercle (1774): Tradition Chinoise et Apport Occidental en Mathématiques.* Paris: Institut des Hautes Études Chinoises, 1990.

Jami, Catherine. Scholars and mathematical knowledge during the late Ming and early Qing, *Historia Scientiarum* **42** (1991), 99–109.

Jami, Catherine. Western mathematics in China, seventeenth century and nineteenth century. In *Science and Empires: Historical Studies about Scientific Development and European Expansion*, ed. Patrick Petitjean, Catherine Jami, and Anne Marie Moulin, pp. 79–88. Dordrecht: Kluwer, 1992.

Jami, Catherine. History of mathematics in Mei Wending's (1633–1721) work, *Historia Scientiarum* **4** (1994), 157–172.

Jami, Catherine. Teachers of mathematics in China: The Jesuits and their textbooks (1580–1723). In *History of Mathematical Sciences: Portugal and East Asia II*, ed. Luis Saraiva, pp. 79–99. Singapore: World Scientific, 2004.

Jami, Catherine. Western learning and imperial scholarship: The Kangxi emperor's study, *East Asian Science, Technology, and Medicine* **27** (2007), 146–172.

Jami, Catherine. Heavenly learning, statecraft, and scholarship: The Jesuits and their mathematics in China. In *The Oxford Handbook of the History of Mathematics*, ed. Eleanor Robson and Jacqueline Stedall, pp. 57–84. Oxford: Oxford University Press, 2009.

Jami, Catherine. *The Emperor's New Mathematics: Western Learning and Imperial Authority During the Kangxi Reign (1662–1722)*. Oxford: Oxford University Press, 2012.

Jami, Catherine. La carrière de Mei Wending (1633–1721) et le statut des sciences mathématiques dans le savoir lettré, *Extrême-Orient Extrême-Occident* **36** (2014), 19–47.

Jami, Catherine; Engelfriet, Peter; and Blue, Gregory, eds. *Statecraft and Intellectual Renewal in Late Ming China: The Cross-Cultural Synthesis of Xu Guangqi (1562–1633)*. Leiden: Brill, 2001.

Jami, Catherine; and Han Qi. The reconstruction of imperial mathematics in China during the Kangxi period (1662–1722), *Early Science and Medicine* **8** (2003), 88–110.

Jesseph, Douglas M. Descartes, Pascal, and the epistemology of mathematics: The case of the cycloid, *Perspectives on Science* **15** (2007), 410–433.

Johnston, S. Mathematical practitioners and instruments in Elizabethan England, *Annals of Science* **48** (1991), 319–344.

Johnston, S. *Making Mathematical Practice: Gentlemen, Practitioners, and Artisans in Elizabethan England*. PhD diss., University of Cambridge, 1994.

Jones, Phillip S. Angular measure—enough of its history to improve its teaching, *Mathematics Teacher* **46** (1953), 419–426.

Jones, William. *Synopsis Palmariorum Matheseos*. London: J. Matthews, 1706.

Joseph, George Gheverghese. *A Passage to Infinity: Medieval Indian Mathematics from Kerala and its Impact*. New Delhi: Sage, 2009.

Jouve, Guillaume. The first works of D'Alembert and Euler about the problem of vibrating strings from the perspective of their correspondence, *Centaurus* **59** (2017), 300–307.

Joyce, W. B.; and Joyce, Alice. Descartes, Newton, and Snell's Law, *Journal of the Optical Society of America* **66** (1976), 1–8.

Kahane, Jean-Pierre; and Lemarié-Rieusset, Pierre-Gilles. *Fourier Series and Wavelets*. Luxembourg: Gordon and Breach, 1995.

Karp, Alexander; and Schubring, Gert, eds. *Handbook on the History of Mathematics Education*. New York: Springer, 2014.

Karpinski, Louis Charles. The place of trigonometry in the development of mathematical ideas, *Scripta Mathematica* **11** (1945), 268–272.

Karpinski, Louis Charles. Bibliographical check list of all works on trigonometry published up to 1700 A.D., *Scripta Mathematica* **12** (1946), 267–283.

Kárteszi, Ferenc, ed. *János Bolyai: Appendix: The Theory of Space*. Amsterdam: North-Holland, 1987.

Kästner, Abraham Gotthelf. Fläche durch Seiten und Winkel zusammen bey einem kleinen Dreyeck ausgedruckt. In *Geometrische Abhandlungen*, Vol. 2, pp. 451–458. Göttingen: Vandenhoef/Ruprecht, 1791.

Katz, Victor. The calculus of the trigonometric functions, *Historia Mathematica* **14** (1987), 311–324.

Katz, Victor. Napier's logarithms adapted for today's classroom. In *Learn From the Masters!* ed. Frank Swetz, John Fauvel, Otto Bekken, Bengt Johansson, and Victor Katz, pp. 49–56. Washington, DC: Mathematical Association of America, 1995.

Katz, Victor. Euler's analysis textbooks. In *Leonhard Euler: Life, Work, and Legacy*, ed. Robert E. Bradley and C. Edward Sandifer, pp. 213–233. Amsterdam: Elsevier, 2007.

Kaunzner, Wolfgang. Zur Mathematik Peter Apians. In *Peter Apian: Astronomie, Kosmographie und Mathematik am Beginn der Neuzeit*, ed. Karl Rötter, pp. 183–216. Buxheim/Eichstätt: Polygon-Verlag, 1997.

Keill, John. *The Elements of Plain and Spherical Trigonometry*. 3rd ed. Dublin: Wilmot/Fuller, 1726.

Keith, Thomas. *An Introduction to the Theory and Practice of Plane and Spherical Trigonometry*. 2nd ed. London: Longman, Hurst, Rees, Orme, and Brown, 1810.

Keith, Thomas. *An Introduction to the Theory and Practice of Plane and Spherical Trigonometry*. 5th ed. London: Longman, Rees, Orme, Brown and Green, 1826.

Kepler, Johannes. *Astronoma nova*. Heidelberg: Voegelinus, 1609.

Kepler, Johannes. *Epitome astronomiae Copernicae*. 3 vols. Linz: Johannes Plancus, 1617–1621.

Kepler, Johannes. *Chilias logarithmorum*. Marburg: Chemlini, 1624. Reprint, Johannes Kepler, *Gesammelte Werke*, Vol. 9., ed. F. Hammer, pp. 277–426, Beck, 1960.

Kepler, Johannes. *Tabulae Rudolphinae*. Ulm: Saurii, 1627.

Kepler, Johannes. *Opera omnia*. 8 vols. Frankfurt/Erlangen: Heyder & Zimmer, 1858–1870.

Kepler, Johannes. *Gesammelte Werke*, ed. Franz Hammer, Walther von Dyck, and Max Caspar. 14 vols. Munich: Beck, 1937–present.

Kepler, Johannes. *Les Mille Logarithmes et le Supplément aux Mille*, transl. Jean Peyroux. Paris: Blanchard, 1993a.

Kepler, Johannes. *New Astronomy*, transl. William H. Donahue. Cambridge, UK: Cambridge University Press, 1993b.

Kleiner, Israel. Evolution of the function concept: A brief survey, *College Mathematics Journal* **20** (1989), 282–300.

Klügel, Georg Simon. *Analytische Trigonometrie*. Braunschweig: Fürstl. Waisenhausbuchhandlung, 1770.

Klügel, Georg Simon. *Mathematisches Wörterbuch*. 3 vols. Leipzig: Schwikert, 1803–1808.

Knott, Cargill Gilston. *Napier Tercentenary Volume*. London: Longmans, Green, 1915.

Komenský, Jan Amos; and Marci, Jan Marcus. Dějiny přírodních věd v českých zemích (19. část), *Elektro* **2008** (5), 60.

Komenský, Jan Amos; and Marci, Jan Marcus. Dějiny přírodních věd v českých zemích (27. část), Český matematik a diplomat—Jakub Kresa, *Elektro* **2009** (2), 60.

Kresa, Jakob. *Analysis speciosa trigonometriae*. Prague: Typis Universitatis Carolo-Ferdinaneae, 1720.

Lacroix, Sylvestre. *Traité Élémentaire de Trigonométrie Rectligne et Sphérique, et d'Application de l'Algèbre à la Géométrie*. Paris: Crapelet, An VII (1798).

Lagrange, Joseph-Louis. Recherches sur la nature, et la propagation du son, *Miscellanea Taurensia* **1** (1759), separately paginated, I–X, 1–112.

Lagrange, Joseph-Louis. De quelques problèmes, relatifs aux triangles sphériques, avec une analyse complète de ces triangles, *Journal de l'École Polytechnique*, Sixième Cahier, Tome 2 (An 7) (1799), 270–296.

Lam Lay-Yong; and Shen Kangsheng. Right-angled triangles in ancient China, *Archive for History of Exact Sciences* **30** (1984), 87–112.

Lam Lay-Yong; and Shen Kangsheng. Mathematical problems on surveying in ancient China, *Archive for History of Exact Sciences* **36** (1986), 1–20.

Lambert, Johann Heinrich. Mémoire sur quelques propriétés remarquables des quantités transcendentes circulaires et logarithmiques, *Mémoires de l'Académie Royale des Sciences et Belles-Lettres* **17** (1768), 265–322.

Lambert, Johann Heinrich. Observations trigonométriques, *Mémoires de l'Académie Royale des Sciences et Belles-Lettres* **17** (1770), 327–354.

Lambert, Johann Heinrich. Anmerkungen und zusätze zur trigonometrie. In *Beiträge zum Gebrauche der Mathematik und deren Anwendung*, Vol. 1, pp. 369–424. Berlin: Konigl. Realschule, 1792.

Lattis, James M. *Between Copernicus and Galileo: Christoph Clavius and the Collapse of Ptolemaic Cosmology*. Chicago, IL: University of Chicago Press, 1994.

Launert, D. *Nova Kepleriana: Bürgis Kunstweg im Fundamentum Astronomiae*. München: Verlag der Bayerischen Akademie der Wissenschaften, 2015.

Le Cointe, I.-L.-A. *Leçons sur al Théorie des Fonctions Circulaires et la Trigonométrie*. Paris: Mallet-Bachelier, 1858.

Le Corre, Loïc. John Neper et la merveilleuse tables des logarithmes. In *Histoires de Logarithmes,* ed. Évelyne Barbin et al., pp. 73–112. Paris: Ellipses, 2006.

Lefort, F. Note sur les erreurs que contient une des tables des logarithmes de Callet, *Comptes Rendus* **44** (1857), 1097–1100.

Lefort, F. Description des grandes tables logarithmiques et trigonmétriques, calculées au Bureau du Cadastre, sous la direction de M. de Prony, et exposition des méthodes et proceeds mis en usage pour leur construction, *Annales de l'Observatoire Impérial de Paris* **4** (1858), Additions, [123]–[150].

Lefort, F. Observations on Mr Sang's remarks relative to the great logarithmic table compiled at the Bureau de Cadastre under the direction of M. Prony, *Proceedings of the Royal Society of Edinburgh* **8** (1875), 563–581.

Legendre, Adrien-Marie. Mémoire sur les operations trigonométriques, dont les résultats dependent de la figure de la terre, *Histoire de l'Académie Royale des Sciences* 1787 (publ. 1789), 352–383.

Legendre, Adrien-Marie. Méthode pour determiner la longueur exacte du quart du méridien, d'après les observations faites pour la mesure de l'arc compris entre Dunkerque et Barcelonne. In J. B. J. Delambre, *Méthodes Analytiques pour la Détermination d'un Arc du Méridien*, An 7, pp. 1–16. Paris: Crapelet, 1798.

Leibniz, Gottfried Wilhelm. Supplementum geometricae practica sese ad problemata transcendentia extendens, ope novae methodi generalissimae per series infinitas, *Acta Eruditorum* (1693), 178–180.

Leibniz, Gottfried Wilhelm. *The Early Mathematical Manuscripts of Leibniz*, transl. J. M. Child. Chicago/London: Open Court, 1920.

l'Hôpital, Guillaume François Antoine de. *Analyse des Infiniment Petits*. Paris: Imprimerie Royale, 1696.

Li Yan and Du Shiran. *Chinese Mathematics: A Concise History,* transl. John N. Crossley and Anthony W.-C. Lun. Oxford: Clarendon Press, 1987.

Libbrecht, Ulrich. *Chinese Mathematics in the Thirteenth Century: The Shu-shu chiu-chang of Ch'in Chiu-shao*. Cambridge, MA: MIT Press, 1973.

List, Martha; and Bialas, Volker. *Die Coss von Joost Bürgi in der Redaktion von Johannes Kepler*. Munich: Verlag der Bayerischen Akademie der Wissenschaften, 1973.

Lobachevsky, Nicolai. Géométrie imaginaire, *Journal für die Reine und Angewandte Mathematik* **17** (1837), 295–320.

Lobachevsky, Nicolai. *Geometrische Untersuchungen zur Theorie der Parallellinien*. 1840. Reprint, Berlin: Mayer and Müller, 1887.

Lobachevsky, Nicolai. *Geometrical Researches on the Theory of Parallels*, new ed., transl. George Bruce Halstead. Chicago: Open Court, 1914.

Lobachevsky, Nicolai. *Pangeometry*, transl. Athanase Papadopoulos. Zurich: European Mathematical Society, 2010.

Loeffel, H. Elementargeometrische integrationem von Pascal, *Elemente der Mathematik* **41** (1986), 83–87.

Lohne, J. A. Thomas Harriot als mathematiker, *Centaurus* **11** (1965/66), 19–45.

Lovett, Edgar Odell. Note on Napier's rules of circular parts, *Bulletin of the American Mathematical Society* **10** (1898), 552–554.

Lü Lingfeng. Eclipses and the victory of European astronomy in China, *East Asian Science, Technology and Medicine* **27** (2007), 127–145.

Lu Yongxiang, ed. *A History of Chinese Science and Technology*, Vol. 1. Heidelberg: Springer, 2015.

Magini, Antonio. *De planis triangulis liber unicus*, Venice: Giovanni Battista Ciotti, 1592.

Magini, Antonio. *Primum mobile* (contains *Trigonometria Sphaericorum* and *Magnus trigonometricus canon*). Bologna, 1609.

Mahoney, Michael S. Barrow's mathematics: Between ancients and moderns. In *Before Newton: The Life and Times of Isaac Barrow*, ed. Mordechai Feingold, pp. 179–249. Cambridge, UK: Cambridge University Press, 1990.

Maierù, Luigi; and Toth, Imre. *Fra Descartes e Newton: Isaac Barrow e John Wallis*. Soveria Mannelli: Rubbettino, 1994.

Malcolm, Noel. *Aspects of Hobbes*. Oxford: Oxford University Press, 2002.

Malcolm, Noel; and Stedall, Jacqueline. *John Pell (1611–1685) and His Correspondence with Sir Charles Cavendish*. Oxford: Oxford University Press, 2005.

Malet, Antoni. James Gregorie on tangents and the "Taylor" rule for series expansions, *Archive for History of Exact Sciences* **46** (1993), 97–137.

Malet, Antoni. Barrow, Wallis, and the remaking of seventeenth century indivisibles, *Centaurus* **39** (1997), 67–92.

Malet, Antoni. Renaissance notions of number and magnitude, *Historia Mathematica* **33** (2006), 63–81.

Maor, Eli. *Trigonometric Delights*. Princeton, NJ: Princeton University Press, 2002.

Marolois, Samuel. *Fortification, ou Architecture Militaire . . . , Reueüe Augmentée et Corrigée par Albert Girard*. Amsterdam: Janssen, 1627.

Marolois, Samuel. *The Art of Fortification . . . , Reviewed, Augmented and Corrected by Albert Girard*, transl. Henry Hexham. Amsterdam: Johnson, 1638.

Marolois, Samuel. *Geometria theoretica ac practica*. Amsterdam: Jansson, 1647.

Martin, Benjamin. *The Young Trigonometer's Compleat Guide*. 2 vols. London: J. Noon, 1736.

Martzloff, Jean-Claude. Le géometrie euclidienne selon Mei Wending, *Historia Scientiarum* **21** (1981a), 27–42.

Martzloff, Jean-Claude. *Recherches sur l'Oeuvre Mathématique de Mei Wending (1633–1721)*. Paris: Presses Universitaires de France, 1981b.

Martzloff, Jean-Claude. Space and time in Chinese texts of astronomy and of mathematical astronomy in the seventeen and eighteenth centuries, *Chinese Science* **11** (1993–94), 66–92.

Martzloff, Jean-Claude. *A History of Chinese Mathematics*. Berlin: Springer, 1997.

Mauduit, A.-R. *Principes d'Astronomie Sphérique; ou Traité Complet de Trigonométrie Sphérique*. Paris: H. L. Guerin & L. F. Delatour, 1765.

Maurice, M. F.; Sprague, Thomas; and Williams, J. Hill. On interpolation, *Journal of the Institute of Actuaries and Assurance Magazine* **14** (1867), 1–36.

Maurolico, Francesco. *Theodosii sphaericorum elementorum libri III*, etc. Messina: Pietro Spira, 1558.

Maurolico, Francesco. *Opuscula mathematica*. Venice: Franciscum Franciscium Senensem, 1575.

Mendoza y Ríos, Joseph de. *A Complete Collection of Tables for Navigation and Nautical Astronomy*. London: Bensley, 1805.

Mercator, Gerard. *Mercator's Map of the World*, introduction by B. van't Hoff, *Imago Mundi* Supplement 2, 1961.

Mercator, Nicolaus. *Logarithmotechnia*. London: Godbid/Pitt, 1668.

Merker, Claude. *Le Calcul Intégrale dans la Dernière Oeuvre Scientifique de Pascal*. Besançon: Presses Universitaires de Franche-Comté, 1995.

Merker, Claude. *Le Chant du Cygne des Indivisibles: Le Calcul Intégrale dans la Dernière Oeuvre Scientifique de Pascal*. Besançon: Presses Universitaires de Franche-Comté, 2001.

Meskens, Ad. Michiel Coignet's contribution to the development of the sector, *Annals of Science* **54** (1997), 143–160.

Meskens, Ad. *Practical Mathematics in a Commercial Metropolis: Mathematical Life in Late 16th-Century Antwerp*. Dordrecht: Springer, 2013.

Mikami, Yoshio. *The Development of Mathematics in China and Japan*. New York: Chelsea, 1913.

Miura, Nobuo. The applications of trigonometry in Pitiscus: A preliminary essay, *Historia Scientiarum* **30** (1986), 63–78.

Miura, Nobuo. The applications to logarithms to trigonometry in Richard Norwood, *Historia Scientiarum* **37** (1989), 17–30.

Möbius, August Ferdinand. Über eine neue Behandlungsweise der analytischen Sphärik, *Abhandlungen bei Begründung der Königlich Sächsischen Gesellschaft der Wissenschaften* 1846, Leipzig, 45–86.

Mollweide, Karl. Zusatze zur ebenen und sphärischen trigonometrie, *Monatlichen Correspondenz zur Beförderung der Erd- und Himmels-kunde* **18** (1808), 394–400.

Mollweide, Karl. Leichte und einfache herleitung der Cagnolischen formeln zue auflösung des vom Hrn. Prof. Gauss im Octbr. Heft der M. C. von 1808 vorgetragenen und aufgelöften problems der sphärischen astronomie, *Monatlichen Correspondenz zur Beförderung der Erd- und Himmels-kunde* **19** (1809), 423–428.

Monmonier, Mark. *Rhumb Lines and Map Wars: A Social History of the Mercator Projection*. Chicago, IL: University of Chicago Press, 2004.

Montelle, Clemency; and Ramasubramanian, K. Determining the Sine of one degree in the *Sarvasiddhāntarāja* of Nityānanda, *SCIAMVS* **19** (2018), 1–52.

Montelle, Clemency; Ramasubramanian, K.; and Dhammaloka, J. Computation of Sines in Nityānanda's *Sarvasiddhāntarāja*, *SCIAMVS* **17** (2016), 1–53.

Moore, Jonas. *A New Systeme of the Mathematicks*. London: Godbid and Playford, 1681.

Moritz, Robert. On Napier's fundamental theorem relating to right spherical triangles, *American Mathematical Monthly* **22** (1915), 220–222.

Moscheo, Rosario. Il *corpus* mauroliciano degli "Sphaerica": problemi editoriali. In *filosofia e Scienze nella Sicilia dei Secoli XVI e XVII*, ed. C. Dollo, pp. 39–84. Catania: Centro di Studi per la Storia della Filosofia in Sicilia, 1992.

Mungello, D. E. *The Great Encounter of China and the West, 1500–1800*. 3rd ed. Lanham, MD: Rowman & Littlefield, 2009.

Nabonnand, Philippe. L'argument de la généralité chez Carnot, Poncelet et Chasles. In *Justifier en Mathématiques*, ed. Dominique Flament and Philippe Nabonnand, pp. 17–47. Paris: Éditions de la Maison des Sciences de l'Homme, 2011.

Nabonnand, Philippe. Utiliser des éléments imaginaires en géométrie: Carnot, Poncelet, von Staudt et Chasles. In *Éléments d'une Biographie de l'Espace Géométrique*, ed. Lise Bioesmat-Martagon, pp. 69–105. Nancy: Univ. Nancy and Univ. Lorraine, 2016.

Nádeník, Zbyněk. Legendre's Theorem on spherical triangles. In *50 Years of the Research Institute of Geodesy, Topography and Cartography: Jubilee Proceedings 1954–2004*, pp. 41–48. Zdiby: Research Institute of Geodesy, Topography and Cartography, 2005.

Nahin, Paul J. *An Imaginary Tale: The Story of $\sqrt{-1}$*. Princeton, NJ: Princeton University Press, 1998.

Napier, John. *A Plain Discoverie of the Whole Revelation of Saint John*. Edinburgh, 1593.

Napier, John. *Mirifici logarithmorum canonis descriptio*. Edinburgh: Hart, 1614.

Napier, John. *A Description of the Admirable Table of Logarithmes*, transl. Edward Wright. London: Okes, 1616.

Napier, John. *Rabdologiae seu numerationis per virgulas*. Edinburgh: Hart, 1617.

Napier, John. *A Description of the Admirable Table of Logarithmes*, transl. Edward Wright. London: Waterson, 1618.

Napier, John. *Mirifici logarithmorum canonis constructio*. Edinburgh: Hart, 1619.

Napier, John. *De arte logistica*, ed. Mark Napier. Edinburgh, 1839.

Napier, John. *The Wonderful Canon of Logarithms*, transl. Herschell Filipowski. Edinburgh: Lizars, 1857.

Napier, John. *The Construction of the Wonderful Canon of Logarithms*, transl. William Rae Macdonald. Edinburgh/London: Blackwood and Sons, 1889. Reprint, London: Dawsons of Pall Mall, 1966; Reprint, New York: Gryphon, 1997.

Napier, Mark. *Memoirs of John Napier of Merchiston*. Edinburgh: Blackwood, and Cadell, London, 1834.

Napoli, Federico. Alla vita ed ai lavori di Francesco Maurolico, *Bullettino di Bibliografia e di Storia delle Scienze Mathematiche e Fisiche* 9 (1876), 1–112.

Nauenberg, Michael. Barrow, Leibniz and the geometrical proof of the Fundamental Theorem of Calculus, *Annals of Science* 71 (2014), 335–354.

Naux, Charles. *Histoire des Logarithmes de Neper à Euler*. 2 vols. Paris: Blanchard, 1966/1971.

Naux, Charles. Le père Christophere Clavius (1537–1612): Sa vie et son oeuvre. III. Clavius mathématicien, *Revue des Questions Scientifiques* 154 (1983), 325–347.

Neal, Katherine. The rhetoric of utility: Avoiding occult associations for mathematics through profitability and pleasure, *History of Science* 37 (1999), 151–178.

Neal, Katherine. *From Discrete to Continuous: The Broadening of Number Concepts in Early Modern England*. Dordrecht: Kluwer, 2002.

Needham, Joseph. *Science and Civilisation in China*. Vol. 3. Cambridge, UK: Cambridge University Press, 1959.

Nell, A. Neue Herleitung des Legendreschen Satzes nebst einer Erweiterung desselben, *Zeitschrift für Mathematik und Physik* 19 (1874), 324–334.

Newton, Isaac. *Arithmetica universalis*. London: Benjamin Tooke, 1707.

Newton, Isaac. *Analysis per quantitatum series, fluxiones, ac differentias.* London: Pearson, 1711.

Newton, Isaac. *The Correspondence of Isaac Newton*, ed. H. W. Turnbull, J. F. Scott, A. Rupert Hall, and Laura Tilling. 7 vols. Cambridge, UK: Cambridge University Press, 1960–1977.

Newton, Isaac. *The Mathematical Works of Isaac Newton*, ed. Derek T. Whiteside. 2 vols. New York: Johnson Reprint Corporation, 1964–1967.

Newton, Isaac. *The Mathematical Papers of Isaac Newton*, ed. Derek T. Whiteside. 8 vols. Cambridge, UK: Cambridge University Press, 1967–1981.

Newton, John. *Institutio mathematica.* London: Leybourn, 1654.

Newton, John. *Astronomia Britannica.* London: Leybourn, 1657.

Newton, John. *Trigonometria Britannica.* London: Leybourn, 1658.

Nobis, Heribert Maria; and Pastori, Anna Maria. *Nicolaus Copernicus Gesamtausgabe*, Band VIII/I: *Receptio Copernicana: Texte zur Aufnahme der Copernicanischen Theorie*, Berlin: Akademie Verlag, 2002.

Norman, Robert. *The Newe Attractive.* London: John Tapp, 1614.

Norwood, Richard. *Trigonometrie, or, The Doctrine of Triangles.* 2 vols. London: Jones, 1631.

Norwood, Richard. *Fortification, or Architecture Military.* London: Thomas Cotes, 1639.

Norwood, Richard. *Epitomie: Being the Application of the Doctrine of Triangles to Certain Problems. . . .* London: Hurlock, 1645.

Oughtred, William. *Trigonometria.* London: Leybourn, 1657.

Panza, Marco. Concept of function, between quantity and form, in the 18th century. In *History of Mathematics and Education: Ideas and Experiences*, ed. N. H. Jahnke et al., pp. 241–274. Göttingen: Vandenhoeck, 1996.

Panza, Marco. *Newton.* Paris: Belles Lettres, 2003.

Panza, Marco. Euler's *Introductio ad analysin infinitorum* and the program of algebraic analysis: Quantities, functions and numerical partitions. In *Euler Reconsidered: Tercentenary Essays*, ed. Roger Baker, pp. 121–169. Heber City, UT: Kendrick Press, 2007.

Panza, Marco. Isaac Barrow and the bounds of geometry. In *Liber Amoricum Jean Dhombres*, ed. Patricia Radelet-de Grave, pp. 365–411. Turnhout: Brepols, 2008.

Papadopoulos, Athanase. On the works of Euler and his followers on spherical geometry, *Gaṇita Bhārati* **36** (2014), 53–108.

Papadopoulos, Athanase; and Théret, Guillaume. Hyperbolic geometry in the work of J. H. Lambert, *Gaṇita Bhārati* **36** (2014), 129–155.

Parsons, E. J. S.; and Morris, W. F. Edward Wright and his work, *Imago Mundi* **3** (1939), 61–71.

Pascal, Blaise. *Lettres de A. Dettonville à M. Carcavi.* Paris: Desprez, 1659.

Pedersen, Kirsti Møller. Roberval's method of tangents, *Centaurus* **13** (1968), 151–182.

Pell, John. *Controversiae de vera circuli mensuri.* Amsterdam: Blaeu, 1647.

Pepper, Jon V. A note on Hariot's method of obtaining meridional parts, *Journal of Navigation* **20** (1967a), 347–349.

Pepper, Jon V. The study of Thomas Harriot's manuscripts. II. Harriot's unpublished papers, *History of Science* **6** (1967b), 17–40.

Pepper, Jon V. Harriot's calculation of the meridional parts as logarithmic tangents, *Archive for History of Exact Sciences* **4** (1968), 359–413.

Pepper, Jon V. Gunter, Edmund. In *Dictionary of Scientific Biography*, Vol. 5, ed. Charles Coulton Gillispie, pp. 593–594. New York: Charles Scribner's Sons, 1972.

Pepper, Jon V. Some clarifications of Harriot's solution of Mercator's problem, *History of Science* **14** (1976), 235–244.

Peurbach, Georg. *Tractatus Georgii Peurbachii super propositiones Ptolemai de sinubus et chordis*, Nuremberg: Petreus, 1541.

Pfister, Louis. *Notices Biographiques et Bibliographiques sur les Jesuites de l'Ancienne Mission de Chine 1552–1773*. 2 vols. Shanghai: Impremerie de la Mission Catholique, 1934.

Phillips, G. M. The development of logarithms, *Bulletin of the Institute of Mathematics and its Applications* **16** (1980), 165–168.

Pingré, Alexandre Guy. La trigonométrie sphérique réduite à quatre analogies, *Mémoires de l'Académie Royale des Sciences* 1756 (publ. 1762), 301–306.

Pingree, David. An astronomer's progress, *Proceedings of the American Philosophical Society* **143** (1999), 73–85.

Pingree, David. Philippe de La Hire at the court of Jayasiṃha. In *History of Oriental Astronomy*, ed. S. M. Razaullah Ansari, pp. 123–131. Dordrecht: Kluwer, 2002a.

Pingree, David. Philippe de La Hire's planetary theories in Sanskrit. In *From China to Paris: 2000 Years Transmission of Mathematical Ideas*, ed. Yvonne Dold-Samplonius, Joseph W. Dauben, Menso Folkerts, and Benno van Dalen, pp. 429–454. Stuttgart: Steiner, 2002b.

Pitiscus, Bartholomew. *Trigonometriae*. Augsburg: Manger, 1600.

Pitiscus, Bartholomew. *Sinuum, tangentium et secantium canon manualis accomodatus ad trigonometriam Barthomomaei Pitisci Grunbergensis Silesii*. Heidelberg, 1613a.

Pitiscus, Bartholomew. *Thesaurus mathematicus*. Frankfurt: N. Hoffmann, 1613b.

Pitiscus, Bartholomew. *Trigonometry, or, The Doctrine of Triangles*, transl. Raphe Handson. London: Edward Allde, 1614.

Plofker, Kim. The "error" in the Indian "Taylor series approximation" to the sine, *Historia Mathematica* **28** (2001), 283–295.

Plofker, Kim. An Indian version of al-Kāshī's method of iterative approximation of Sin 1°, *Ganita Bhāratī* **39** (2017), 95–106.

Pogo, A. Gemma Frisius, his method of determining differences of longitude by transporting timepieces (1530), and his treatise on triangulation (1533), *Isis* **22** (1935), 469–506.

Prékopa, András. The revolution of János Bolyai. In *Non-Euclidean Geometries: János Bolyai Memorial Volume*, ed. András Prékopa and Emil Molnár, pp. 3–59. New York: Springer, 2006.

Price, Michael Haydn. *The Reform of English Mathematical Education in the Late Nineteenth and Early Twentieth Centuries*. PhD diss., University of Leicester, 1981.

Pritchard, Kailyn. Determining the sine tables underlying early European tangent tables. In *Editing and Analyzing Historical Astronomical Tables*, ed. M. Husson, C. Montelle, and B. van Dalen. Turnhout: Brepols, Forthcoming.

Pund, Otto. Über substitutionsgruppen in der sphärischen trigonometrie, insbesondere die Neperschen regeln für die rechtwinkligen spharischen dreiecke, *Mitteilungen der Mathematischen Gesellschaft in Hamburg* **3** (7) (1897), 290–301.

Qu Anjing. Reconstruction of a difference table of solar shadow in the *Dayan Calendar* (AD 724), *Studies in the History of Natural Science* **16** (1997), 233–244.

Ransom, W. R. Napier's rules, *American Mathematical Monthly* **45** (1938), 34–36.

Rashed, Roshdi. A pioneer in anaclastics. Ibn Sahl on burning mirrors and lenses, *Isis* **81** (1990), 464–491.

Rathborne, Aaron. *The Surveyor.* London: Stansby, 1616.

Ravetz, Jerome R. Vibrating strings and arbitrary functions. In *The Logic of Personal Knowledge: Essays Presented to Michael Polanyi on His Seventieth Birthday*, ed. Polanyi Festschrift Committee, pp. 71–88. London: Routledge and Paul, 1961.

Regiomontanus, Johannes. *Tabulae directionum*. Impressum Uenetiis: Ingenio ac impensa Petri Liechtenstein, 1504.

Regiomontanus, Johannes. *De triangulis omnimodis libri quinque*. Nuremberg: Johannes Petreius, 1533.

Regiomontanus, Johannes. *De triangulis planis et sphaericis*, ed. Daniel Santbeck. Basel, 1561.

Regiomontanus, Johannes. *On Triangles*, transl. Barnabas Hughes. Madison: University of Wisconsin Press, 1967.

Reich, Karin. Quelques remarques sur Marinus Ghetaldus et François Viète. In *Actes du Symposium International "La Géometrie et l'Algèbre au Début du XVIIe Siècle,"* pp. 171–174. Zagreb: Institut d'Histoire des Sciences Naturelles, Mathématiques et Médicales de l'Académie Yougoslve des Sciences et des Arts, 1969.

Reich, Karin. Leonhard Euler, "Introduction" to Analysis (1748). In *Landmark Writings in Western Mathematics, 1640–1940*, ed. Ivor Grattan-Guinness, pp. 181–190. Amsterdam: Elsevier, 2005.

Reich, Karin; and Gericke, Helmuth. *François Viète: Einführung in die Neue Algebra.* Munich: Fritsch, 1973.

Reidt, Friedrich. *Sammlung von Aufgaben und Beispielen aus der Trigonometrie und Stereometrie.* Leipzig: Teubner, 1872.

Reinhold, Erasmus. *Tabularum directionum.* Tubingen: Ulrici Morhardi, 1554.

Rheticus, Georg. *Canon doctrinae triangulorum.* Leipzig: Wolfgang Gunter, 1551.

Rheticus, Georg; and Otho, Valentin. *Opus palatinum de triangulis.* Neustadt: Matthaeus Harnisch, 1596.

Riccati, Vincenzo. *Opusculorum ad res physicas, et mathematicas pertinentium.* 2 vols. Bologna: Laelium a Vulpe Instituti Scientiarum Typographum, 1757/1762.

Rice, Adrian. In search of the "birthday" of elliptic functions, *Mathematical Intelligencer* **30** (2) (2008), 48–56.

Richeson, A. W. Additions to Karpinski's trigonometry check list, *Scripta Mathematica* **18** (1952), 94.

Rickey, Fred; and Tuchinsky, Philip. An application of geography to mathematics: History of the integral of the secant, *Mathematics Magazine* **53** (1980), 162–166.

Ritter, Frédéric. Introduction à l'art analytique par François Viète, *Bullettino di Bibliografia e de Storia delle Scienze Matematiche e Fisische* **1** (1868), 223–276.

Ritter, Frédéric. *François Viète: Inventeur de l'Algèbre Moderne*, Paris: Revue Occidentale, 1895.

Robusto, C. C. The cosine-haversine formula, *American Mathematical Monthly* **64** (1957), 38–40.

Roegel, Denis. Bürgi's *"Progress Tabulen"* (1620): Logarithmic tables without logarithms, 2010a. http://hal.archives-ouvertes.fr/docs/00/54/39/36/PDF/buergi1620 doc.pdf.

Roegel, Denis. A reconstruction of Briggs' *Logarithmorum chilias prima* (1617), 2010b. http://hal.archives-ouvertes.fr/docs/00/54/39/35/PDF/briggs1617doc.pdf.

Roegel, Denis. A reconstruction of Smogulecki and Xue's table of logarithms of numbers (ca. 1653), 2011a. http://locomat.loria.fr/smogulecki1653/smogulecki1653 doc1.pdf.

Roegel, Denis. A reconstruction of Smogulecki and Xue's table of trigonometrical logarithms (ca. 1653), 2011b. http://locomat.loria.fr/smogulecki1653/smogulecki1 653doc2.pdf.

Roegel, Denis. A reconstruction of the tables of Pitiscus' *Thesaurus mathematicus* (1613), 2011c. http://locomat.loria.fr/pitiscus1613/pitiscus1613doc.pdf.

Roegel, Denis. A reconstruction of the tables of Rheticus' *Canon doctrinae triangulorum* (1551), 2011d. http://locomat.loria.fr/rheticus1551/rheticus1551doc.pdf.

Roegel, Denis. A reconstruction of the tables of Rheticus' *Opus palatinum* (1596), 2011e. http://locomat.loria.fr/rheticus1596/rheticus1596doc.pdf.

Roegel, Denis. A reconstruction of the tables of the *Shuli Jingyun* (1713–1723), 2011f. http://locomat.loria.fr/shulijingyun1723/shuli1723intro.pdf.

Roegel, Denis. A reconstruction of Viète's *Canon mathematicus* (1579), 2011g. http://locomat.loria.fr/viete1579/viete1579doc1.pdf.

Roegel, Denis. A reconstruction of Viète's *Canonion triangulorum* (1579), 2011h. http://locomat.loria.fr/viete1579/viete1579doc2.pdf.

Roegel, Denis. A reconstruction of Henri Andoyer's trigonometric tables (1915–1918). Vol. 1, 2012a. http://locomat.loria.fr/andoyer1915/andoyer1915doc.pdf.

Roegel, Denis. A reconstruction of Henri Andoyer's trigonometric tables (1915–1918). Vol. 2, 2012b. http://locomat.loria.fr/andoyer1915/andoyer1916doc.pdf.

Roegel, Denis. A reconstruction of Henri Andoyer's trigonometric tables (1915–1918). Vol. 3, 2012c. http://locomat.loria.fr/andoyer1915/andoyer1918doc.pdf.

Romanus, Adrianus. *Ideae mathematicae pars prima*. Antwerp: Keerburgium, 1593.

Romanus, Adrianus. *Canon triangulorum sphaericorum*. Mainz: Albini, 1609.

Rose, Paul Lawrence. *The Italian Renaissance of Mathematics*. Geneva: Librairie Droz, 1975.

Rosen, Edward. Maurolico's attitude toward Copernicus, *Proceedings of the American Philosophical Society* **101** (1957), 177–194.

Rosenfeld, Boris A. *A History of Non-Euclidean Geometry: Evolution of the Concept of a Geometric Space*, transl. Abe Shenitzer. New York: Springer, 1988.

Rosenfeld, Boris A.; and Hogendijk, Jan P. A mathematical treatise written in the Samarkand observatory of Ulugh Beg, *Zeitschrift für Geschichte der Arabisch-Islamischen Wissenschaften* **15** (2002–03), 25–65.

Roseveare, W. N. On "circular measure" and the product forms of the sine and cosine, *Mathematical Gazette* **3** (49) (1905), 129–137.

Rosińska, Grażyna. Don't give to Rheticus what is Regiomontanus's [in Polish], *Kwartalnik Historii Nauki i Techniki* **28** (1983), 615–619.

Ross, Richard P. Oronce Fine's *De sinibus libri II*: The first printed trigonometric treatise of the French Renaissance, *Isis* **66** (1977), 379–386.

Röttel, Karl, ed. *Peter Apian: Astronomie, Kosmographie und Mathematik am Beginn der Neuzeit*, Buxheim: Polygon-Verlag, 1997.

Roy, Ranjan. *Sources in the Development of Mathematics: Infinite Series and Products from the Fifteenth to the Twenty-First Century*. Cambridge, UK: Cambridge University Press, 2011.

Russo, François. Pascal et l'analyse infinitésimale, *Revue d'Histoire des Sciences et de leurs Applications* **15** (1962), 303–320.

Sabra, Abdelhamid I. *Theories of Light from Descartes to Newton*. London: Oldbourne, 1967. Reprint, Cambridge, UK: Cambridge University Press, 1981.

Sachs, J. M. A curious mixture of maps, dates, and names, *Mathematics Magazine* **60** (1987), 151–158.

Sandifer, C. Edward. *The Early Mathematics of Leonhard Euler*. Washington, DC: Mathematical Association of America, 2007a.

Sandifer, C. Edward. Foundations of calculus. In C. Edward Sandifer, *How Euler Did It*, Washington, DC: Mathematical Association of America, 2007b, pp. 147–151.

Sandifer, C. Edward. *How Euler Did It*. Washington, DC: Mathematical Association of America, 2007c.

Sandifer, C. Edward. *How Euler Did Even More*. Washington, DC: Mathematical Association of America, 2015.

Sang, Edward. On the theory of commensurables, *Transactions of the Royal Society of Edinburgh* **23** (1864), 721–760.

Sang, Edward. *A New Table of Seven-Place Logarithms of All Numbers from 20 000 up to 200 000*. London: Layton, 1871.

Sang, Edward. Remarks on the great logarithmic and trigonometrical tables computed in the Bureau de Cadastre under the direction of M. Prony, *Proceedings of the Royal Society of Edinburgh* **8** (1875a), 421–436.

Sang, Edward. Reply to M. Lefort's observations, *Proceedings of the Royal Society of Edinburgh* **8** (1875b), 581–587.

Sang, Edward. On the need for decimal subdivisions in astronomy and navigation, and on tables requisite therefor, *Proceedings of the Royal Society of Edinburgh* **12** (1884), 533–544.

Sasaki, C. The acceptance of the theory of proportion in the sixteenth and seventeenth centuries: Barrow's reaction to the analytic mathematics, *Historia Scientiarum* **29** (1985), 83–116.

Schmidt, Robert; and Black, Ellen, transl. *The Early Theory of Equations: On Their Nature and Constitution: Translations of Three Treatises by Viète, Girard, and De Beaune.* Annapolis, MD: Golden Hind Press, 1986.

Schneider, Ivo. Der mathematiker Abraham de Moivre (1667–1754), *Archive for History of Exact Sciences* **5** (1968), 177–317.

Schönbeck, Jürgen. Thomas Fincke und die *Geometria rotundi*, *NTM* **12** (2004), 80–99.

Schubring, Gert. *Conflicts Between Generalization, Rigor, and Intuition: Number Concepts Underlying the Development of Analysis in 17–19th Century France and Germany.* New York: Springer, 2005.

Schulze, Johann Carl. *Neue und Erweiterte Sammlung Logarithmischer, Trigonometrischer und anderer zum Gebrauch der Mathematik unentbehrlicher Tafeln.* 2 vols. Berlin: August Mylius, 1778.

Schuster, John A. Descartes *opticien*: The construction of the law of refraction and the manufacture of its physical rationales, 1618–1629. In *Descartes' Natural Philosophy*, ed. Stephen Gaukroger, John A. Schuster, and John Sutton, pp. 258–312. London: Routledge, 2000.

Scriba, Christoph. James Gregory's frühe schriften zur infinitesimalrechnung, *Mitteilungen aus dem Mathematischen Seminar der Universität Giessen* **55** (1957).

Scriba, Christoph. Gregory's converging double sequence: A new look at the controversy between Huygens and Gregory over the "analytical" quadrature of the circle, *Historia Mathematica* **10** (1983), 274–285.

Seller, John. *Practical Navigation.* London: Darby, 1669.

Serret, Joseph Alfred. *Traité de Trigonométrie.* Paris: Bachelier, 1850.

Serret, Joseph Alfred. *Traité de Trigonométrie.* 5th ed. Paris: Gauthier-Villars, 1875.

Serret, Paul. *Des Méthodes en Géométrie.* Paris: Mallet-Bachelier, 1855.

Service Géographique de l'Armée. *Tables des Logarithmes a Huit Décimales des Nombres Entiers de 1 a 120 000 et des Sinus et Tangentes de Dix Secondes en Dix Secondes d'Arc dans le Système de la Division Centésimale du Quadrant.* Paris: Imprimerie Nationale, 1891.

Shah, Jayant. Absence of geometric models in medieval Chinese astronomy, 2012a. http://www.northeastern.edu/shah/papers/Yixing.pdf.

Shah, Jayant. Absence of Indian astronomy in *Ta-yrn li* of I-Hsing, 2012b. http://citeseerx.ist.psu.edu/viewdoc/download?doi=10.1.1.221.5275&rep=rep1&type=pdf.

Shen Kangshen; Crossley, John; and Lun, Anthony. *The Nine Chapters on the Mathematical Art: Companion and Commentary.* Oxford: Oxford University Press, 1999.

Shennan, Francis. *Flesh and Bones: The Life, Passions and Legacies of John Napier.* Edinburgh: Napier Polytechnic, 1989.

Sherwin, Henry, ed. *Mathematical Tables.* London: Richard Mount/Thomas Page, 1705.

Shi Yunli. Nikolaus Smogulecki and Xue Fengzuo's *True Principles of the Pacing of the Heavens*: Its production, publication, and reception, *East Asian Science, Technology, and Medicine* **27** (2007), 63–126.

Shi Yunli. Islamic astronomy in the service of Yuan and Ming monarchs, *Suhayl* **13** (2014), 41–61.

Shirley, John W., ed. *Thomas Harriot: Renaissance Scientist.* Oxford: Clarendon Press, 1974.

Shirley, John W. *A Source Book for the Study of Thomas Harriot.* New York: Arno Press, 1981.

Shirley, John W. *Thomas Harriot, a Biography.* Oxford: Clarendon Press, 1983.

Silverberg, Joel. Circles of illumination, parallels of equal altitude, and le calcul du point observe: Nineteenth century advances in celestial navigation. In *Proceedings of the Canadian Society for History and Philosophy of Mathematics 32nd Annual Meeting*, Vol. 20, ed. Antonella Cupillari, pp. 272–296. 2007.

Silverberg, Joel. Napier's rules of circular parts, *Proceedings of the Canadian Society for History and Philosophy of Mathematics* **21** (2008), 160–174.

Silverberg, Joel. Nathaniel Torporley and his *Diclides coelometricae* (1602)—a preliminary investigation, *Proceedings of the Canadian Society for History and Philosophy of Mathematics* **22** (2009), 142–154.

Simpson, Thomas. *A New Treatise of Fluxions.* London: Gardner, 1737.

Simpson, Thomas. *Trigonometry, Plane and Spherical.* London: Nourse, 1748.

Sivin, Nathan. *Granting the Seasons: The Chinese Astronomical Reform of 1280, With a Study of Its Many Dimensions and a Translation of Its Records.* New York: Springer, 2009.

Sleight, E. R. John Napier and his logarithms, *National Mathematics Magazine* **18** (1944), 145–152.

Smith, A. Mark. Descartes' theory of light and refraction: A discourse on method, *Transactions of the American Philosophical Society* **77** (1987), 1–92.

Smith, David Eugene. *A Source Book in Mathematics.* New York/London: McGraw-Hill, 1929.

Snell, Willebrord. *Cyclometricus.* Leiden: Elsevier, 1621.

Snell, Willebrord. *Doctrinae triangulorum canonicae.* Leiden: Maire, 1627.

Sobel, Dava. *Longitude: The True Story of a Lone Genius who Solved the Greatest Scientific Problem of His Time.* London: Fourth Estate, 1995.

Sommerville, D. M. Y. Napier's rules and trigonometrically equivalent polygons. In *Napier Tercentenary Memorial Volume*, ed. C. G. Knott, pp. 169–176. London: Longmans Green, 1915.

Sonar, Thomas. *Der Fromme Tafelmacher: Die Frühen Arbeiten des Henry Briggs.* Berlin: Logos, 2002.

Sørenson, Henrik; and Kragh, Helge. Longomontanus and the quadrature of the circle [in Danish], *Normat* **55** (2007), 97–118.

Sorlin, A. N. J.; and Gergonne, Joseph Diaz. Recherches de trigonométrie sphérique, *Annales de Mathématiques Pures et Appliquées* **15** (1825), 273–305.

Speidell, John. *New Logarithmes*. London, 1619.

Standaert, Nicolas. Xue Fengzuo's and Smogulecki's translation of Cardano's commentaries on Ptolemy's *Tetrabiblos, Sino-Western Cultural Relations Journal* **23** (2001), 51–80.

Steggall, J. E. A. A short account of the treatise *De arte logistica*. In *Napier Tercentenary Memorial Volume*, ed. C. G. Knott, pp. 145–161. London: Longmans Green, 1915.

Stevin, Simon. *Wisconstighe Ghedachtenissen*. Leiden: Bouvvensz, 1608a.

Stevin, Simon. *Hypomnemata mathematica*, transl. Willebrord Snell. 5 vols. Leiden: Ioannis Patii, 1608b.

Stevin, Simon. *Oeuvres Mathématiques de Simon Stevin de Bruges*, transl. Albert Girard. Leiden: Bonaventure & Elsevier, 1634.

Struik, Dirk Jan. *The Principal Works of Simon Stevin*. 6 vols. Amsterdam: C. V. Swets & Zeitlinger, 1958.

Struik, Dirk Jan. *A Source Book in Mathematics, 1200–1800*. Cambridge, MA: Harvard University Press, 1969.

Study, Eduard. *Sphärische Trigonometrie, Orthogonale Substitutionem, und Elliptische Functionen*. Leipzig: Hirzel, 1893.

Sultan, Alan. CORDIC: How hand calculators calculate, *College Mathematics Journal* **40** (2009), 87–92.

Swerdlow, Noel. Kepler's iterative solution to Kepler's equation, *Journal for the History of Astronomy* **31** (2000), 339–341.

Swerdlow, Noel; and Neugebauer, Otto. *Mathematical Astronomy in Copernicus's De Revolutionibus*. 2 parts. New York: Springer, 1984.

Swetz, Frank. *The Sea Island Mathematical Manual: Surveying and Mathematics in Ancient China*. University Park, PA: Pennsylvania State University Press, 1992.

Swetz, Frank; and Ang Tian Se. A brief chronological and bibliographic guide to the history of Chinese mathematics, *Historia Mathematica* **11** (1984), 39–56.

Sylla, Edith. Compounding ratios: Bradwardine, Oresme, and the first edition of Newton's *Principia*. In *Transformation and Tradition in the Sciences: Essays in Honor of I. Bernard Cohen*, ed. Everett Mendelsohn, pp. 11–43. Cambridge, UK: Cambridge University Press, 1984.

Taha, Abdel-Kaddous; and Pinel, Pierre. Sur les sources de la version de Maurolico des *Sphériques* de Ménélaos, *Bollettino di Storia della Scienze Matematiche* **17** (1997), 149–198.

Taha, Abdel-Kaddous; and Pinel, Pierre. La version de Maurolico des Sphériques de Ménélaos et ses sources. In *Medieval and Classical Traditions and the Renaissance of Physico-Mathematical Sciences in the 16th Century*, ed. P. Souffrin and Pier Napolitani, pp. 23–32. Turnhout: Brepols, 2001.

Tanner, R. C. H. Nathaniel Torporley's "Congestor analyticus" and Thomas Harriot's "De triangulis laterum rationalium," *Annals of Science* **34** (1977), 393–428.

Taurinus, Franz Adolph. *Die Theorie der Parallellinien*. Cologne: Johann Peter Bachem, 1825.

Taurinus, Franz Adolph. *Geometriae prima elementa*. Cologne: Johann Peter Bachem, 1826.

Taylor, Andrew. *The World of Gerard Mercator: The Mapmaker Who Revolutionized Geography*. New York: Walker, 2004.

Taylor, E. G. R. The earliest account of triangulation, *Scottish Geographical Magazine* **43** (1927), 341–345.

Taylor, E. G. R. *The Mathematical Practitioners of Tudor & Stuart England*. Cambridge, UK: Cambridge University Press, 1954.

Taylor, E. G. R. John Dee and the nautical triangle, *Journal of Navigation* **8** (1955), 318–325.

Taylor, E. G. R. *The Haven-Finding Art: A History of Navigation from Odysseus to Captain Cook*. New York: Abelard-Schuman, 1957.

Taylor, E. G. R. *A Regiment for the Sea by William Bourne*. London: Cambridge University Press, 1963.

Taylor, E. G. R. *The Mathematical Practitioners of Hanoverian England, 1714–1840*. Cambridge, UK: Cambridge University Press, 1966.

Taylor, E. G. R. *The Haven-Finding Art: A History of Navigation from Odysseus to Captain Cook*. London: Hollis & Carter, new aug. ed., 1971.

Taylor, E. G. R.; and Sadler, D. H. The doctrine of nauticall triangles compendious, *Journal of the Institute of Navigation* **6** (1953), 131–147.

Taylor, Katie. Reconstructing vernacular mathematics: The case of Thomas Hood's sector, *Early Science and Medicine* **18** (2013), 153–179.

Thiele, Rüdiger. *Leonard Euler*. Leipzig: Teubner, 1982.

Thiele, Rüdiger. The mathematics and science of Leonhard Euler (1707–1783). In *Mathematics and the Historian's Craft*, ed. Glen Van Brummelen and Michael Kinyon, pp. 81–140. New York: Springer, 2005.

Thiele, Rüdiger. The rise of the function concept in analysis. In *Euler Reconsidered: Tercentenary Essays*, ed. Roger Baker, pp. 366–405. Heber City, UT: Kendrick Press, 2007a.

Thiele, Rüdiger. What is a function? In *Euler at 300: An Appreciation*, ed. Robert C. Bradley, Lawrence D'Antonio, and C. Edward Sandifer, pp. 63–83. Washington, DC: Mathematical Association of America, 2007b.

Thomas, W. R. John Napier, *Mathematical Gazette* **19** (1935), 192–205.

Thompson, A. J. Henry Briggs and his work on logarithms, *American Mathematical Monthly* **32** (1925), 129–131.

Thompson, A. J. *Logarithmetica Britannica*. Cambridge, UK: Cambridge University Press, 1952.

Thomson, James. *Plane and Spherical Trigonometry*. Belfast: Joseph Smyth, 1825.

Thomson, W. On deep-sea sounding by pianoforte wire, *Journal of the Society of Telegraph Engineers* **3** (1874a), 206–228.

Thomson, W. On the perturbations of the compass produced by the rolling of the ship, *Philosophical Magazine* (4) **48** (1874b), 363–368.

Thoren, Victor E. Prosthaphaeresis revisited, *Historia Mathematica* **15** (1988), 32–39.

Todhunter, Isaac. *Spherical Trigonometry*. Cambridge/London: Macmillan, 1859.

Todhunter, Isaac. *A History of the Mathematical Theories of Attraction and the Figure of the Earth*. 2 vols. London: Macmillan, 1873a.

Todhunter, Isaac. Note on the history of certain formulae in spherical trigonometry, *London, Edinburgh, and Dublin Philosophical Magazine* (4) **45** (1873b), 98–100.

Todhunter, Isaac; and Leathem, J. G. *Spherical Trigonometry*. London: Macmillan, 1901.

Torporley, Nathaniel. *Diclides Coelometricae*. London: Kingston, 1602.

Tropfke, Johannes. *Geschichte der Elementar-Mathematik*. Vol. 2. Leipzig: Verlag von Veit & Comp., 1903.

Tropfke, Johannes. *Geschichte der Elementar-Mathematik*. 7 vols. Leipzig: De Gruyter, 1923.

Truesdell, Clifford. *The Rational Mechanics of Flexible or Elastic Bodies, 1638–1788, Leonhardi Euleri Opera Omnia* (2) **11**. Zurich: Birkhäuser, 1960.

Turnbull, Herbert Westren, ed. *James Gregory Tercentenary Memorial Volume*. London: Bell & Sons, 1939.

Tweddle, Ian. John Machin and Robert Simson on inverse-tangent series for π, *Archive for History of Exact Sciences* **42** (1991), 1–14.

Udias, Agustin. Jesuit astronomers in Beijing, 1601–1805, *Quarterly Journal of the Royal Astronomical Society* **35** (1994), 463–478.

Ursinus, Benjamin. *Cursus mathematici practici*. Guthij, 1618.

Ursus, Nicolaus Raymarus. *Fundamentum astronomicum*. Strasbourg: Bernardus Iobin, 1588.

Valette, Siméon. *La Trigonométrie Sphérique, Resolue par le Moyen de la Règle et du Compas*. Bourges: Jacques Boyer; Paris: J. Barbou, 1757.

Van Brummelen, Glen. *The Mathematics of the Heavens and the Earth: The Early History of Trigonometry*. Princeton, NJ: Princeton University Press, 2009.

Van Brummelen, Glen. *Heavenly Mathematics: The Forgotten Art of Spherical Trigonometry*. Princeton, NJ: Princeton University Press, 2013.

Van Brummelen, Glen. The end of an error: Bianchini, Regiomontanus, and the conversion of stellar coordinates, *Archive for History of Exact Sciences* **72** (2018), 547–563.

Van Brummelen, Glen; and Byrne, James. Maurolico, Rheticus, and the birth of the secant function. Forthcoming.

van Ceulen, Ludolph. *Arithmetische en Geometrische Fondamenten*. Leiden: Colster, 1615.

van Dalen, Benno. Tables of planetary latitude in the *Huihui li* (II). In *Current Perspectives in the History of Science in East Asia*, ed. Yung Sik Kim and Francesca Bray, pp. 316–329. Seoul: Seoul National University Press, 1999.

van Dalen, Benno. A non-Ptolemaic Islamic star table in Chinese. In *Sic Itur ad Astra*, ed. Menso Folkerts and Richard Lorch, pp. 147–175. Wiesbaden: Harrassowitz, 2000.

van Dalen, Benno. Islamic astronomical tables in China: The sources for the *Huihui li*. In *History of Oriental Astronomy*, ed. S. M. Razaullah Ansari, pp. 19–31. Dordrecht: Kluwer, 2002a.

van Dalen, Benno. Islamic and Chinese astronomy under the Mongols: A little-known case of transmission. In *From China to Paris: 2000 Years Transmission of Mathematical Ideas*, ed. Yvonne Dold-Samplonius, Joseph W. Dauben, Menso Folkerts, and Benno van Dalen, pp. 327–356. Stuttgart: Franz Steiner, 2002b.

van Dalen, Benno; and Yano, Michio. Islamic astronomy in China: Two new sources for the *Huihui li* ("Islamic Calendar"), *Highlights of Astronomy* **11B** (1998), 697–700.

Van Egmond, Warren. A catalog of François Viète's printed and manuscript works. In *Mathemata: Festschrift für Helmuth Gericke*, ed. Menso Folkerts and Uta Lindgren, pp. 359–396. Wiesbaden: Franz Steiner, 1985.

Van Haaften, M. Ce n'est pas Vlacq, en 1628, mais De Decker, en 1627, qui a publié une table de logarithmes étendue et complète, *Nieuw Archief voor Wiskunde* **15** (1928), 49–54.

van Hee, P. L. Le *Hai-tao souan-king* de Lieou, *T'oung Pao* **20** (1920), 51–60.

van Hee, P. L. Le classique de l'île maritime, ouvrage chinois du IIIe siècle, *Quellen und Studien zur Geschichte der Mathematik* **2** (1932), 255–280.

van Lansberge, Philip. *Triangulorum geometriae*. Médécin: F. Raphelengium, 1591.

van Lansberge, Philip. *Cyclometriae novae*. Middelburg: Schilders, 1616.

van Lansberge, Philip. *Triangulorum geometriae*. 2nd ed. Amsterdam: Blaevw, 1631.

van Maanen, Jan. The refutation of Longomontanus's quadrature by John Pell, *Annals of Science* **43** (1986), 315–352.

van Poelje, Otto E. Adriaen Vlacq and Ezechiel de Decker: Dutch contributors to the early tables of Briggsian logarithms, *Journal of the Oughtred Society* **14** (2005), 30–40.

van Roomen, Adriaan. *Canon triangulorum sphaericorum*. Mainz: Albini, 1609. (Publ. 1607.)

Van Sickle, Jenna. *A History of Trigonometry Education in the United States: 1776–1900*. PhD diss., Columbia University, 2011a.

Van Sickle, Jenna. The history of one definition: Teaching trigonometry in the US before 1900, *International Journal for the History of Mathematics Education* **6** (2) (2011b), 55–70.

Vanvaerenbergh, Michel; and Ifland, Peter. *Line of Position Navigation: Sumner and Saint-Hilaire, the Two Pillars of Modern Celestial Navigation*. Bloomington, IN: Unlimited, 2003.

Vega, Georgio. *Thesaurus logarithmorum completus*. Leipzig: Weidmann, 1794.

Vekerdi, László. Infinitesimal methods in Pascal's mathematics [in Hungarian], *Magyar Tudományos Akadémia Matematikai és Fizikai Tudományok Osztályának Közleményei* **13** (1963), 269–285.

Vermij, Rienk, ed. *Mercator und seine Welt*. Duisburg: Mercator, 1997.

Vetter, Quido; and Archibald, Raymond Claire. The development of mathematics in Bohemia, *American Mathematical Monthly* **30** (1923), 47–58.

Victor, Stephen K. *Practical Geometry in the High Middle Ages: Artis cuiuslibet consummatio and the Pratike de geometrie*. Philadelphia, PA: American Philosophical Society, 1979.

Viète, François. *Canon mathematicus seu ad triangula*. Paris: Jean Mettayer, 1579.

Viète, François. *Variorum de rebus mathematicis responsorum, liber VIII*. Tours: Mettayer, 1593.

Viète, François. *Ad problema, quod omnibus mathematicis totius orbis construendum proposuit Adrianus Romanus, Francisci Vitae responsum*. Paris: Mettayer, 1595.

Viète, François. *De numerosa potestatum purarum, atque adfectarum ad exegesin resolutione tractatus*, ed. Marino Ghetaldi. Paris, 1600.

Viète, François. *Ad angularium sectionum analyticen*. Paris: Oliver de Varennes, 1615.

Viète, François. *Opera mathematica*, ed. Franz van Schooten. Leiden: Bonaventura and Abraham Elzevier, 1646.

Viète, François. *The Analytic Art*, transl. T. Richard Witmer. Kent, OH: Kent State University Press, 1983.

Viète, François; Albert Girard; and de Beaune, Florimond. *The Early Theory of Equations: On Their Nature and Constitution*, ed. Robert Schmidt and Ellen Black. Annapolis, MD: Golden Hind, 1986.

Vita, Vincenzo. Gli indivisibli di Roberval, *Archimede* **25** (1973), 38–46.

Vlacq, Adrian. *Arithmetica logarithmorum*. 2nd ed. Gouda: Rammaseijn, 1628.

Vlacq, Adrian. *Trigonometria artificialis*. Gouda: Rammaseijn, 1633.

Volder, Jack E. The CORDIC trigonometric computing technique, *IRE Transactions on Electronic Computers* **EC-8** (1959), 330–334.

Volder, Jack E. The birth of CORDIC, *Journal of VLSI Signal Processing* **25** (2000), 101–105.

Vollgraff, J. A. Snellius' notes on the reflection and refraction of rays, *Osiris* **1** (1936), 718–725.

von Braunmühl, Anton. Zur geschichte der sphärischen Polardreiecke, *Bibliotheca Mathematica* (2) **3** (1898), 65–72.

von Braunmühl, Anton. Die Entwickelung der Zeichen- und Formelsprache in der Trigonometrie, *Bibliotheca Mathematica* (3) **1** (1900), 64–74.

von Braunmühl, Anton. *Vorlesungen über Geschichte der Trigonometrie*. 2 vols. Leipzig: Teubner, 1900/1903.

von Braunmühl, Anton. Zur geschichte der entstehung des sogenannten Moivreschen satzes, *Bibliotheca Mathematica* (3) **2** (1901a), 97–102.

von Braunmühl, Anton. Zur geschichte der Trigonometrie im achtzehnten Jahrundert, *Bibliotheca Mathematica* (3) **2** (1901b), 103–110.

von Braunmühl, Anton. Beiträge zur geschichte der integralrechnung bei Newton und Cotes, *Bibliotheca Mathematica* (3) **5** (1904), 355–365.

von Oppel, Friedrich Wilhelm. *Analysis triangulorum.* Dresden/Leipzig: Georgii Conradi Waltheri, 1746.

von Schubert, Friedrich Theodor. Trigonometria sphaerica e Ptolemaeo, *Nova Acta Academiae Scientiarum Imperialis Petropolitanae* **12** (1796), 165–175.

Wagner, Roy; and Hunziker, Samuel. Jost Bürgi's methods of calculating sines, and possible transmission from India, *Archive for History of Exact Sciences* **73** (2019), 243–260.

Walker, Evelyn. *A Study of the Traité des Indivisibles of Gilles Personne de Roberval.* New York: Columbia University, 1932.

Walker, Helen M. Abraham de Moivre, *Scripta Mathematica* **2** (1934), 316–333.

Wallis, John. *Arithmetica infinitorum.* Oxford: Lichfield/Robinson, 1656.

Wallis, John. *Mechanica: sive, De motu, Tractatus geometricus.* London: Godbid/Pitt, 1670.

Wallis, John. *Opera mathematica.* Vol. 1. Oxford, 1695.

Wallis, John. *The Arithmetic of Infinitesimals. John Wallis 1656*, transl. Jacqueline Stedall. New York: Springer-Verlag, 2004.

Wang Yusheng. Li Shanlan: Forerunner of modern science in China. In *Chinese Studies in the History and Philosophy of Science and Technology*, ed. Fan Dainian and Robert S. Cohen, pp. 345–368. Dordrecht: Kluwer, 1996.

Ward, John. *The Lives of the Professors of Gresham College.* London: Moore, 1740.

Wardhaugh, Benjamin. A "lost" chapter in the calculation of π: Baron Zach and MS Boldeian 949, *Historia Mathematica* **42** (2015), 343–351.

Wardhaugh, Benjamin. Filling a gap in the history of π: An exciting discovery, *Mathematical Intelligencer* **38** (1) (2016), 6–7.

Waters, David Watkin. *The Art of Navigation in England in Elizabethan and Early Stuart Times.* New Haven, CT: Yale University Press, 1958.

Weiss, Ad. *Handbuch der Trigonometrie.* Fürth: J. Ludwig Schmid, 1851.

Westfall, Richard S. *Never at Rest: A Biography of Isaac Newton.* Cambridge, UK: Cambridge University Press, 1980.

Wheeler, Gerald F.; and Crummett, William P. The vibrating string controversy, *American Journal of Physics* **55** (1987), 33–37.

Whiteside, Derek. Henry Briggs: The binomial theorem anticipated, *Mathematical Gazette* **45** (1961a), 9–12.

Whiteside, Derek. Newton's discovery of the general binomial theorem, *Mathematical Gazette* **45** (1961b), 175–180.

Whiteside, Derek. Patterns of mathematical thought in the later 17th century, *Archive for History of Exact Sciences* **1** (1961c), 179–388.

Whitman, E. A. Some historical notes on the cycloid, *American Mathematical Monthly* **50** (1943), 309–315.

Wildberger, N. J. *Divine Proportions: Rational Trigonometry to Universal Geometry.* Sydney: Wild Egg, 2005.

Williams, J. H. Briggs's method of interpolation; being a translation of the 13th chapter and part of the 12th of the preface to the "Arithmetica Logarithmica," *Journal of the Institute of Actuaries and Assurance Magazine* **14** (1868), 73–88.

Williams, Michael R. Difference engines: From Müller to Comrie. In *The History of Mathematical Tables: From Sumer to Spreadsheets*, ed. Martin Campbell-Kelly, Mary Croarken, Raymond Flood, and Eleanor Robson, pp. 123–142. Oxford: Oxford University Press, 2003.

Williams, Michael R.; and Tomash, Erwin. The sector: Its history, scales, and uses, *IEEE Annals of the History of Computing* **25** (1) (2003), 34–47.

Willmoth, Frances. *Sir Jonas Moore: Practical Mathematics and Restoration Science*. Woodbridge, UK: Boydell Press, 1993.

Wilson, Curtis A. Perturbations and solar tables from Lacaille to Delambre: The rapprochement of observation and theory, part I, *Archive for History of Exact Sciences* **22** (1980), 53–188.

Wilson, Henry. *Trigonometry Improv'd, and Projection of the Sphere Made Easy*. London: J. Senex, 1720.

Wingate, Edmond. *Arithmétique logarithmétique*. Paris, 1625.

Wolf, Rudolf. *Astronomische Mitteilungen*. Vols. 31–40. Zürich: Zürcher und Furrer, 1872–1876.

Wolf, Rudolf. *Handbuch der Astronomie: Ihrer Geschichte und Litteratur*. 2 vols. Zürich: F. Schulthess, 1890/1892.

Wolfson, Paul. The crooked made straight: Roberval and Newton on tangents, *American Mathematical Monthly* **108** (2001), 206–216.

Wright, Edward. *Certaine Errors in Navigation*. London: Sims, 1599.

Wu, Rex H. The story of Mollweide and some trigonometric identities, 2007. http://www.geocities.ws/galois_e/pdf/mollweide%20article.pdf.

Wylie, Alexander. *Chinese Researches*, Shanghai, 1897.

Wylie, Alexander. *Notes on Chinese Literature*, new ed. Shanghai: American Presbyterian Mission Press, 1901.

Yabuuti, Kiyosi. The Chiuchih-li: An Indian Astronomical Book in the T'ang Dynasty, *Chugoku Chisei Kagaku Gijutsushi no Kenkyu* (1963), 493–538.

Yabuuti, Kiyosi. Researches on the *Chiu-chih li*—Indian astronomy under the T'ang dynasty, *Acta Asiatica* **36** (1979), 7–48.

Yabuuti, Kiyosi. The influence of Islamic astronomy in China. In *From Deferent to Equant: A Volume of Studies in the History of Science in the Ancient and Medieval Near East in Honor of E. S. Kennedy*, ed. David A. King and George Saliba, pp. 547–559. New York: New York Academy of Sciences, 1987.

Yabuuti, Kiyosi. Islamic astronomy in China during the Yuan and Ming dynasties, *Historia Scientiarum* **7** (1997), 11–43.

Yano, Michio. Tables of planetary latitude in the *Huihui li* (I), in *Current Perspectives in the History of Science in East Asia*, ed. Yung Sik Kim and Francesca Bray, pp. 307–315. Seoul: Seoul National University Press, 1999.

Yoshida, Haruyo; and Takata, Seiji. The growth of Fourier's theory of heat conduction and his experimental study, *Historia Scientiarum* **1** (1991), 1–26.

Youschkevitch, A. P. The concept of function up to the middle of the 19th century, *Archive for History of Exact Sciences* **16** (1976–77), 37–85.

Youschkevitch, A. P. L. Euler's unpublished manuscript *Calculus Differentialis.* In *Leonhard Euler 1707–1783: Beitrage zu Leben und Werk*, ed. Emil A. Fellmann, pp. 161–170. Springer, 1983.

Youschkevitch, A. P. Chinese mathematics: Some bibliographic comments, *Historia Mathematica* **13** (1986), 36–38.

Zeller, Mary Claudia. *The Development of Trigonometry from Regiomontanus to Pitiscus*, PhD diss., University of Michigan, Ann Arbor, MI, 1944.

Index

Arabic names are alphabetized after the initial "al-" if it appears. Thus, for instance, "al-Bīrūnī" appears in the Bs. Historical works are listed under their author's name, if known.

Abel, Niels Henrik, 315
Abraham bar Ḥiyya, 47
abstract algebra, 255
Abū Naṣr Manṣūr ibn ʿIrāq, 37
Abū'l-Wafā, 206
Academy of Sciences (Berlin), 164, 176
Acta Eruditorum, 133, 140
Alfonsine Tables, 2
altimetry, 46–51, 303
altitude triangle, 270
An Qingqiao, *Yi xianbiao*, 235–236
analemma, 43, 156
al-Andalus, 2
Anderson, Alexander, 25
Andoyer, Henri, *Nouvelles Tables Trigono-métriques Fondamentales*, 173, 279
Andrew, James, *Astronomical and Nautical Tables*, 268
angle of parallelism, 302
Annales de Gergonne. See *Annales de Mathématiques Pure et Appliqués*
Annales de Mathématiques Pure et Appliqués, 254
anomaly, 111
Apian, Peter: *Instrumentum sinuum seu primi mobilis*, 12, 18, 54–55; *Introductio geographica*, 17–18; *Cosmographia*, 51
arc cosecant, series for, 237
arc cosine, derivative of, 145
arc cotangent, series for, 237
arc secant, series for, 237
arc sine, series for, 129–131, 139, 228, 239–241; table of, 18
arc tangent, series for, 132, 139–142, 170, 236–237
Archimedean spiral, 147
Archimedes, 110–111, 113, 126, 140, 197–198, 207, 241; *On the Sphere and Cylinder*, 111
architecture, 48, 106–109, 303, 315
area of hyperbolic triangle, 299–300
area of spherical triangle, 61, 92–94, 244
Āryabhaṭa, 143
ascensional difference, 65–66
astrolabe, 43–45, 255
astronomical triangle, 266–271

astronomy, 99–102, 185–191, 198–202
Aubrey, John, 55
auxiliary circle, 111

Babbage, Charles, 273, 279
Barrow, Isaac, 121–126, 128, 133, 143–144; *Lectiones geometricae*, 122–126
Bartels, Martin, 301
barycentric coordinates, 244
Baum, Simon, 275
Beltrami, Eugenio, 303
Bernoulli, Daniel, 161, 162, 283
Bernoulli, Jacob, 132, 173, 178
Bernoulli, Johann, 132, 161–164, 178, 294
Bernoulli, Nicholas, 171
Bianchini, Giovanni, 3, 5, 7, 16, 21, 78
Biaodu shuo, 204
bienao, 215–216
Bili duishu biao, 211
bili shu, 211
Biligui jie, 204
binomial coefficients, 80
binomial theorem, 129–130, 136, 169, 174
Blagrave, John, *The Mathematical Jewel*, 53
Blundeville, Thomas, *Exercises*, 53
Board of Longitude, 265
Bolyai, Farkas, *Tentamen juventutem studiosam in elementa matheseos purae*, 301
Bolyai, János, 301, 302
Bond, Henry, 104
Borda, Jean Charles de, 277
Borough, William, 53; *Discourse on the Variation of the Cumpas*, 54
Boscovich, Roger, 156
brachistochrone problem, 177–178
Brahe, Tycho, 24, 32, 35, 51, 63, 210
Brahmagupta, *Khaṇḍakhādyaka*, 186
Bressieu, Maurice, *Metrices astronomicae*, 17, 39–40, 47–48, 109
Briggs, Henry, 53, 55, 59, 68, 75, 76, 78, 80–81, 91, 100, 129, 135, 136, 220, 274; *Arithmetica logarithmica*, 68, 73–74, 81; *Logarithmorum chilias prima*, 68, 72–73, 80; *Trigonometria Britannica*, 28, 72, 74, 76, 83–84, 91, 211

British East India Company, 261
Burckhardt, Johann Karl, 263
Bureau de Cadastre, 276
Bureau des Longitudes, 276
Bureau of Astronomy, 204, 207, 213
Bürgi, Joost, 24, 27, 69–71, 84, 100;
    *Aritmetische und Geometrische Progress
    Tabulen*, 69; *Fundamentum astronomiae*,
    69–71
Buzengeiger, Karl, 263

Cadastre Tables. See *Tables du Cadastre*
Cagnoli, Antonio, *Trigonometria piana e
    sferica*, 249
calculus of variations, 178–184
Callet, François, 172, 273–274
Cantor, Georg, 290
Cardano, Gerolamo, 25, 29; *Ars magna*, 25;
    De subtilitate, 111
Carnot, Lazare, 255, 277, 290–294; *De la
    Corrélation des Figures de Géométrie*,
    291–293; *Géométrie de Position*, 291–293
Carroll, Lewis, *Euclid and His Modern
    Rivals*, 305
Cartesian coordinates, 95, 126, 147, 311
Caswell, John, 156
catenary. See hyperbolic cosine
Cavalieri, Bonaventura, 67, 112–113, 115,
    142; *Directorium generale uranometri-
    cum*, 93–94
Cavendish, Charles, 95
Celiang quanyi, 204, 207–208
center of gravity, 116
Cesàro, Ernesto, 259
Cesàro, Giuseppe, 258–260
Ceyuan haijing, 237
Chasles, Michel, 293
Chen Jixin, 230; *Geyuan milü jiefa*, 230
chong cha, 192–197
Chongzhen, 204
Chongzhen lishu, 204–208, 210
chord, 1, 4–5, 18, 83–84, 223; series for,
    230–234
chordal triangle, 262–264
chouren, 186, 229–230
Chousuan, 204
circular parts. See Napier's rules
Clairaut, Alexis, *Élémens de Géométrie*, 304
Clavius, Christoph, 14–15, 39–40, 158, 203,
    205; *Astrolabium*, 43–45, 159; *Geometria
    practica*, 49–50; *Triangula rectilinea*, 30
clearing the distance, 265
Coignet, Michiel, 55

Collins, John, 128, 132–133, 143, 267, 268
colunar triangle, 260
complex numbers, 152–153, 163–169, 176,
    242, 294, 296–303
Condamine, Charles Marie de la, 263–264
Condorcet, Nicolas de, 261
Connaissance des Temps, 249
continuity, 283, 287–290
Copernicus, Nicolaus, 6–8, 12–13, 43; *De
    lateribus et angulis triangulorum*, 8, 17;
    De revolutionibus, 8
CORDIC, 280–281
correlative system, 292–293
Corvair, 280
cosecant, 9, 15, 99, 205, 224, 270; abbrevia-
    tions of, 16; series for, 237
cosine, 9, 73–74, 97, 99, 205, 224; abbrevia-
    tions of, 16, 165–166; curve, 115; Taylor
    series for, 131, 144–145, 169, 175–176, 273
cosine-haversine formula. See haversine
    formula
cotangent, 9, 12, 15, 73–74, 97, 99, 205, 224;
    abbreviations of, 16; series for, 170, 237
Cotes, Roger, 142, 143, 147, 153, 294, 312;
    Aestimatio errorum in mixta mathesi,
    149–150; *De methodo differentiali
    Newtoniana*, 142; *Harmonia mensuram*,
    147–152
Craig, John, 63
Crelle's Journal, 254, 301
curtain, 107
cycloid, 113–115, 117, 120–121, 178
cyclometry, 46

Da tang kaiyuan suanjing, 186
Dace, 204–207, 212, 221
Dai Xu, *Waiqie milü*, 237
Dai Zhen, 222–227; *Gougu geyuan ji*, 223,
    237; *Mengzi ziyi shuzheng*, 222–223
d'Alembert, Jean le Rond, 164, 282–283,
    291; *Encyclopédie*, 290
Datong li, 204
Davis, Percy, *Requisite Tables*, 270
de Beaugrand, Jean, 27
De Decker, Ezechiel, 73, 76; *Nieuw Telkonst*,
    75
De Gus de Malves, Jean Paul, 237–248
de La Caille, Nicolas-Louis, 150
de Lagny, Thamas Fantet, 132, 142, 147
de Moivre, Abraham, 132, 152–153; *De
    sectione anguli*, 153; *The Doctrine of
    Chances*, 152; *Miscellanea analytica*,
    153

de Moivre's formula, 153, 169
De Morgan, Augustus, 8, 88, 308
de Ursis, Sabatino, 204
de Witt, Jan, 95
decimal division of degrees, 74–75, 83, 211, 220, 235–236
decimal numeration, 16, 78, 109, 276–278
declination, 56–58, 100–102, 199, 201–202
Dee, John, 61
defect, 299–300
del Ferro, Scipione, 29
Delambre, Jean Baptiste Joseph, 248–249, 260, 277
Delambre's analogies, 248–250, 259
derivatives, 143–145, 149, 154, 174–176
derived set, 290
Descartes, René, 114, 120, 121; *Discourse on Method*, 104; *Géometrie*, 95, 114
Dettonville, A. *See* Pascal, Blaise
development, 157–158, 216–218, 224
difference formula, cosine, 16, 167, 206; sine, 16, 19, 21, 167, 206, 236, 251, 277; tangent, 140
difference-to-product identities (sine, cosine), 31
differential equations, 145, 162–164, 282–289, 294, 315
differential triangle, 124, 144–145, 154
Digges, Leonard, *Prognostication*, 53
dijia fa, 231
direct methods, 270
Dirichlet, Gustav Peter Lejeune, 288
Dong Youcheng, *Geyuan lianbili shu tujie*, 236–237
Donnay, J. D. H., 259
double altitude problem, 264
double-angle formula: cosine, 268; sine, 206; tangent, 94, 95, 140
du, 188–190, 207, 211, 223, 235–236
Du Mei. *See* Jartoux, Pierre
duality, 37–38, 182–183, 254–255
duoji shu, 236, 238

e, 148, 297
eccentric anomaly, 112
École Polytechnique, 256, 276
elliptic integrals, 315
Emerson, William, 249
equation of time, 267
error analysis, 149–150
Euclid, 185, 299, 300; *Elements*, 89, 95, 107, 154–155, 191, 203–204, 206, 237, 304–305, 307

Euler, Leonhard, 150, 161–184, 243, 247, 254–255, 277, 281–283, 289, 290, 294, 297, 304, 309; *Calculus differentialis*, 161–162, 165; *Institutiones calculi differentialis*, 174–177; *Institutionum calculi integralis*, 176–177; *Introductio in analysin infinitorum*, 158, 164–173; *Methodus inveniendi lineas curvas maximi minimive proprietate gaudentes*, 178
Euler-Lagrange equation, 178–180
Euler's formula, 170

Fabri, Honoré: *Opusculum geometricum de linea sinuum et cycloide*, 116; *Synopsis geometrica*, 116
fangzhi yi, 216
Faulhaber, Johann, *Zehntausend Logarithmi*, 75
fen, 211–212
Fermat, Pierre de, 121, 122, 241
Fibonacci. *See* Leonardo of Pisa
Filipowski, Herschell, 273
Fincke, Thomas, 13–16, 39, 97, 249; *Geometriae rotundi*, 13–17, 22, 30, 39–40, 48
Fine, Oronce, *De sinibus*, 18
finite differences. *See* interpolation
Firmin Didot, 278
fix, 272
fixed point iteration, 112
Flamsteed, John, 137
flank, 108
Foncenex, Daviet de, 296
Foster, Samuel, 107
Fourier, Jean-Baptiste Joseph, 243, 284–289; *Description de l'Égypte*, 284; *Théorie Analytique de la Chaleur*, 284–289
Fourier series, 243, 281–290
fractional trigonometry, 315
Frederick the Great, 164, 176
French Revolution, 276, 284, 290–291
Frisius, Gemma, 51, 58, 61
function, 1, 133, 143, 162, 164–166, 283, 287–290, 311
Fundamental Theorem of Algebra, 92–93
Fundamental Theorem of Calculus, 119–122

Galileo Galilei, 55, 204
Gardiner, William, *Tables for Logarithms for All Numbers from 1 to 102100*, 274
Gauss, Carl Friedrich, 293, 300, 301; *Theoria motus corporum coelestium*, 248

Gautama Siddhārtha. *See* Qutan Xita
Geber, 32–33
Geber's Theorem, 32–33, 181
geodesic, 178
geodesy, 48, 51, 214, 243
geographic position, 272
geography, 48, 104–106
Gergonne, Joseph Diaz, 254–255
geyuan, 197–198, 232, 238
Geyuan baxianbiao, 207, 212
Girard, Albert, 41–42, 92, 106; *Invention Nouvelle en l'Algèbre*, 92–94; *Tables des Sinus, Tangentes & Secantes*, 97
Girard's Theorem, 300
Glaisher, J. W. L., *Report of the Committee on Mathematical Tables*, 273–274
gnomometry. *See* sundials
Goldbach, Christian, 161
goniometry, 46
Goodwin, H. B., 270
gorge, 108
gougu. See Pythagorean Theorem
grades. *See* gradians
gradians, 275–279, 313
Great Trigonometrical Survey of India, 261
greenhouse effect, 284
Gregory, James, 121, 126, 132–133, 136, 139, 143, 294; *Exercitationes geometricae*, 116, 142–143; *Geometriae pars universalis*, 126; *Vera circuli et hyberbolae quadratura*, 126–128
Gregory-Newton formula, 136
Gresham College, 59, 68
Grunert, Johann August, 263
Gunter, Edmund, 55–59, 68, 97; *Canon triangulorum*, 72–74; *De sector et radio*, 55–58; *Description and Use of the Sector, Crosse-Staffe and Other Instruments*, 56, 106–107
Guo Shoujing, 198–202, 216–217, 223, 224; *Shoushi li cao*, 198
gyrotrigonometry, 315

half-angle formulas: planar, 16, 19, 21–22, 168, 206, 212, 236; spherical, 88–89
Halley, Edmund, 140, 143, 148, 152, 265
Handson, Raphe, *Trigonometrie: Or, The Doctrine of Triangles*, xii, 40, 45, 53–54
Harriot, Thomas, 23, 61, 71, 79–81, 92, 94, 104, 128, 135, 136, 158; *Magisteria magna*, 79–81
Harris, John, 159
Harrison, John, 265–266

Hassler, Ferdinand Rudolph, *Elements of Analytic Trigonometry*, 305–308, 311, 313
haversine, 268–270
haversine formula, 269–270
Hawney, William: *Complete Measurer*, 160; *Doctrine of Plain and Spherical Trigonometry*, 159–160
heat equation, 285–287
Heine, Eduard, 289–290
Helmert, F., 263
Henri, M., 248
Henrion, Dénis, *Traicté des Logarithmes*, 75
Herigone, Pierre, *Cursus mathematicus*, 104
Hermann, Jacob, 132, 140, 161
Hind, John, 314
Hipparchus of Rhodes, 1, 2, 16, 273, 279, 308
Hood, Thomas, 55
Horner's method, 200
hour angle, 267
Huihui li, 191
Hutton, Charles, *A Course of Mathematics*, 310–313
Huygens, Christiaan, 95, 126, 133, 294
Hymers, John, *Treatise on Plane and Spherical Trigonometry*, 242
hyperbolic cosine, 171, 294, 296–299; series for, 296–297
hyperbolic geometry/trigonometry, 244, 294–303, 315
hyperbolic sine, 171, 294, 296–299; series for, 296–297

Ibn Sahl, 104
imaginary numbers. *See* complex numbers
Imperial Board of Astronomy, 229
infinite series (polynomial), 126–143, 154, 169–177, 227–242, 243, 273, 277, 279, 282, 297, 304
infinite series (trigonometric). *See* Fourier series
Institut de France, 284
integration, 111, 113–120, 126–129, 133–134, 150–154, 176–177, 238–241
integration by parts, 133
Internal Angle Bisector Theorem, 95
interpolation, 24, 61, 69–71, 78–84, 129, 135–137, 142, 188–190, 211–212, 221, 236, 273, 277–279
Islamic Astronomical Bureau, 191

Jābir ibn Aflaḥ. *See* Geber
Jacobi, Carl Gustav Jacob, 315
Jai Singh, xiv

Jartoux, Pierre, 229–230
jia shu, 211
jian zhui, 238–241
Jiang Yong, 212
jiefa, 229, 230
Jiuzhang suanshu, 192, 197, 203, 208, 223, 237
John of Murs, 47
Jones, William, *Synopsis Palmariorum Matheseos*, 139–140
ju, 223

Kangxi, 213–214, 220, 222, 227
kaozheng, 222
kardajas, 18
al-Kāshī, Jamshīd, 21, 23, 25, 27, 83, 173, 198, 221, 236
Kästner, Abraham, 263
Kazan Messenger, 301
Keill, John: *Elements of Plain and Spherical Trigonometry*, 154, 245; *Introductio ad veram physicam*, 245
Keith, Thomas, *Introduction to the Theory and Practice of Plane and Spherical Trigonometry*, 160, 308
Kelvin, Lord. *See* Lord Kelvin
Kepler, Johannes, 69, 75–76, 99–100, 104; *Astronomia nova*, 110–112; *Chilias logarithmorum*, 75–76; *Epitome astronomiae Copernicae*, 112; *Rudolphine Tables*, 75–76, 100
Kepler equation, 112
Kepler's Laws, 110
Kerala school. *See* Madhava school (Kerala)
Khublai Khan, 191, 198
Klein, Felix, 255, 293
Klügel, Georg Simon: *Analytische Trigonometrie*, 309–313; *Mathematisches Worterbuch*, 309
Kovalevskaya, Sophia, 255
Kresa, Jakob, *Analysis speciosa trigonometriae*, 155–156

Lacroix, Sylvestre, *Traité Élémentaire de Trigonométrie Rectiligne et Sphérique*, 313
Lagrange, Joseph-Louis, 178, 247, 253, 283–284, 288, 296, 307, 314
Lambert, Johann Heinrich, 88, 250–253, 256, 275, 296–300, 303, 309; *Theory of Parallel Lines*, 299
Laplace, Pierre-Simon, 76
Law of Cosines: hyperbolic, 300, 303; planar, 250, 258–259, 303; spherical,

35–37, 40, 88–89, 106, 181, 182, 247, 250, 251, 254, 258–259, 262, 267, 269–270, 300, 303, 307
Law of Cosines for angles, 35–39, 88–89, 182, 254
Law of Sines: hyperbolic, 303; planar, 103, 211, 224, 259, 303; spherical, 35, 40, 158, 181, 247, 303
Law of Tangents: planar, 30–31, 39, 224, 305–307; spherical, 40, 249
Lefort, F., 277–278
Legendre, Adrien-Marie, 263–264, 277, 315
Legendre's Theorem, 262–264
Leibniz, Gottfried Wilhelm, 110, 119–121, 133–134, 136, 143–145, 154, 162, 164, 294
Leonardo of Pisa, 47
Levi ben Gerson, 206
Lexell's Theorem, 244
l'Hôpital, Guillaume de, *Analyse des Infiniment Petits*, 154
L'Huilier's Theorem, 244
li, 223
Li Shanlan, 237–242; *Fangyuan chanyou*, 238; *Hushi qimi*, 237
Li Tianjing, *Chongzhen lishu*, 204
Li Xinmei, 239–241
Li Zijin, *Tianhu xiangxian biao*, 212
line of position navigation, 272
linear algebra, 71, 214, 221, 280–281
Liu Hui, 192–198; *Haidao suanjing*, 192–197
liuzong sanyao erjianfa, 206
Lobachevsky, Nicolai, 301; *Geometrische Untersuchen zur Theorie der Parallellinien*, 301–303; *Pangeometry*, 301
logarithms, 32, 35, 52, 59, 62–109, 147–148, 150–152, 172, 208–213, 220, 250, 252, 267–271, 273, 277, 291, 294, 296, 303, 314; base 10, 68, 73, 148; of complex numbers, 164; logistic, 75–76, 100; Napierian, 64–68, 73, 77–78, 100; natural, 104, 128, 142, 143, 148, 176, 274–275; tables of, 66–67, 71–78, 90–91, 172, 250–251, 273–279
longest line of defense, 109
Longomontanus, Christian, 94
Loomis, Elias, *Elements of Analytical Geometry and of the Differential and Integral Calculus*, 237
Lord Kelvin, 314
loxodrome. *See* rhumb line
lü, 208, 231–232
lunar distances method, 265
Luo Shilin, 236

Machin, John, 139–142
Maclaurin series for sine and cosine. *See* cosine: Taylor series for; sine: Taylor series for
Madhava school (Kerala), 144, 230, 234, 273
Magini, Antonio, 97, 205; *De planis triangulis*, 15, 22, 41; *Trigonometricae sphaericorum*, 208
Maier, Friedrich Christian, 161; *Trigonometria*, 156, 161
map projections, 58–59
Marolois, Samuel: *Fortification*, 106–107; *Geometria theoretica ac practica*, 106
Martin, Benjamin, *Young Trigonometer's Compleat Guide*, 45, 159–160, 305–307
Maseres, Francis, 140
Maskelyne, Nevil, 265; *Nautical Almanac*, 265
mathematical practitioners, 52, 303, 314, 315
Mathematics Subject Classification, 314
Maty, Matthew, 152
Mauduit, René, 249
Maurolico, Francesco, 12–13, 22; *De sphaera*, 12; *Theodosii sphaericorum*, 12–13, 17
Mayer, Tobias, 265
mean anomaly, 111–112
Méchain, Pierre, 260
Mei Juecheng, 215, 227–229; *Chishui yizhen*, 228–229; *Meishi congshu jiyao*, 228–229; *Yuzhi shuli jingyun*, 215, 227–229
Mei Wending, 214–220, 227, 228, 231; *Bilishu jie*, 220; *Husanjiao juyao*, 215; *Jie baxian geyuan zhi gen*, 218–220; *Lisuan quanshu*, 218–219; *Qiandu celiang*, 215; *Sanjiaofa juyao / Pingsanjiao juyao*, 214–215
Mendoza y Ríos, Josef de, 268
Menelaus configuration, 224–226
Menelaus's theorem: planar, 255; spherical, 14, 32, 39, 248
mensuration, 46
Mercator, Gerard, 53, 58–59, 61, 79, 102–103, 110–111, 143, 148
Mercator, Nicholas, 104, 129, 136; *Logarithmotechnia*, 128
meridional parts, 58–61, 79, 103–104
Mersenne, Marin, 95, 113, 116, 117
Metius, Adrian, 140; *De astrolabio catholico*, 44
metric system, 275
Ming dynasty, 198, 203
Minggatu, 229–238, 241
Möbius, August Ferdinand, 244
Mollweide, Karl, 248–249

Mollweide's formulas, 249
Monge, Gaspard, 254
Moore, Jonas, *A New Systeme of the Mathematicks*, 267
Muir, Thomas, 149, 314
multiple-angle formulas (sine, cosine, tangent, secant), 25–28, 131–132, 147, 152–153, 173

Napier, John, 39, 52, 54, 62–68, 69, 75, 76, 114, 158, 226–227, 253; *De arte logistica*, 63; *Mirifici logarithmorum canonis constructio*, 75, 77–78, 90–91; *Mirifici logarithmorum canonis descriptio*, 59, 63–68, 72, 77, 84–90, 97–98, 181; *Plaine Discoverie of the Whole Revelation of Saint John*, 62–63; *Rabdologiae*, 63
Napier, Mark, 63
Napier's analogies, 90–91, 182, 248, 250, 256
Napier's rods, 63, 204, 222
Napier's rules, 86–88, 90, 245–247, 256
Napoleon, 284, 290–291
National Convention (France), 291
navigation, 53–61, 75, 79, 99, 102–104, 143, 148, 243–244, 264–272, 303, 308, 314, 315
negative magnitudes, 290–293
Nell, A., 263
Newton, Isaac, 104, 110, 120, 121, 128–137, 139–140, 143–144, 152, 154, 173, 178, 184, 228, 241, 282; *Arithmetica universalis*, 249; *De analysi per aequationes numero terminorum infinitas*, 128–131; *epistola posterior*, 133–136; *Methodus differentialis*, 136–137, 142; *Principia Mathematica*, 136, 147; *Regula differentiarum*, 136
Newton, John, 75; *Astronomia Britannica*, 100–102; *Institutio mathematica*, 99; *Trigonometria Britannica*, 99, 100, 102
Newton-Cotes formulas, 142
Newton's laws of motion, 284
Nikitin, Basil, 249
Nityānanda, xiv
non-Euclidean geometry, 243, 296, 299–303
Norman, Robert, *The Newe Attractive*, 54
Norwood, Richard: *Epitomie*, 75–76, 99, 102–106; *Fortification, or Architecture Military*, 106–109; *Trigonometrie*, 72, 99, 105
Nuñez, Pedro, 58, 61

Ogura, Sinkiti, 270–272
Oldenbourg, Henry, 131, 133

Opium Wars, 242
optics, 104
Otho, Valentin, 17, 24, 33, 279
Oughtred, William, 78, 91, 97–98;
    Trigonometria, 99
out-in complementarity principle, 195–197

Pañcasiddhāntikā, 186
Pascal, Blaise, 114–115, 117, 144; *Traité des
    sinus du quart de cercle*, 117–120
Pascal's triangle, 236
Pell, John, 94, 95; *Controversiae de vera
    circuli mensura*, 94–95
pentagramma mirificum, 85–88, 226, 253,
    256
permutation groups, 88
philology, 222–227
Philosophical Transactions, 152
physics, mathematical, 281–289
π, 94–95, 128, 132–135, 138–142, 165–166,
    170–172, 192, 197–201, 207, 223, 228,
    230, 241, 296–297
Pingré, Guy, 244–247
Pitiscus, Bartholomew, 24, 39, 52–53,
    75–76, 206, 221; *Sinuum, tangentium et
    secantium canon manualis accomodatus
    ad trigonometriam*, 207; *Thesaurus
    mathematicus*, 24; *Trigonometriae*, xii,
    17, 36, 40, 48, 97, 106, 205
Poincaré disk, 303
point-set topology, 290
polar sine, 184
polar triangle, 36–39, 183, 253–254
practical geometry, 46–50
primitive circle, 45
primitive system, 292–293
Principal Triangulation of Great Britain and
    Ireland, 261, 264
product-to-sum and product-to-difference
    formulas (sine, cosine), 32, 62, 168, 173
projections (spherical), 89–90; orthographic,
    158; stereographic, 43–45, 88, 92, 158–160,
    255–260
projective geometry, 254, 291
Prony, Gaspard de, 276–279
prosthaphairesis, 32, 54, 62–63, 69, 88, 168
pseudosphere, 303
Ptolemy, Claudius, *Almagest*, 1, 5, 18, 20–21,
    23, 32–33, 39, 43, 69, 84, 138, 167, 187, 206,
    307
Pythagorean Theorem: planar, 18, 46,
    130–131, 155, 167, 192, 199, 206, 212,
    214, 221, 223, 224; spherical, 33, 270

qi, 223
qiandu, 215–217
Qianlong, 227
qibla, 104–106
Qing dynasty, 207, 213–214, 242
quadrant, 55
quadrantal spherical triangles, 85–86
quadrature. See integration
Qutan Xita, 186; *Jiuzhi li*, 186, 190

radian measure, 111, 138–139, 148–149, 166,
    275, 312–314
radius, base circle, 16, 78, 165–166, 186, 205
Raleigh, Sir Walter, 61, 79
Ramus, Peter, 9; *Geometria*, 13
Rathborne, Aaron, *The Surveyor*, 52–53
rational trigonometry, 23–24, 275–276, 315
Recorde, Robert, 98
recurrence relations, 25–28
Regiomontanus, 3, 7, 11–12, 17, 21, 45, 53,
    88, 205; *Compositio tabularum sinuum
    rectorum*, 18–21; *De triangulis omni-
    modis*, 3–5, 11, 30, 35, 39–40; *Tabulae
    directionum*, 5, 17
Reidt, Friedrich, *Sammlung von Aufgaben
    und Beispielen der Trigonometrie und
    Stereometrie*, 249–250
Reidt's analogies, 249–250, 252
Reinhold, Erasmus, 5–7, 17, 53; *Prutenic
    Tables*, 6; *Tabularum directionum*, 6–7
Rheticus, Georg, 3, 7–9, 11, 13, 24, 25,
    33–34, 53, 97, 206, 221; *Canon doctrinae
    triangulorum*, 8–9, 11, 17, 21–22, 33, 53;
    Narratio prima, 8; *Opus palatinum*, 8–9,
    17, 21, 24, 25, 33, 61, 75–76, 221, 279
Rho, Giacomo, 204; *Celiang quanyi*, 204,
    207–208, 215, 216; *Geyuan baxianbiao*,
    207, 212
rhumb line, 58
Riccati, Jacopo, 294
Riccati, Vincenzo, 294–297; *Opusculum ad
    res physicas, et mathematicas pertinen-
    tium*, 294–296
Ricci, Matteo, 203–204, 237; *Jihe yuanben*,
    203
Riemann, Bernhard, 303
Riemann sum, 104, 111
right ascension, 199
Roberval, Gilles, 94, 113, 117, 120–122;
    *Observations sur la Composition des
    Mouvemens*, 121; *Traité des Indivisibles*,
    113–116
Roe, Nathaniel, *Tabulae logarithmicae*, 72

Romanus, Adrianus, 24, 28, 46, 97; *Canon triangulorum*, 11, 17, 42–43; *Ideae mathematicae pars prima*, 46
Roseveare, W. N., 314
Royal Mathematical School, 267
Royal Observatory, 137
Ruggieri, Michele, 203
Rule of Four Quantities, 33, 40

Saint Hilaire, Marcq, 272
Sang, Edward, 273, 276–279
Sanjufīnī Zīj, 191
Savile, Henry, 55
Schall von Bell, Adam, 204, 207, 214; *Geyuan baxianbiao*, 207, 212
Schreck, Johann, 204; *Geyuan baxianbiao*, 207, 212
Schulze, Johann Carl, *Neue und Erweiterte Sammlung Logarithmischer, Trigonome-trischer und anderer zum Gebrauch der Mathematik unentbehrlicher Tafeln*, 274–276
Schweikart, F. K., 300
Scultetus, Abraham, *Sphaericorum*, 40
Sea Island Mathematical Manual. See Liu Hui: *Haidao suanjing*
secant, 9, 11, 14–17, 30, 205, 224, 270; abbreviations of, 16, 97–99; curve, 116–117, 147; derivative of, 149; integral of, 60–61, 103–104, 110–111, 143, 148; series for, 132, 237; tables of, 12–13, 15, 22, 31
second flank, 109
sector, 55–58, 62, 204, 221
Seller, John, *Practical Navigation*, 99
separation of variables, 286
Serret, Joseph Alfred, 256
Serret, Paul: *Des Méthodes en Géométrie*, 256–260; *Traité de Trigonometrie*, 249, 313
set theory, 290
sexagesimal numeration, 16
shadow lengths, 187–190, 204, 213
Sharp, Abraham, 136–139, 169, 273; *Geometry Improved*, 137
Shen Kuo, *Mengxi bitan*, 199–200
Sherwin, Henry, 137
shishu, 216
short methods, 270
sight reduction, 266–271
Simpson, Thomas, 249; *New Treatise of Fluxions*, 154; *Trigonometry, Plane and Spherical*, 154

Simpson's Three-Eighths Rule, 142
sine, 2, 4–5, 9, 205, 224; abbreviations of, 16, 97–99, 165–166; curve, 115–116; derivative of, 149, 175–176; integral of, 118–120; tables of, 18, 69–71, 75, 83, 106, 135–143, 167–172, 186–187, 224, 235; tangent to curve, 121–122; Taylor series for, 131, 137, 144–145, 169, 175, 229, 230, 273
Smith, John, 135–136
Smith, Robert, 148
Smogulecki, Nikolaus, 209; *Suan sanjiao fa*, 211; *Tianbu zhenyuan*, 209–211
Snell, Willebrord, 51, 104, 140, 253; *Doctrinae triangulorum canonicae*, 37–38, 41
Snell's Law, 104
solar timekeeping, 46
Song dynasty, 203, 227
Sorlin, A. N. J., 255
Souvoroff, Prochor, 249
Speidell, John, *New logarithmes*, 72–73
spherical excess, 92, 244, 262–264
spheroid, 183–184, 243, 263
St. Petersburg Academy of Sciences, 161
step function, 287–288
stereographic projection. *See* projections (spherical): stereographic
stereometry, 46
Stevin, Simon: *Driehouckhandel*, 34, 41–42; *Hypomnemata mathematica*, 39, 212–213; *Mémoires Mathématiques*, 205; *Wisconstighe Ghedachtenissen*, 41
Stirling, James, 132
Study, Eduard, *Spherical Trigonometry, Orthogonal Substitutions, and Elliptic Functions*, 255
Suan sanjiao fa, 211
Suan shu shu, 185
Suanjing shishu, 227
Suanxue guan, 214–215
subtangent, 124
sum formula, cosine, 16, 167; hyperbolic cosine, 296, 298–299; hyperbolic sine, 296, 298–299; sine, 16, 19, 21, 167, 175–176, 212, 236, 251, 277; tangent, 140
Sumner, Thomas Hubbard, 272
sum-to-product formulas, 168, 249
sundials, 12, 48
surveying, 51–52, 260–264, 303, 308, 315
Suzzio, Josepho, 296

tables, collections of trigonometric, 8–9, 11,
    13, 53, 135–143, 154, 172–173, 220–221,
    267, 273–281
Tables du Cadastre, 273, 276–278
tangent, 2, 9, 11–12, 14–17, 30, 205, 224;
    abbreviations of, 16, 97–99, 166; curve,
    116, 147; derivative of, 122–126, 149;
    integral of, 116, 142; series for, 132, 170,
    236–237; tables of, 5–8, 15–24, 31,
    187–190, 204, 235
tangent line, 117, 120–126
Tapp, John, 54
Tartaglia, Niccolò Fontana, 29
Taurinus, Franz Adolph, 302; *Die Theorie
    der Parallellinien*, 300–301; *Geometriae
    prima elementa*, 300
Taylor series, 2, 133, 307, 311
Thacker, Anthony, 249
theodolite, 51
Theodosius, *Spherics*, 15, 30, 40
Thomson, James, 149, 314
Thomson, William. *See* Lord Kelvin
three-body problem, 174
Tianbu zhenyuan, 209–211
time triangle, 270
Todhunter, Isaac, 249; *Spherical Trigonom-
    etry*, 41
Toledan Tables, 2
Torporley, Nathaniel, 23, 80; *Diclides
    coelometricae*, 43, 88
Torricelli, Evangelista, 113, 121
transfinite numbers, 290
triangulation, 261–262
trigonometer, 51
triple angle formula (sine), 29, 221–222
al-Ṭūsī, Naṣīr al-Dīn, *Treatise on the
    Quadrilateral*, 34
Twysden, John, 107

unit circle, 165–166
United States Naval Academy, 272
Ursinus, Benjamin, 72; *Cursus mathematici
    practici*, 75
Ursus, Nicolaus Raymarus, 24, 32, 69;
    Fundamentum astronomicum, 24

van Ceulen, Ludolph, 126, 140; *Arithme-
    tische en Geometrische Fondamenten*,
    46
van Lansberge, Philip, 140, 210, 253;
    Cyclometriae novae, 46; *Triangulorum
    geometriae*, 30, 35, 38, 39–41

van Roomen, Adriaan. *See* Romanus,
    Adrianus
van Schooten, Frans, 95–96
Vega, Georgio, *Thesaurus logarithmorum
    completus*, 274
Verbiest, Ferdinand, 213
versed cosine, 205, 224
versed sine, 2, 111, 137, 199–202, 205, 223,
    224, 267–268; series for, 229, 230; tables
    of, 267
vibrating string problem. *See* wave equation
Viète, François, 9–11, 15, 25–29, 30–31, 83,
    84–85, 97, 132, 135, 140, 168, 170, 173,
    198, 221, 236, 253; *Ad angularium
    sectionum analyticen*, 25–28; *Ad
    logisticem speciosam notae priores*,
    27–28; *Ad problema quod omnibus
    mathematicis totius orbis construendum
    proposuit Adrianus Romanus*, 28–29;
    Canon mathematicus seu ad triangula,
    10–11, 17, 22–24, 31, 33–34, 39, 53; *De
    aequationum recognitione*, 28; *De
    numerosa potestum purarum*, 28;
    In artem analyticam isagoge, 10;
    Supplementum geometriae, 28; *Variorum
    de rebus mathematicis responsorum*,
    30–31, 34–35, 37; *Zeteticorum*, 28
Vlacq, Adriaan, 74–75, 76, 221; *Arithmetica
    logarithmica*, 72–74, 211, 273–274;
    Trigonometria artificialis, 72, 273–274
Volder, Jack, 280–281
volume of revolution, 116
von Oppel, Friedrich Wilhelm, 249; *Analysis
    triangulorum*, 156–158
von Schubert, Friedrich Theodor, 248

Waddington, Robert, 265
Wallis, John, 115, 121, 136, 171–172;
    Arithmetica infinitorum, 129; *Mechanica*,
    146–147; *Tractatus de motu*, 116–117
Wang Xun, 198
wave equation, 282–284
Werner, Johann, 32
Wildberger, Norman, 315
Wilson, Henry, 159
Wingate, Edmund, *Arithmétique Logarith-
    métique*, 72, 75
Wolfram, Isaac, 274–276
Wright, Edward, 53, 55, 59, 63, 68, 97–98,
    110–111; *Certaine Errors in Navigation*,
    59–61, 102–104
Wylie, Alexander, 237

Xu Guangqi, 237; *Chongzhen lishu*, 204; *Jihe yuanben*, 203–204
Xu Youren: *Ceyuan milü*, 236–237; *Geyuan baxian zhuishu*, 237
Xue Fengzuo, 209–212, 220; *Bili sixian biao*, 211–212; *Lixue huitong*, 209–213; *Suan sanjiao fa*, 211; *Tianbu zhenyuan*, 209–211; *Zhenxian bu*, 212–213; *Zhongfa sixian*, 211

Yang Zuomei, *Jie baxian geyuan zhi gen*, 220
yangma, 215–217

yicheng tongchu, 224
Yixing, 187; *Dayan li*, 187–190
Yongzhen, 227
Young, Grace Chisholm, *Algebraic-Group Theoretic Investigations of Spherical Trigonometry*, 255
Yuan dynasty, 191, 198
Yuzhi shuli jingyun, 220–222, 236

Zhao Youqin, 198
Zhoubi suanjing, 223–224